P9-CKY-513

PROCEEDINGS

Fourth FORUM on GEOLOGY of INDUSTRIAL MINERALS

March 14 and 15, 1968
Austin, Texas

Themes:

I. Geology of Chemical Raw Materials

II. Depositional Models in Economic Stratigraphy

L. F. Brown, Jr., Editor

Bureau of Economic Geology
The University of Texas at Austin
December 1968

EDITOR'S NOTE

Papers in this volume were presented at the Fourth Forum on Geology of Industrial Minerals sponsored by the Bureau of Economic Geology, The University of Texas at Austin, on March 14 and 15, 1968. Program and meetings were organized and directed by co-chairmen W. L. Fisher and P. U. Rodda, Bureau of Economic Geology. Papers in this volume are arranged in order of presentation at the meetings; contributions have been edited only as necessary to present a standard format. Mr. James Macon, Cartographer, and Mrs. Elizabeth T. Moore, Technical Reports Editor, Bureau of Economic Geology, were chiefly responsible for prompt and efficient processing of manuscripts and illustrations.

L. F. Brown, Jr.

CONTENTS

INTRODUCTION

Peter T. Flawn
Director, Bureau of Economic Geology
The University of Texas at Austin
Austin, Texas

In this day when words are more commonly used for their sound than their meaning and labels do not reflect substance, it is indeed a pleasure to introduce this Forum on Geology of Industrial Minerals. It is truly a FORUM--a place where science and practice meet to apply theory, a place where geologists try to put their science to work on the practical problems of finding, evaluating, and producing industrial minerals. This Fourth Forum continues the fine traditions started four years ago in Ohio.

In Texas, as in other states, the value of the industrial mineral increment is rising more rapidly than the total value of mineral production. In 1967 the value of nonmetallic industrial minerals produced in Texas was nearly 425 million dollars. With the exception of salt and brine, which are mined underground and through well fields, and sulfur which is extracted through bore holes, industrial minerals in Texas, as elsewhere, are produced from open holes in the ground or by dredges operating in bays, estuaries, and rivers. So in Texas, as elsewhere, it is the industrial mineral producer who bears the brunt of the rising concern about environmental quality--concern with air and water pollution and with land reclamation. Texas already has 23 metropolitan centers, more than any other state. By the year 2000, it is estimated that Texas will have 50 metropolitan centers. Metropolitan centers are by their very nature resource-consumption centers and waste-producing centers. It is in and around these centers that industry will have to find the low unit value mineral and rock materials (construction materials and clays) needed to supply the center. It is here that industry will have to produce the deposits so that the environment will be protected during production and reclaim the land so that the depleted site can be restored to beneficial use. It is in and around these centers that wastes from mineral beneficiation and refining will be produced and where their dispersal will be a problem. It is here that industry will have to establish contacts and exchange information with planners and citizen groups that control land use. It is significant that the first violation and fine assessed under the Texas air pollution laws came about as the result of air pollution caused by a mineral producer.

What is the role of geology in this effort? More precisely, what has the theme "Depositional Models in Economic Geology" to do with the problem? In simple language, an understanding of the earth processes responsible for the concentration of minerals and elements in certain parts of the earth's crust and a study in three dimensions of the distribution of different rock types enable the geologist to reconstruct the depositional system, and from this reconstruction, to predict the occurrence of facies likely to contain valuable mineral deposits. Beyond the finding of such deposits, the same model enables the geologist to assist the engineer in the design of waste disposal systems, drainage systems, and the formulation of reclamation plans, and to assist the planner in development of land-use plans compatible with the geology of the terrain and the geologic processes operating therein.

The nonmetallic industrial mineral industry has only recently emerged from the era of small individually-owned operations that seldom benefited from engineering and scientific information. Now these individual enterprises have been joined into larger corporations with engineering departments. But full use of geology in exploration, production, and reclamation is yet to be realized. This Forum and those to follow will do much toward making geology an indispensable part of the nonmetallic mineral industry.

THE VIEW FROM THE FORUM

John B. Patton
State Geologist, Indiana Geological Survey, and
Chairman, Department of Geology, Indiana University
Bloomington, Indiana

ABSTRACT

The Forum on the Geology of Industrial Minerals originated through widespread recognition that the principles of economic geology have been less systematized for the nonmetallic mineral commodities than for the ores and the fossil fuels, and that the industrial minerals have received less than their due in geologic literature.

Textbooks pointed toward the nonmetals have been few, and most of the general economic texts have been written by ore geologists who viewed all mineral occurrences as "deposits" in the sense of localized bodies. Some industrial minerals, notably those of igneous, hydrothermal, or metamorphic origin, lend themselves to this approach, but more are sedimentary, residual, or otherwise supergene, and factors other than localization and tenor determine their exploitability.

A host of new ideas and data, especially from the fields of sedimentology and geochemistry, are waiting to be incorporated into our studies of the industrial minerals. Numerous instruments and investigative methods developed within only the last few years have yet to be applied broadly to the nonmetallic minerals. Predictable market demands for these materials assure a coming era of prosperity for the mineral industries, and the opportunity is here for the geology of the industrial minerals to become a sophisticated field of scientific endeavor.

Our view from this, the Fourth Forum on the Geology of Industrial Minerals, may be as clear as any that we are likely to have soon of geologic education and the state of the art for the nonmetals. The industrial minerals have not been a neglected subject, as volumes have been written about their mining, quarrying, processing, and uses, but their geology has not received the professional consideration accorded to the ores and to the fossil fuels.

By looking back over the training that most of us received and that the preceding one or two generations received, we can, I believe, see both the evidence of the neglect and some of the reason for it. It is interesting to note that some of the early English-language texts in economic geology emphasized the nonmetals. David Page's ECONOMIC GEOLOGY, published in 1874, gave only 37 pages of attention to the metals and ores but 118 to the industrial minerals and their products, 76 percent of the little volume thus being pointed toward the nonmetals. His example was nearly lost, however, in the swarm of nineteenth-century literature that emanated from western Europe concerning magmatic and hydrothermal processes, and directed largely toward the ores.

In reviewing certain other texts released during the ensuing 90-odd years, I shall use the overly simplified method of comparing pages of coverage, realizing with you that other factors are also part of an appraisal.

George P. Merrill's text THE NONMETALLIC MINERALS, published in 1905, would arouse our hopes because of its title, but it approached the materials through a chemical classification ranging from the elements to the sulfates, with the hydrocarbons and a miscellaneous classification at the end for good measure. We find the title mildly misleading when we discover arsenic to be among the nonmetallic elements, cobaltite and molybdenite among the arsenides and sulfides, and bauxite among the oxides, all of which shows that Merrill recognized the artificiality of a classification based on whether metal was or was not recovered from a mineral, as most minerals contain one or more metallic elements, and many of the ore minerals have some use for purposes other than recovery of metal.

In 1920 appeared Volume I (and only) of Grabau's GEOLOGY OF THE NONMETALLIC MINERAL DEPOSITS OTHER THAN THE SILICATES. The preface began "This book is essentially a treatise on applied stratigraphy," an accurate description that set the tone for the work and revealed the fact that Grabau was separating economic mineral deposits not so much on the basis of whether they were metallic or nonmetallic as upon whether their genesis was sedimentary or related to igneous activity and/or metamorphism. His approach would have worked as well for the ores of sedimentary origin. The text made nodding acknowledgment toward economic applications by including paragraphs concerning the geographic distribution of known commercial nonmetallic deposits.

In the next year (1921) appeared C. K. Leith's ECONOMIC ASPECTS OF GEOLOGY, which gave 38 percent of its attention, in terms of pages, to the industrial minerals.

W. S. Bayley's GUIDE TO THE STUDY OF NONMETALLIC MINERAL PRODUCTS, published in 1930, was perhaps the first to deal with the industrial minerals from the viewpoint of the systemics of genesis rather than as a handbook of commodities, but even it managed only twenty-four pages of general principles against 480 pages of encyclopaedic treatment presented commodity by commodity.

In the same year, Heinrich Ries' ELEMENTARY ECONOMIC GEOLOGY appeared, and as we might expect from the author's standing as a specialist in certain of the nonmetallic commodities, particularly industrial sands and clays, the treatment accorded the industrial minerals was respectful. He gave 46 percent of the attention accorded commodities to the nonmetals.

Four editions of Lindgren's MINERAL DEPOSITS, spanning a period from 1913 to 1933, covered the genesis of mineral deposits comprehensively. Lindgren's chapter headings included such rolling perorations as "Deposits formed by mechanical processes of transportation and concentration; detrital deposits," and surely the person whose interest is industrial minerals cannot feel that Lindgren ignored their genesis. Yet the manner of presentation relegated most of the industrial minerals to second class citizenship. Only the nonmetallic materials that resulted from igneous activity, hydrothermal deposition, or metamorphism were viewed as serious concerns of the economic geologist.

W. A. Tarr, in first and second editions of his INTRODUCTORY ECONOMIC GEOLOGY published in 1930 and 1938, gave 80 pages to the topic "History and Origin of Earth Materials" and in them touched lightly on such nonmetallic minerals as the evaporites. In the second edition he gave 39 percent of the nonfuel attention to the nonmetallic earth materials. It should be noted that his coverage of some industrial minerals was as well done, for the space available, as <u>any</u> text literature.

Ernest R. Lilley's ECONOMIC GEOLOGY OF MINERAL DEPOSITS was issued in 1936. He gave fairly brief treatment to principles, genesis, classification, exploration, and technology, but extensive coverage to commodities. The metals were sandwiched between various categories of industrial minerals but still received somewhat more than half the attention of the pages on nonfuel commodities. The nonmetallic minerals were accorded 47 percent of the space.

Alan Bateman's first and second editions of ECONOMIC MINERAL DEPOSITS, published in 1942 and 1950, respectively, have dominated the teaching of economic geology from the time of the first edition nearly to the present day. In the second edition, Part I, "Principles and Processes," occupied 420 pages and touched obliquely on the industrial minerals as examples of certain processes. Nonmetallic minerals were accorded 43 percent of the commodity coverage.

Midway in the Bateman period McKinstrey's MINING GEOLOGY appeared, and despite the breadth implied by the name it is written largely as if mining were restricted to the ores. Its principles and procedural instructions, nevertheless, <u>are</u> applicable to most nonmetallic minerals that are related to igneous and metamorphic activity.

A popular treatment of economic geology was offered in 1959 in Riley's OUR MINERAL RESOURCES. The volume gave 26 percent of the space on nonfuels to the nonmetallic minerals. The coverage of the industrial minerals was whimsical, as exemplified by the fact that 16 pages were given to gems and only 13 to rock and mineral building materials. In the latter category dimension stone received 3-1/2 pages of attention and the vastly more important crushed stone only 1/3 of a page.

The recent economic geology text by Park and McDiarmid is refreshingly and accurately titled ORE DEPOSITS, but despite a statement in the preface that it is "designed to present the principles and data basic to understanding the genesis and localization of the metallic ores <u>and the nonmetallic minerals associated with them</u>," the authors presented all the material on barite and fluorite in a section entitled "The Nature of the Gangue," which puts industrial minerals in their place fairly forthrightly.

In 1960, Robert L. Bates, the godfather of this Forum, initiated a development that we hope will become a trend by issuing a true textbook on the geology of the industrial rocks and minerals. It approached the nonmetals as if they are understandable from the viewpoint of genesis. It contains information on uses but does not confuse uses with materials. Laying no claim to being a handbook, it does <u>not</u> attempt to be comprehensive in the sense of covering <u>all</u> industrial minerals. It is a textbook that presents information in a coordinated manner and offers a basis for a sound course. Unfortunately few undergraduate curricula include a course in geology of industrial minerals, and most graduate courses on the subject have small enrollment. The availability of this text <u>should</u> have stimulated training in the subject, and possibly did, although a formal course may be less necessary when the course content is so well organized within a single volume. The text, now out of print, should be reissued.

The most recent of the full-blown texts in economic geology is Lamey's METALLIC AND

INDUSTRIAL MINERAL DEPOSITS, published in 1966. Part I deals for 105 pages with the characteristics and origin of mineral deposits and makes fleeting acknowledgment to certain types of nonmetallic minerals through specific reference to pegmatites and brief sections on sedimentary beds and residual deposits. The sections on metallic and nonmetallic deposits give 26 percent of the coverage to the industrial minerals.

In summary of this review of texts, Lindgren and other authors who wrote concurrently or subsequently supplied the educational base for training in economic geology for most persons still active in the profession today. The field includes a notable dearth of texts pointed toward the nonmetals. Most of the classical textbooks in economic geology during this period were written by ore geologists, who looked upon all useful minerals as "deposits," in the sense of localized concentrations. Can a stratigraphic unit such as the Mississippian Ste. Genevieve Limestone, 100 feet or more thick, and lying within reach of open-pit quarrying or shallow mining through a region of several thousand square miles, be looked upon as a "deposit"? The parameters that determine its economic utilization at any given location are likely to be urbanization, land ownership, quarrels with adjacent property owners, and whether the county decides to improve the road. Such considerations do not yield to the traditional approach of the economic geologist. Questions concerning the potential value can be answered only partially, at best, by diamond drill coring, chemical analysis, and physical testing.

For the most part, a different attitude has existed toward the industrial minerals than toward the ores. The economic geology textbooks have been written largely by persons who approached the field of economic geology from the view of magmatic activity and associated phenomena. Some industrial minerals, such as the pegmatites, it is true, are of igneous origin and yield to this approach. They are "deposits" in the classical meaning of the term.

Some of my audience will wonder how I can speak thus critically of the period preceding the Bates text when we have had available such volumes as Ladoo's NONMETALLIC MINERALS and the Ladoo and Myers second edition, or the three editions of the AIME Sealey Mudd Memorial Volume, INDUSTRIAL MINERALS AND ROCKS. My answer is that these are not textbooks of economic geology but handbooks concerning mineral commodities, extremely useful, as handbooks tend to be, but lacking the comprehensive approach required of a text that deals with the principles of a subject. The two editions of Ladoo's work showed the desirable homogeneity that results from a single author, or in the case of the second volume, a pair of co-authors. The AIME volumes lack this homogeneity but have the advantage that each chapter is written by a specialist in the field. This assures a heterogeneity of approach which in the case of these volumes is intensified by the fact that the authors of the various chapters are not all geologists. Some are, in which case the chapter is likely to be a substantial work on the geology of an industrial mineral commodity. Other chapters are written by chemical engineers, or by managers, and these variations in authorship are reflected in the coverage. One chapter offers sound coverage of clay mineralogy, another is a thorough analysis of pressure and temperature relationships or of processing methods, and another is a shrewd analysis of profit and loss. The result is that a geology-oriented course in the nonmetals can use some chapters almost alone for coverage of a mineral resource but must depend entirely on outside references for geologic coverage of other materials.

To turn now from individual texts and handbooks on economic geology to broader works, we gain some impression of the state of our art in 1941 by scanning the 50th anniversary volume of the Geological Society of America. The preface has a section entitled "The Field of Geology as Recognized by the Geological Society of America." Under economic geology it listed ore deposits, engineering geology, and petroleum geology. Each of these three subjects was treated in a chapter by a specialist in the field, and the volume lacks any reference to industrial minerals.

The 50th anniversary volume of ECONOMIC GEOLOGY, published in 1951, got through the first 500-odd pages that constituted Part I without reference to industrial minerals, but Part II, a collection of rather unrelated chapters, included three contributions bearing in various ways on the nonmetallic minerals. One was "Recent Developments in Clay Mineralogy and Technology" by Ralph Grim, another dealt with calcium and magnesium carbonates and was written by Don Graf and J. E. Lamar, and a third, by Richard C. Jahns, was a most thorough state-of-the-art paper entitled "The Study of Pegmatites." The three industrial mineral topics together occupied about 200 pages, or nearly 18 percent of the anniversary volume.

A survey of recent professional literature is no more encouraging. One might turn first and most hopefully to ECONOMIC GEOLOGY, but a quick review of the last five-year run reveals a decline from 29 percent nonmetals coverage in 1963 to 9 percent in 1967, exclusive of those papers that could not clearly be assigned to either category or that spanned both.

The journal MINING ENGINEERING was once a fairly fertile source of information on the nonmetals, but in 1962 a change in publishing procedure carried technical articles to the

TRANSACTIONS, where they are immortal but not accessible to that majority of the membership that receives the journal but does not buy the TRANSACTIONS of the Society of Mining Engineers. The coverage of MINING ENGINEERING is now topical rather than technical, and about one-third of the articles that can be classified fall into the nonmetals category.

We may ask ourselves whether the picture revealed by a survey of the literature is a cause for a poorly developed discipline of industrial minerals or whether it is an effect attributable to the fact that study of the nonmetals is not a discipline in the same sense that studies of ore deposits or petroleum geology are. One relevant observation is that the geologists working in nonmetals have not contributed to the written body of knowledge to a degree comparable to workers in other fields, whether through diffidence or lack of expertese. It seems unlikely that the imbalance reflects any prejudice in the groups that choose the papers to be published in various appropriate journals.

Another possible answer is that the geologic literature of industrial minerals is appearing but is not universally recognized because it is not labelled. I suggest that numerous papers appearing in SCIENCE, the JOURNAL OF SEDIMENTARY PETROLOGY, the JOURNAL OF PALEONTOLOGY, various journals in geochemistry, geophysics, and oceanography, and even such general purpose periodicals as the JOURNAL OF GEOLOGY, transactions of unspecialized societies, and a host of monographs published as memoirs, special papers, and proceedings are the literature of the new economic geology without ever the mention of a deposit or a practical application.

If I may draw a parallel from the field of petroleum geology, let me point out that as recently as the 1940's a high proportion of the literature in that field appeared in the AAPG BULLETIN or in special volumes directed toward the oil industry. Consult a paper of today that is agreed to be significant to petroleum geology, and you will find that its citations are largely to fundamental works. May we hope that the field of industrial minerals can telescope (to borrow a word from the geology of ore deposits) its scientific development from the stage of youth to that of maturity without going through a period of adolescence? If this idea has merit, I suggest that a host of recent and current works in modern sedimentation, low temperature geochemistry, and geophysical determination of materials' properties are waiting to be incorporated into our studies of the industrial minerals, and that we should be adding to this segment of the literature instead of describing quarries, mines, and pits, or the materials that come from them, in terms of their present economic potential.

In a similar vein we can rely upon industry, and the governmental agencies concerned with minerals, to try out new instruments and investigative procedures that are described as applicable to mineral resource problems and needs, but we are less prompt in seeing the applicability of a device or a method that is developed for obtaining basic information, particularly when it originates in an apparently unrelated field of science or technology.

With all this said, the most fundamental need that faces us in the geology of the industrial minerals is to show beyond doubt that geology can not only contribute to the field but is essential. I firmly believe that the majority of current nonmetals operations, at least in terms of numbers, are proceeding without any geologic advice, a situation as absurd as random drilling is recognized to be in oil exploration. On the hopeful side I can report an evident change in the attitude toward new ventures, at least in my own State. I believe that 20 years ago not a single major quarry, pit, or mine for industrial minerals in Indiana had been located with the benefit of geologic advice. Within the last five years not a single substantial new minerals operation has been initiated without geologic advice.

All forecasts of market demands during the coming decades predict enormously increased use of most industrial minerals, and particularly those used in construction and the chemical industries. These materials will be produced in huge amounts whether or not geology is utilized in exploration or development. The efficiency will be immeasurably increased if geologic direction at all stages comes to be considered necessary. Moreover, the demand for minerals, and the financial support that will be available during the assured era of prosperity, offer the opportunity for the geology of industrial minerals to become a sophisticated field of endeavor.

RESOURCE AND ECONOMIC IMPORTANCE OF GULF COASTAL SALT DEPOSITS

M. E. Hawkins and S. O. Wood, Jr.
Office of Mineral Resources
U. S. Bureau of Mines
Dallas, Texas

ABSTRACT

In 1966 Gulf Coastal salt production was 16.6 million short tons, valued at approximately $79 million. This production amounted to about 45 percent of the U. S. total of 36.5 million tons and also more than 13 percent of the world output of 122.8 million tons. Salt was mined from 23 salt domes in Alabama, Louisiana, and Texas by fresh-water solution or by room-and-pillar methods. In the 23 domes, the estimated resource of salt down to a depth of 2 miles totals about 104 cubic miles. Assuming a recovery of 50 percent, the producible salt would be almost 517 billion tons, enough to supply the United States for several thousand years based upon the 1966 rate. In addition to the domes now being exploited, 108 other inland domes having salt at depths less than 3,000 feet offer good possibilities for salt extraction or for underground storage sites of liquefied petroleum gases (LPG).

As of December 1967, 23 inland domes were being used for underground storage of LPG. The storage capacity in these 23 domes was about 80 million barrels, or approximately 57 percent of the U. S. total underground LPG storage of 141 million barrels.

During the past decade the annual salt consumption per capita in the United States increased 31 percent from about 280 to 367 pounds. The growth of the chemical complex along the Gulf Coast contributed greatly to the increased demands for Gulf Coastal salt production from 8.2 million tons in 1957 to 16.6 million tons in 1966. Chemical uses for salt have altered the end-use patterns considerably. There are more than 34 major chemical derivatives of salt and more than 14,000 uses for these derivatives. As technology advances, additional derivatives and uses are expected to be developed, thus increasing future demand for salt. Projection of recent production trends indicates that the U. S. demand for salt in 1980 will be about 75 million short tons. The Gulf Coast States' share of the 1980 demand could reach 50 million tons.

INTRODUCTION

Salt, known as halite in the crystalline form and as sodium chloride chemically, is a necessity for human life. Virtually everything that man consumes is prepared, produced, or processed by direct or indirect usages of sodium compounds, of which salt is the most common. Historically, salt has been used as a medicine and a medium of exchange, as well as a symbol of purity and of friendship. During the days of the Roman Empire, Caesar's soldiers received salt as part of their pay, and this was known as "salarium" which is the predecessor of our word "salary." Among natives of some of the South Sea islands and in some areas of Africa, salt is still used as a medium of exchange.

In addition to salt contained in ocean waters (an average of about 1 pound of salt for 4 gallons of water), large deposits of this mineral occur in many basins throughout the world. Figure 1, adapted from Lang (1957), shows the principal areas of rock salt deposits in the continental United States. Structurally, these deposits range from flat-bedded salt to almost vertical columns of salt that have ruptured or pierced overlying formations. Although deposits of rock salt are fairly common, only two areas in the United States are known to contain salt domes. The principal area of occurrence is in the Gulf Coast region which includes parts of Texas, Louisiana, Mississippi, Alabama, and adjacent offshore tidelands. The other area is the Paradox basin in Colorado and Utah.

The mineral salt has been a factor in the Gulf Coast economy for almost 200 years. White settlers in Louisiana began recovering salt by evaporating natural brines emanating from saline springs during the late 1700's. In later years, and in particular during the Civil War, wells were drilled or hand dug in these saline spring areas to increase the flow and concentration of the brine. It was during the digging of a brine well on Avery Island, Iberia Parish, Louisiana, in 1862 that the first important deposit of rock salt was found in North America. This finding

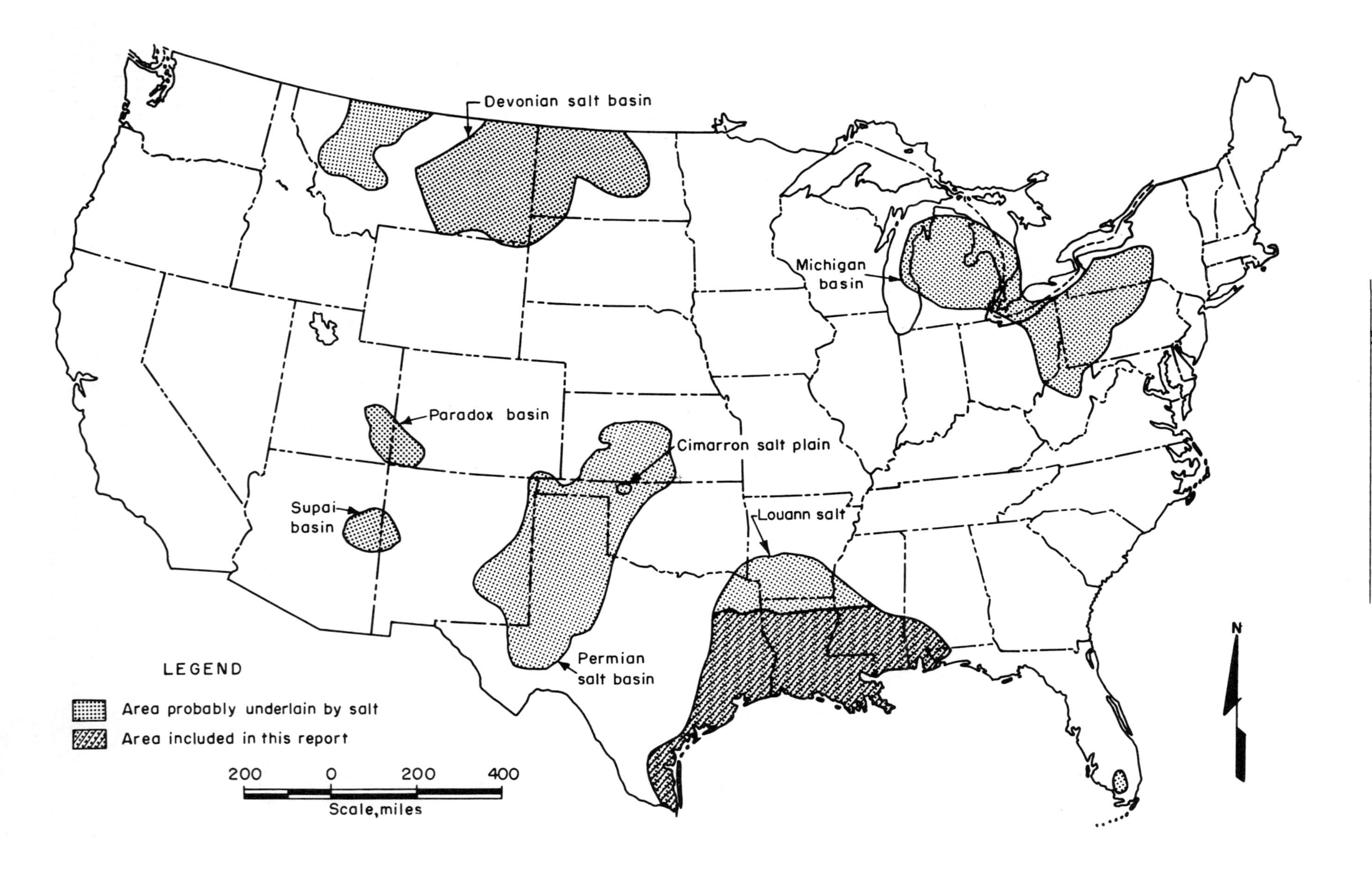

FIG. 1. Continental rock salt deposits in the United States (adapted from Lang, 1957).

also marked the first discovery of a salt dome, although several years passed before this fact became known. In addition to being the first discovery of an important supply of rock salt, Gulf Coast domes were the scene of the birthplace of the Frasch sulfur industry, and the location of the first large oil gusher that ushered in the liquid fuel age.

This report provides information pertaining to the Gulf Coastal salt resource and economic importance of selected domes that offer good possibilities for salt extraction or underground storage sites. Parameters used for selection of domes are that (1) the domes must be inland and (2) subsurface depth to the salt must be less than 3,000 feet; 130 domes meet the selection criteria (Fig. 2). Statistical data related to the domes are presented in tables A-1, A-2, and A-3.

SALT PRODUCTION METHODS

Solution and room-and-pillar mining operations are the principal methods of recovering salt in the Gulf Coast region of the United States. Solution mining can be performed by drilling a well to the desired depth into the salt. Casing can then be set into the cap rock or salt and cemented, and two strings of tubing installed concentrically into the well in such a manner that water can be pumped into the well through the annular space between tubing strings and into the upper part of the cavity. Because water dissolves salt that it contacts, the specific gravity of the resulting brine increases, and the heavier liquid gravitates toward the deeper part of the cavity. The near-saturated solution is then flowed or pumped to the surface through the inner tubing string. In theory, about 800 gallons of fresh water is required to recover a ton of salt. To protect the casing and casing seat, a blanket of gas is sometimes maintained between the casing and outer tubing string and in the upper portion of the cavity.

Room-and-pillar mining consists of removing salt in a checkerboard pattern, leaving solid pillars of rock salt for roof support. After developing a room, a new working level is started by vertical mining of the salt from the room floor to a depth of about 30 to 50 feet. Mining then proceeds by undercutting the new salt face, drilling shot holes, and blasting salt onto the lower floor where it is loaded and transported to either a crusher or the mine shaft. Because of repeated vertical mining operations, rooms sometimes reach heights of 75 to 90 feet. Salt recovery in a room-and-pillar mining operation normally ranges from 50 to 80 percent.

As of January, 1968, salt was being recovered from 23 domes in the Gulf Coastal region (Fig. 3). In three domes--Hockley in Texas, Cote Blanche Island and Jefferson Island in Louisiana--salt is being recovered exclusively by room-and-pillar mining. There are 16 domes where salt is being produced exclusively by solution-mining methods. Seven of these domes are in Texas: Barbers Hill, Blue Ridge, Bryan Mound, Palangana, Pierce Junction, Spindletop, and Stratton Ridge. The eight domes in Louisiana are Anse La Butte, Bayou Choctaw, Darrow, West Hackberry, Napoleonville, Sorrento, Starks, and Sulphur Mines. The other dome undergoing solution-mining operations exclusively is the McIntosh Dome in Alabama. Both solution mining and room-and-pillar mining operations are being conducted on four domes--Avery Island, Belle Isle, and Weeks Island in Louisiana, and Grand Saline in Texas. Table 1 presents a listing of operators, recovery methods, and names of the domes being exploited for salt.

In 1966, solution mining accounted for about 95 percent of the 7.7 million tons of salt produced in Texas and almost 57 percent of the 8.7 million tons produced in Louisiana; the average was 75 percent for the entire Gulf Coast region.

SALT RESOURCE

Of the 130 inland domes having a depth to top of salt less than 3,000 feet, the depth is less than 1,000 feet in 47 domes, from 1,000 to 2,000 feet in 52 domes, and from 2,000 to 3,000 feet in 31 domes. Depth to the salt is less than 1,000 feet in 13 of the domes being exploited for salt, between 1,000 and 2,000 feet in nine domes, and greater than 4,000 feet in one dome.

In the 23 domes currently being mined, the estimated volume of salt to a depth of 2 miles totals about 104 cubic miles (Fisher, 1965; Hawkins and Jirik, 1966). Assuming that 50 percent of this salt volume could be recovered, the producible volume would be about 517 billion tons. At the 1966 production rate of 36.5 million short tons in the United States, the calculated producible volume would be adequate for several thousand years. Using the same assumptions, the other domes listed in tables A-1, A-2, and A-3 contain an estimated producible volume of about 2,150 billion tons of salt.

SALT CONSUMPTION

During the decade ending in 1967 the per capita annual salt consumption in the United States increased about 31 percent, from about 280 to 367 pounds. During the same period Gulf Coast annual salt production increased from 8.2 million to 16.6 million tons, an increase of about 102 percent.

As shown in Figure 4, salt production in the

Table 1. Gulf Coast salt domes where salt is mined.

Number	Dome	County or parish	Company	Address	Mining method
			ALABAMA		
1	McIntosh	Washington	Mathieson Alabama Chemical Corp.	McIntosh, Alabama	Solution.
			LOUISIANA		
2	Anse la Butte	St. Martin	Gordy Salt Co.	P. O. Box 638, New Iberia, La.	Solution.
3	Avery Island	Iberia	International Salt Co.	Clarks Summit, Pa.	Solution, room-and pillar.
4	Bayou Choctaw	Iberville	Allied Chemical Corp. (Solvay Process Div.).	40 Rector St., New York, N. Y.	Solution.
	...do	...do	Ethyl Corp.	100 Park Ave., New York, N. Y.	Do.
5	Belle Isle	St. Mary	Cargill, Inc.	Cargill Bldg., Minneapolis, Minn.	Solution, room-and-pillar.
6	Cote Blanche Island	...do	Carey Salt Co. (with Monsanto Chemical Co.).	1800 Carey Blvd., Hutchinson, Kan.	Room-and-pillar.
7	Darrow1/	Ascension	Wyandotte Chemicals Corp.	1609 Diddle Ave., Wyandotte, Mich.	Solution.
8	Hackberry, West	Cameron	Olin Mathieson Chemical Co.	P. O. Box 896, Lake Charles, La.	Do.
9	Jefferson Island	Iberia	Diamond Crystal Salt Co. (Jefferson Island Div.).	916 Riverside, St. Clair, Mich.	Room-and-pillar.
10	Napoleonville	Assumption	Dow Chemical Co.	Midland, Mich.	Solution.
	...do	...do	Hooker Chemical Co.	666 Fifth Ave., New York, N. Y.	Do.
11	Sorrento	Ascension	Kaiser Aluminum and Chemical Corp.	300 Lakeside Drive, Oakland, Calif.	Do.
12	Starks	Calcasieu	Pittsburgh Plate Glass Co.	Chemical Div., Box 1000, Lake Charles, La.	Do.
13	Sulphur Mines	...do	...do	...do	Do.
14	Weeks Island	Iberia	Morton Salt Co.	110 N. Wacker Drive, Chicago, Ill.	Solution, room-and-pillar.
15	Winnfield2/	Winn	Carey Salt Co.	1800 Carey Blvd., Hutchinson, Kan.	Do.
			TEXAS		
16	Barbers Hill	Chambers	Diamond Alkali Co.	300 Union Commerce Bldg., Cleveland, Ohio.	Solution.
17	Blue Ridge	Fort Bend	United Salt Corp.	4614 Montrose Blvd., Houston, Tex.	Do.
18	Bryan Mound	Brazoria	Dow Chemical Co.	Midland, Mich.	Do.
19	Grand Saline	Van Zandt	Morton Salt Co.	110 N. Wacker Drive, Chicago, Ill.	Solution, room-and-pillar.
20	Hockley	Harris	United Salt Corp.	4614 Montrose Blvd., Houston, Tex.	Room-and-pillar.
21	Palangana	DuVal	Pittsburgh Plate Glass Co.	Box 4026, Corpus Christi, Tex.	Solution.
22	Pierce Junction	Harris	Texas Brine Corp.	11650 Alameda Road, Houston, Tex.	Do.
23	Spindletop	Jefferson	...do	...do	Do.
24	Stratton Ridge	Brazoria	Dow Chemical Co.	Midland, Mich.	Do.

1/ Only dome being mined for salt with depth to top of salt more than 3,000 ft.
2/ The mine filled with water, and operations were discontinued in November 1965.

Table A-1. Louisiana onshore domes with depth to top of salt 3,000 feet or less.

Ref-erence No.	Map index	Dome	Location: Parish	Location: Sec. T. R.	Location: Miles from parish seat	Depth to cap rock, feet	Depth to salt, feet	Salt volume, cu mi 1/	Year first known 2/	Discovery well and year 3/	Remarks	Ref-erence No.
1	119	Anse la Butte....	St. Martin......	117-9S-5E........	11 NW St. Martinville.	(4/)	160	3.1	1899	Anse la Butte Oil Co. well 2 (salt) - 1901.	LPG storage; brine production.	1
2	56	Arcadia..........	Bienville.......	29-18N-5W........	1 SE Arcadia.......	1,282	1,400	-	1922	Arcadia Syndicate (Smitherman) White well 1 (salt) - 1922.	LPG storage.	2
3	111	Avery Island.....	Iberia..........	39-13S-5E........	10 SW New Iberia...	(4/)	6	4.0	1862	Hand-dug well found salt at 16 feet - 1862.	Salt mine. First discovery of rock salt deposits in North America.	3
4	101	Bay Ste. Elaine..	Terrebonne......	17-22S-18E.......	31 SSE Houma.......	710	1,200	6.7	1927	The Texas Co., State-Bay Ste. Elaine well 5 (salt) - 1928.	Sulfur operations abandoned.	4
5	115	Bayou Bleu.......	Iberville.......	74-9S-10E........	9 W Plaquemine....	2,793	2,801	4.6	1926	Standard Oil Co. of Louisiana, Wilbert well 1-A (salt) - 1928.	-	5
6	117	Bayou Bouillon...	St. Martin and Iberville.	24-9S-8E.........	17 NE St. Martinville.	1,260	1,375	4.5	1901	Crowley Oil & Mineral Co., well 2 (salt) - 1914.	-	6
7	116	Bayou Choctaw....	Iberville.......	52-9S-11E........	6 NW Plaquemine....	237	629	1.3	1926	Texas Gulf Sulphur Co., Wilbert's Myrtle Grove well 5 (salt) - 1930.	LPG storage. Brine production.	7
8	104	Belle Isle.......	St. Mary........	28-17S-10E.......	19 SSE Franklin....	110	137	1.9	1896	Captain A. F. Lucas well 2 (salt) - 1897.	Salt mine.	8
9	124	Big Lake.........	Cameron.........	12-12S-9W........	15 NNE Cameron.....	(4/)	1,295	-	1934	Humble Oil & Refining Co., Miami Corp. well 2-L (salt) - 1956.	-	9
10	52	Bistineau........	Webster.........	35-18N-10W.......	9 SW Minden........	1,375	1,500	-	1902	Gulf Refining Co., Bistineau well 2 (salt) - 1913.	Salt produced from brine wells on dome during Civil War.	10
11	130	Black Bayou......	Cameron.........	7-12S-12W........	24 NW Cameron......	881	1,035	2.8	1927	Shell Oil Co. (Roxana), Watkins well 1 (salt) - 1928.	-	11
12	98	Bully Camp.......	Lafourche.......	1-19S-20E........	37 SE Thibodaux....	1,256	1,296	2.7	1942	Gulf Oil & Refining Co., Delta Securities well 1 (salt) - 1942.	Sulfur production expected in 1968.	12
13	123	Calcasieu Lake...	Cameron.........	13S-9W...........	6 NNE Cameron......	1,446	2,345	3.1	1927	Louisiana Land & Exploration Co.-Texas Co., state well 1 (salt) - 1928.	-	13
14	64	Cedar Creek......	Winn............	30-11N-2W........	2 SE Winnfield.....	500	750	-	1907	Winnfield Oil Co., Butler well 1 (cap rock) - 1907.	-	14
15	105	Chacahoula.......	Lafourche.......	71-15S-15E.......	10 SW Thibodaux....	875	1,100	9.1	1926	Gulf Petroleum Corp., Starks well 25 (salt) - 1930.	Brine production. Sulfur operations reopened in 1967.	15
16	59	Chestnut.........	Natchitoches....	23-13N-6W........	24 NE Natchitoches.	(4/)	2,450	-	1926	Ohio Oil Co., S. F. Thomas well 1 (salt - 1927.	-	16
17	99	Clovelly.........	Lafourche.......	32-18S-22E.......	42 SE Thibodaux....	349	1,168	.6	1951	Texas Gulf Sulphur Co., Humble-Shell fee well 1 (salt) - 1951.	-	17
18	62	Coochie Brake....	Winn............	31-10N-4W........	13 SW Winnfield....	Surface	2,500	-	1899	Ohio Oil Co., Edenborn well 1 (salt) - 1927.	-	18
19	109	Cote Blanche Island.	St. Mary........	23-15S-7E........	12 WSW Franklin....	(4/)	298	6.8	1901	Southern Salt Syndicate fee well 9 (salt) - 1921.	Salt mine.	19
20	103	Dog Lake.........	Terrebonne......	6-22S-16E........	30 SSW Houma.......	1,439	1,725	3.2	1927	The Texas Co., State-Dog Lake well 1-B (salt) - 1929.	-	20
21	61	Drakes...........	Natchitoches and Winn.	21-12N-5W........	16 WNW Winnfield...	200	850	-	1902	Savage Bros. fee well 1 (salt) - 1906.	Salt produced from brine wells on dome during Civil War.	21
22	114	Fausse Pointe....	Iberia and St. Martin.	26-11S-8E........	14 NE New Iberia...	792	823	7.9	1926	Union Sulphur Co., Shaw fee well 1 (salt) - 1926.	-	22
23	102	Four Isle Bay....	Terrebonne......	23-21S-16E.......	26 S Houma.........	498	1,305	3.1	1927	The Texas Co., Louisiana Land & Exploration Co.-Four Isle well 1 (salt) - 1929.	-	23
24	93	Garden Island Bay	Plaquemines.....	24-23S-20E.......	51 SE Point a la Hache.	1,691	2,014	6.3	1928	The Texas Co., State-Garden Island Bay well 1 (salt) - 1929.	Sulfur productive.	24
25	55	Gibsland.........	Bienville.......	16-18N-7W........	9 W Arcadia........	612	885	-	1925	L. R. Hawks et al., Calhoun well 1 (salt) - 1926.	LPG storage.	25
26	65	Gilbert..........	Franklin........	22-13N-8E........	7 SE Winnsboro.....	1,425	1,770	-	1937	Continental Oil Co., Sherrouse well 1 (salt) - 1939.	-	26
27	126	Hackberry, East..	Cameron.........	13-12S-10W.......	17 N Cameron.......	2,749	2,950	4.7	1926	Calcasieu Oil Co., Watkins well 2 (salt) - 1928.	LPG storage.	27
28	125	Hackberry, West..	...do...........	21-12S-10W.......	16 NW Cameron......	1,234	1,960	11.5	1902	Yount-Lee Oil Co., Gulf Land Co. well 5 (salt) - 1924.	Brine production.	28
29	113	Iberia (Little Bayou).	Iberia..........	51-12S-7E........	5 E New Iberia.....	5/ 1,078	805	2.1	1917	New Iberia Oil Co., Bolivar well 1 (salt) - 1917.	-	29
30	112	Jefferson Island.	...do...........	59-12S-5E........	9 WSW New Iberia...	5/ 525	31	2.4	1895	Joseph Jefferson water well (salt) - 1895.	Salt mine; sulfur operations abandoned.	30
31	122	Jennings.........	Acadia..........	47-9S-2W.........	14 NW Crowley......	2,000	2,512	1.5	1901	The Texas Co., Latrielle well B-2 (salt) - 1926.	-	31
32	57	Kings............	Bienville.......	35-15N-8W........	25 SW Arcadia......	160	172	-	1902	The Pardee Co. fee well 1 (salt) - 1914.	Salt produced from brine wells on dome during Civil War.	32
33	100	Lake Barre.......	Terrebonne......	37-21S-19E.......	30 SSE Houma.......	721	758	1.3	1928	The Texas Co., State-Lake Barre well 1 (salt) - 1929.	-	33
34	97	Lake Hermitage...	Plaquemines.....	11-18S-25E.......	6 SW Point a la Hache.	905	1,400	.9	1928	Texas Gulf Sulphur Co., Lafourche Basin Levee District well 1 (cap rock) - 1933.	Sulfur production expected in 1968.	34
35	95	Lake Washington (Grande Ecaille)	...do...........	24-20S-26E.......	16 S Point a la Hache.	1,070	1,500	14.5	1928	Humble Oil & Refining Co., Cockrell-Moran well 1 (salt) - 1930.	Sulfur productive.	35
36	53	Minden...........	Webster.........	20-19N-8W........	4 E Minden.........	1,372	1,912	-	-	Gulf Refining Co., Bridgeman well 1 (salt) - 1933.	First commercial oil production from an interior dome in Louisiana.	36

1/ Salt volume estimated from top of salt to a depth of 10,560 feet.
2/ Salt dome indicated by surface expression or geophysical test.
3/ Year first well penetrated salt, cap rock, or discovered oil or gas.
4/ No cap rock reported or information not available.
5/ Cap rock not on dome top.
6/ Depth indicated by geophysical or drilling data.

Table A-1. Louisiana onshore domes with depth to top of salt 3,000 feet or less (continued).

Reference No.	Map index	Dome	Location: Parish	Location: Sec. T. R.	Location: Miles from parish seat	Depth to cap rock, feet	Depth to salt, feet	Salt volume, cu mi 1/	Year first known 2/	Discovery well and year 3/	Remarks	Reference No.
37	108	Napoleonville....	Assumption......	41-12S-13E.......	10 NW Napoleonville	350	650	7.7	1926	Union Sulphur Co., Clifton Land well 1 (salt) - 1926.	Brine production.	37
38	121	Pine Prairie.....	Evangeline......	35-3S-1W.........	10 NW Ville Platte.	Surface	516	2.9	1908	Myles Mineral Co. fee well 1 (salt) - 1909.	LPG storage; limestone quarrying operations abandoned.	38
39	120	Plumb Bob........	St. Martin......	21-8S-7E.........	17 NNE St. Martinville.	1,030	1,189	.2	1933	The Texas Co., St. Martin Land Co. well 9 (salt) - 1937.	-	39
40	96	Potash...........	Plaquemines.....	11-18S-27E.......	8 SE Point a la Hache.	678	1,300	1.0	1929	Humble Oil & Refining Co., Texas Gulf Sulphur Co. well 1 (salt) - 1937.	-	40
41	60	Prices...........	Winn............	30-13N-4W........	15 NW Winnfield....	6/ 1,000	6/ 1,300	-	1902	Surface geology; geological literature describes area in detail in 1902.	Salt produced from brine wells on dome during Civil War.	41
42	58	Rayburns.........	Bienville.......	31-15N-5W........	21 S Arcadia.......	Surface	115	-	1902	J. L. Dorman well (cap rock) - 1910.	Salt produced from brine wells on dome during Civil War.	42
43	118	Section 28.......	St. Martin......	33-9S-7E.........	9 NNE St. Martinville.	951	1,190	2.3	1917	Martin-Hager-Simms fee well 1 (salt) - 1918.	-	43
44	106	Sorrento.........	Ascension.......	15-10S-4E........	14 NE Donaldsonville.	1,568	1,717	5.1	1926	Gulf Refining Co., United Land Co. well 2 (cap rock) - 1927.	LPG storage; brine production.	44
45	128	Starks...........	Calcasieu.......	19-9S-12W........	19 NNW Lake Charles.	1,202	1,925	2.2	1925	Gulf Refining Co., Industrial Lumber Co. well 1 (salt) - 1926.	Brine production; sulfur operations abandoned.	45
46	127	Sulphur Mines....	...do...........	29-9S-10W........	13 W Lake Charles..	315	1,460	1.1	1868	Louisiana Petroleum & Coal Oil Co. well (cap rock) - 1868.	LPG storage, brine production; Frasch process proved.	46
47	54	Vacherie.........	Webster and Bienville.	16-17N-8W........	12 SSE Minden......	658	777	-	1921	Standard Oil Co. of Louisiana, Jordan well 1 (salt) - 1922.	-	47
48	94	Venice...........	Plaquemines.....	26-21S-30E.......	31 SE Point a la Hache.	(4/)	1,328	3.7	1928	Gulf Oil Corp., Buras Levee District well 2-B (salt) - 1931.	-	48
49	129	Vinton...........	Calcasieu.......	33-10S-12W.......	16 WSW Lake Charles.	384	925	-	1902	Vinton Oil & Sulphur Co., well 2 (cap rock) - 1902-04.	-	49
50	66	Walnut Bayou.....	Madison.........	50-17N-13E.......	6 E Tallulah.......	2,650	2,740	-	-	Vasser-Brown, Webb well 1 (salt) - 1953.	-	50
51	110	Weeks Island.....	Iberia..........	38-14S-6E........	16 S New Iberia....	Surface	88	6.1	1897	F. F. Myles fee well 4 (salt) - 1897.	Salt mined since 1902.	51
52	107	White Castle.....	Iberville.......	1-11S-12E........	11 S Plaquemine....	1,693	2,313	1.3	1926	Shell Petroleum Corp., Wilberts Sons well A-1 (salt) - 1928.	-	52
53	63	Winnfield........	Winn............	19-11N-3W........	5 W Winnfield......	Surface	200	-	1907	Cady Petroleum Co., Southern Mineral Co. well 1 (salt) - 1920.	Salt mine became flooded and was abandoned November 1965.	53

1/ Salt volume estimated from top of salt to a depth of 10,560 feet.
2/ Salt dome indicated by surface expression or geophysical test.
3/ Year first well penetrated salt, cap rock, or discovered oil or gas.
4/ No cap rock reported or information not available.
5/ Cap rock not on dome top.
6/ Depth indicated by geophysical or drilling data.

Table A-2. Mississippi and Alabama onshore domes with depth to top of salt 3,000 feet or less.

Reference No.	Map index	Dome	Location: County	Location: Sec. T. R.	Location: Miles from county seat	Depth to cap rock, feet	Depth to salt, feet	Salt volume 1/ cu mi	Year first 2/ known	Discovery well and year 3/	Remarks	Reference No.
		Mississippi										
54	70	Allen..........	Copiah..........	5-9N-6E.......	15 SW Hazlehurst....	2,445	2,774	-	1944	Sun Oil Co., Case Lumber Co. well 2 (salt) - 1944.	-	54
55	76	Arm............	Lawrence........	17-6N-20W......	5 SE Monticello.....	1,218	1,930	-	1945	Humble Oil & Refining Co., S. M. Nelson well 1 (salt) - 1945.	Slight oil stain in upper Tuscaloosa.	55
56	67	Bruinsburg......	Claiborne........	13-11N-1E......	10 WSW Port Gibson..	1,629	2,016	-	1944	Freeport Sulphur Co., W. R. Hammet well 2 (salt) - 1944.	Gas productive; shut in.	56
57	90	Byrd............	Greene..........	16-3N-7W.......	10 NW Leaksville....	1,440	2,058	-	1943	Gulf Refining Co., School Land well 1 (salt) - 1943.	-	57
58	71	Carmichael......	Hinds...........	27-3N-3W.......	14 SSW Raymond......	2,685	2,966	-	1949	Southeastern Drilling Co., Lewis-Ervin well 1 (salt) - 1949.	-	58
59	91	County Line.....	Greene and Wayne.	1-5N-6W........	19 N Leaksville.....	1,239	2,170	-	1948	Sun Oil Co., David Gaines Estate well 1 (salt) - 1948.	Showings of oil in sands from 1,085 feet to 1,270 feet.	59
60	75	D'Lo...........	Simpson.........	17-2N-4E.......	3 NNW Mendenhall....	2,090	4/ 2,400	-	1937	Gulf Refining Co., Blalock-Nichols well 1 (cap rock) - 1942.	-	60
61	81	Dont...........	Covington.......	7-8N-14W.......	7 ENE Collins.......	2,032	4/ 2,300	-	1940	Sun Oil Co., W. W. Speed well D-1 (cap rock) - 1940.	-	61
62	78	Dry Creek.......	...do...........	21-8N-17W......	9 W Collins.........	1,986	4/ 2,300	-	1946	Sippiala Corp., Bertha McRaney well 1 (cap rock) - 1946.	-	62
63	86	Eminence........	Covington and Jones.	5-7N-14W.......	9 SE Collins........	1,964	2,440	-	1947	Humble Oil & Refining Co., Leroy Rogers well 1 (salt) - 1947.	-	63
64	83	Lampton........	Marion..........	21-3N-17W......	6 ESE Columbia......	1,365	1,647	-	1943	Gulf Refining Co., E. H. Bradshaw well 1 (cap rock) - 1943.	-	64
65	68	Leedo..........	Jefferson.......	19-8N-4E.......	12 ESE Fayette......	1,359	2,065	-	1943	Gulf Refining Co., Sarah B. Cupit well 1 (salt) - 1943.	-	65
66	69	McBride........	...do...........	10-9N-4E.......	15 ENE Fayette......	1,865	2,205	-	1946	The California Co., W. G. Greer, et al. well 2 (salt) - 1946.	Gas productive; shut in.	66
67	88	McLaurin.......	Forrest.........	10-2N-13W......	12 S Hattiesburg....	1,705	1,933	-	1948	Danciger Oil & Refining Co., Love Petroleum Co. well 1 (salt) - 1948.	-	67
68	85	Midway.........	Lamar...........	28-4N-15W......	10 NNW Purvis.......	1,613	2,522	-	1932	Sun Oil Co., Scanlan-Semmes well 1 (salt) - 1937.	First proved salt dome in Mississippi. Asphaltic oil shows in cap rock.	68
69	74	Monticello.....	Lawrence........	35-7N-10E......	3 SW Monticello.....	2,256	2,757	-	1943	Gulf Refining Co., J. A. Cox well 1 (salt) - 1943.	Shows of asphaltic oil in flank well.	69
70	80	New Home.......	Smith...........	5-10N-13W......	16 SE Raleigh.......	1,520	2,595	-	1943	Gulf Refining Co., J. F. Dykes well 1 (cap rock) - 1943.	Asphaltic oil shows in cap rock.	70
71	72	Oakley.........	Hinds...........	27-5N-3W.......	4 WSW Raymond.......	2,600	2,634	-	1949	Sun Oil Co., H. T. Shuff well 1 (cap rock) - 1949.	-	71
72	77	Oakvale........	Jefferson Davis..	32-6N-19W......	11 SSW Prentiss.....	1,836	2,696	-	1940	Sun Oil Co., Ed Taylor well 1 (cap rock) - 1940.	-	72
73	87	Petal..........	Forrest.........	25-5N-13W......	2 NNE Hattiesburg...	1,235	1,739	-	1946	Sippiala Corp., J. E. Wilson well 1 (salt) - 1946.	LPG storage.	73
74	79	Raleigh........	Smith...........	17-2N-8E.......	6 SE Raleigh........	1,490	2,140	-	1964	Central Oil Co., Unit 17-14 well 1 (salt) - 1964.	-	74
75	82	Richmond.......	Covington.......	20-6N-15W......	11 SSE Collins......	1,610	1,954	-	1944	Freeport Sulphur Co., Mrs. M. Beasley well 1 (cap rock) - 1944.	-	75
76	89	Richton........	Perry...........	26-5N-10W......	10 NNE New Augusta..	497	722	-	1944	Exploro Corp., W. E. Carter well 1 (cap rock) - 1944.	Largest salt dome known in Mississippi.	76
77	73	Ruth...........	Lincoln.........	15-5N-9E.......	13 SE Brookhaven....	2,208	2,700	-	1942	Freeport Sulphur Co., C. C. Clarke well 2 (cap rock) - 1942.	Showings of asphaltic oil in Wilcox and Cretaceous sands.	77
78	84	Tatum..........	Lamar...........	14-2N-16W......	9 WSW Purvis........	872	1,503	-	1940	Tatum Lumber Co., W. S. F. Tatum well 1 (salt) - 1940.	-	78
		Alabama										
79	92	McIntosh.......	Washington......	29-3N-1E.......	20 SE Chatom.......	270	400	11.8	-	Humble Oil & Refining Co., Washington Lumber and Turpentine Corp. well 1 (salt) - 1948.	Brine production.	79

1/ Salt volume estimated from top of salt to a depth of 10,560 feet.
2/ Salt dome indicated by surface expression or geophysical test.
3/ Year first well penetrated salt, cap rock, or discovered oil or gas.
4/ Depth indicated by geophysical or drilling data.

Table A-3. Texas onshore domes with depth to top of salt 3,000 feet or less.

Reference No.	Map index	Dome	Location: County and district	Location: Survey	Location: Miles from county seat	Depth to cap rock, feet	Depth to salt, feet	Salt volume, cu mi 1/	Year first known 2/	Discovery well and year 3/	Remarks	Reference No.
80	44	Allen........	Brazoria 3	Wm. & Thos. Alley..	17 SSW Angleton...	760	1,380	0.8	1925	Shell Oil Co., Allen well 1 (cap rock) - 1927.	Structure is circular with overhang.	80
81	31	Barbers Hill..	Chambers 3	H. Griffith, A-12..	14 WNW Anahuac....	350	1,000	5.1	1902	Patillo Higgins, Old River well 1 (cap rock) - 1902.	LPG storage; brine production. First important oil found below a salt overhang.	81
82	25	Batson.......	Hardin 3	J. Milhome.........	16 WSW Kountze....	1,080	2,050	3.2	1903	Paraffine Oil Co. fee well 3 (cap rock) - 1903.	-	82
83	10	Bethel.......	Anderson 6	J. N. Acosta, A-1..	19 NW Palestine...	1,440	1,600	8.0	1927	Pure Oil Co., J. R. Cook well 1 (salt) - 1927.	-	83
84	19	Big Creek.....	Fort Bend 3	Michael Young......	9 SSE Richmond....	450	635	1.8	1922	Gulf Production Co.-McSweeney and Snowden, Wheat well 1 (salt) - 1922.	-	84
85	34	Big Hill......	Jefferson 3	S. Eaton..........	26 SSW Beaumont...	200	1,300	2.6	1900	J. M. Guffey, Broussard well 1 (cap rock) - 1901.	LPG storage.	85
86	20	Blue Ridge....	Fort Bend 3	J. Poitevent 3.....	17 E Richmond.....	143	230	1.3	1903	Rio Bravo Oil Co., well 1 (salt) - 1903.	LPG storage; brine production.	86
87	9	Boggy Creek...	Anderson and Cherokee 6	M. Kays, A-459.....	17 NE Palestine...	(4/)	1,829	11.2	1924	Humble Oil & Refining Co., Earl and Ragsdale well 1 (salt) - 1926.	First oil productive interior salt dome.	87
88	38	Boling.......	Wharton and Fort Bend 3	S. F. Austin, A-2..	13 ESE Wharton....	383	975	31.3	1923	Gulf Production Co., Missouri Land Co. well 1 (cap rock) - 1923.	Sulfur productive.	88
89	16	Brenham.......	Washington and Austin 3	S. M. Williams, A-110.	8 SW Brenham......	800	1,150	1.3	1917	Gulf Production Co., Schuerenberg well 2 (salt) - 1917.	-	89
90	8	Brooks.......	Smith 6	Pedro Elias Bean, A-2.	17 SW Tyler.......	195	220	5.5	1903	Salt exploration well hit salt - 1903.	Salt produced from brine wells on dome during Civil War. Oil wells, abandoned.	90
91	43	Bryan Mound...	Brazoria 3	C. Arriola, A-141..	16 S Angleton.....	680	1,112	1.5	1901	Several wells drilled salt - early 1900's.	Brine production; sulfur operations abandoned.	91
92	7	Bullard.......	Smith 6	V. Moore, A-15.....	13 S Tyler........	(4/)	527	3.0	1927	Gulf Production Co., Annie Morgan well 1 (salt) - 1927.	-	92
93	13	Butler........	Freestone 5	Simon Sanchez......	20 E Fairfield....	(4/)	312	1.5	-	Llewellyn Gin well (salt) - 1918-22.	LPG storage, rock quarry; oil well, abandoned.	93
94	15	Clay Creek....	Washington 3	J. F. Perry........	10 NNW Brenham....	1,750	2,400	2.2	1926	Sun Oil Co., Grote well 2 (salt) - 1928-30.	-	94
95	45	Clemens.......	Brazoria 3	John McNeel, A-92..	14 SW Angleton....	530	1,380	1.9	1925	Shell Oil Co., Kropp well 1 (cap rock) - 1926.	LPG storage. Gas well, shut in; sulfur operations abandoned.	95
96	37	Damon Mound...	Brazoria 3	J. P. Mills........	24 NW Angleton....	Surface	529	4.3	1901	J. M. Guffey Petroleum Co., Herndon well 1 (salt) - 1901.	Sulfur operations abandoned.	96
97	24	Davis Hill....	Liberty 3	J. D. Martinez.....	18 N Liberty......	800	1,200	5.5	1905	Drilling proved structure - 1905.	Oil wells, abandoned.	97
98	23	Dayton, North.	Liberty 3	H & T C RR, sec.119	12 NW Liberty.....	580	800	1.7	1901	Taylor-Dayton Co. well (salt) - about 1902.	-	98
99	33	Fannett.......	Jefferson 3	W. H. Smith, A-198.	16 SW Beaumont....	741	2,200	1.0	1925	Gulf Production Co., Thomas well 1 (cap rock) - 1925.	Sulfur productive, LPG storage.	99
100	1	Grand Saline..	Van Zandt 5	Samuel Bell, A-46..	10 NE Canton......	188	212	7.6	1888	Grand Saline Salt Co., S. Q. Richardson well 1 (salt) - 1888.	Salt mine and brine production. Brine springs used by Indians before 1845.	100
101	47	Gulf..........	Matagorda 3	Wm. Simpson, A-89..	18 SSE Bay City...	825	1,100	.7	1901	W. Cash and W. T. Goode, Mayes well 1 (cap rock) - 1901-02.	Sulphur productive. Oil wells, abandoned.	101
102	51	Gyp Hill......	Brooks 4	G B & C N G RR, Blk. 331.	2 SE Falfurrias...	Surface	1,175	2.3	1911	Sun Oil Co., Mary Lasater well B-4 (oil) - 1946.	No oil production reported, 1966. Condensate reported from gas wells.	102
103	2	Hainesville...	Wood 6	Wesley Tollett, A-515.	10 SSE Quitman....	(4/)	1,555	8.3	1927	British American Oil Producing Co., Wisenhunt well 1 (oil) - 1956.	LPG storage; reportedly largest interior Texas dome. Oil wells, abandoned.	103
104	46	Hawkinsville..	Matagorda 3	T. Williams, A-107.	19 ESE Bay City...	95	450	4.4	1925	Gulf Production Co., Craig well 1 (cap rock) - 1925.	Oil wells, abandoned.	104
105	35	High Island...	Galveston 3	Martin Dunman......	36 NE Galveston...	150	1,300	2.1	1901	Chase and Newsome et al. drilled cap rock - 1901.	-	105
106	17	Hockley.......	Harris 3	Thos. Coghill......	31 NW Houston.....	74	1,000	5.9	1902	Patillo Higgins, John Warren well 2 (salt) - 1906.	Salt mine.	106
107	41	Hoskins Mound.	Brazoria 3	Henry Austin.......	10 ENE Angleton...	574	1,150	.8	1904	Mound Oil Co., well drilled salt - 1904-07.	Sulfur operations abandoned. Oil wells, abandoned.	107
108	29	Hull..........	Liberty 3	J. Devore..........	10 NE Liberty.....	260	595	2.6	1908	Sun Oil Co. fee well 1 (salt) - 1908.	LPG storage.	108
109	22	Humble........	Harris 3	W. B. Adams........	18 NNE Houston....	700	1,200	9.8	1902	Beatty fee well 2 (cap rock) - 1905.	-	109
110	11	Keechi........	Anderson 6	W. A. Cook, A-225..	6 NW Palestine....	(4/)	1,361	1.2	1891	Producers Oil Co., Barrett and Greenwood well 1 (salt) - 1916.	Showings of oil in discovery well.	110
111	30	Liberty, South	Liberty 3	M. G. White, A-117.	3 S Liberty.......	275	480	5.4	1901	Liberty-Marble Falls Oil Co., Bullard-Wilson fee well 3 (cap rock) - 1905.	-	111
112	36	Long Point....	Fort Bend 3	H & T C RR, sec. 66	15 S Richmond.....	550	930	2.8	1924	Gulf Production Co., Davis well 1 (cap rock) - 1924.	Sulfur productive. First salt dome discovered on coast by torsion balance.	112
113	48	Markham.......	Matagorda 3	Wm. Hadden, A-194..	12 W Bay City.....	1,380	1,417	1.9	1908	Structure proved 1908 (17 wells drilled).	LPG storage	113
114	32	Moss Bluff....	Liberty and Chambers 3	J. McFadden, A-76..	10 SSE Liberty....	591	1,160	11.2	1926	Union Sulphur Co., Kyle well 1 (cap rock) - 1926.	Sulfur productive.	114
115	4	Mount Sylvan..	Smith 6	Philip Lively, A-578.	8 WNW Tyler.......	(4/)	613	2.9	1927	Humble Oil & Refining Co., L. V. Reese well A-1 (salt) - 1931.	-	115

1/ Salt volume estimated from top of salt to a depth of 10,560 feet.
2/ Salt dome indicated by surface expression or geophysical test.
3/ Year first well penetrated salt, cap rock, or discovered oil or gas.
4/ No cap rock reported or information not available.

Table A-3. Texas onshore domes with depth to top of salt 3,000 feet or less (continued).

Reference No.	Map index	Dome	Location: County and district		Location: Survey	Location: Miles from county seat	Depth to cap rock, feet	Depth to salt, feet	Salt volume, cu mi 1/	Year first known 2/	Discovery well and year 3/	Remarks	Reference No.
116	40	Nash..........	Fort Bend and Brazoria	3	H. N. Cleveland, A-150.	21 SSE Richmond....	620	950	2.0	1924	Rycade Petroleum Co., Wisdom well (cap rock) - 1924.	Sulfur operations abandoned. First important oil discovery by geophysics.	116
117	14	Oakwood.......	Freestone and Leon	5	T. J. Teal, A-774..	17 SE Fairfield....	703	800	2.9	1927	Roxana Petroleum Corp., Marshall well 1 (cap rock) - 1928.	-	117
118	18	Orchard.......	Fort Bend	3	James Frazier......	11 W Richmond......	285	375	1.2	1924	Gulf Oil Corp., J. M. Moore well 1 (cap rock) - 1924.	Sulfur productive. First salt dome on Gulf Coast discovered by seismic survey.	118
119	49	Palangana.....	Duval	4	G H & N RR, A-249..	11 SW San Diego....	350	500	3.6	1916	Sinclair Oil & Gas Co., Schallert well 1 (salt) - 1917-18.	Brine production; sulfur operations abandoned. Oil wells, abandoned.	119
120	12	Palestine.....	Anderson	6	Elias G. Myers, A-48.	6 SW Palestine.....	120	122	3.1	1890	Several wells drilled shallow rock salt - about 1900.	Salt produced from brine wells on dome during Civil War.	120
121	50	Piedras Pintas	Duval	4	A. J. Bryant.......	14 SSW San Diego...	385	1,350	2.6	1900	Producers Oil Co., Mabee well 1 (salt) - 1904-05.	-	121
122	21	Pierce Junction	Harris	3	I & G N RR, A-933..	4 S Houston........	630	950	1.3	1901	Patillo Higgins wells proved structure - 1902.	LPG storage; brine production.	122
123	26	Saratoga.......	Hardin	3	J. F. Cotton.......	12 WSW Kountze.....	1,500	1,900	2.1	1901	Structure proved early 1900's (cap rock).	-	123
124	28	Sour Lake......	Hardin	3	S. Jackson, A-34...	15 S Kountze.......	660	719	1.8	1901	Salt found in Davis and Gilbert wells about 1903.	LPG storage; minor oil production before 1900. First Tex. oil refinery - 1896.	124
125	27	Spindletop.....	Jefferson	3	J. A. Veatch.......	3 SE Beaumont......	700	1,200	1.8	1890	Higgins Fuel & Oil Co. fee well 3 (salt) - 1901-02.	Sulfur productive. First major oil production from a salt dome in Texas.	125
126	3	Steen..........	Smith	6	M. Carmona, A-3....	10 N Tyler.........	75	300	1.9	1874	Lestergett, Scott heirs wells (salt) - 1902.	Salt produced from brine wells on dome during Civil War.	126
127	42	Stratton Ridge.	Brazoria	3	J. E. Groce........	8 S Angleton.......	850	1,250	9.6	1913	J. Dannenbaum, Mrs. J. J. Stratton well 1 (cap rock) - 1913.	LPG storage, brine production.	127
128	5	Tyler, East....	Smith	6	John Lucky, A-584..	2 NE Tyler.........	800	890	4.3	1927	Humble Oil & Refining Co., Lassiter well 1 (salt) - 1928.	LPG storage.	128
129	39	West Columbia..	Brazoria	3	Tennille League....	17 W Angleton......	650	750	.8	1901	Exploration wells drilled cap rock - 1902.	-	129
130	6	Whitehouse.....	Smith	6	Nancy Childs, A-209	6 S Tyler..........	485	2,000	2.5	1927	Humble Oil & Refining Co., Van Hovenberg et al., well 1-A (cap rock) - 1928.	-	130

1/ Salt volume estimated from top of salt to a depth of 10,560 feet.
2/ Salt dome indicated by surface expression or geophysical test.
3/ Year first well penetrated salt, cap rock, or discovered oil or gas.

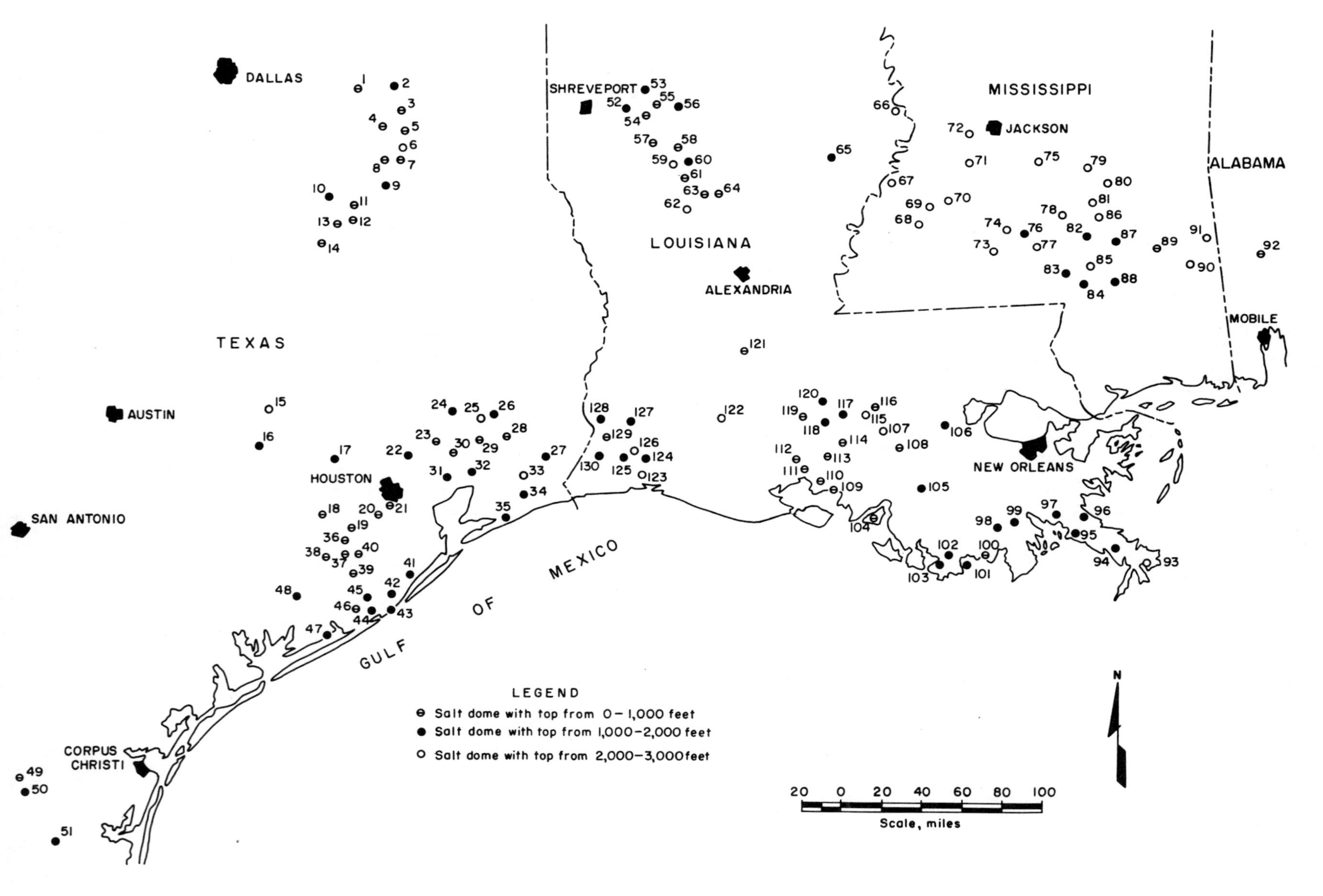

FIG. 2. Onshore domes with depth to top of salt less than 3,000 feet.

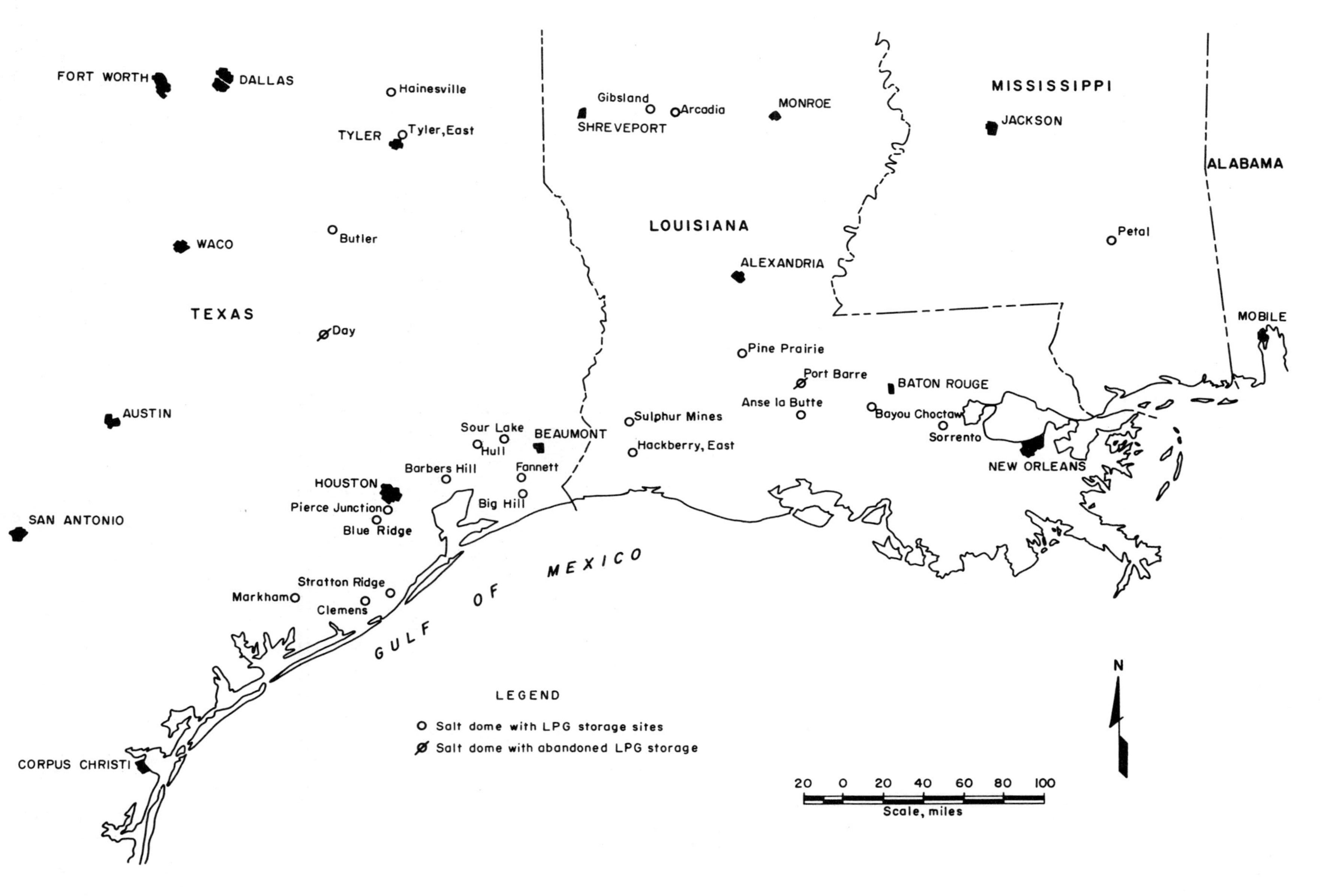

FIG. 3. Domes with salt mining operations.

Gulf Coast region has increased substantially in recent years. Statistics pertaining to salt in storage or stock on hand are not assembled; thus, statistics of salt used or sold by producers have been used to indicate produced volumes. Because of the difficulty and the expense of salt storage, salt stocks are small when compared to total produced volumes. Also, many chemical companies (the largest consumers of salt) produce their own supply and can alter production rates to conform to changing requirements.

There are more than 34 major chemical derivatives of salt, and the estimated number of uses for these derivatives exceeds 14,000 (Watt, 1962). Nationwide, about 66 percent of the 1966 salt production was used for chemical feedstock. About 40 percent was used as a raw material for the manufacture of chlorine gas and the coproduct caustic soda (sodium hydroxide). The manufacture of soda ash (sodium carbonate) by the Solvay process consumed about 18 percent of the U. S. salt production.

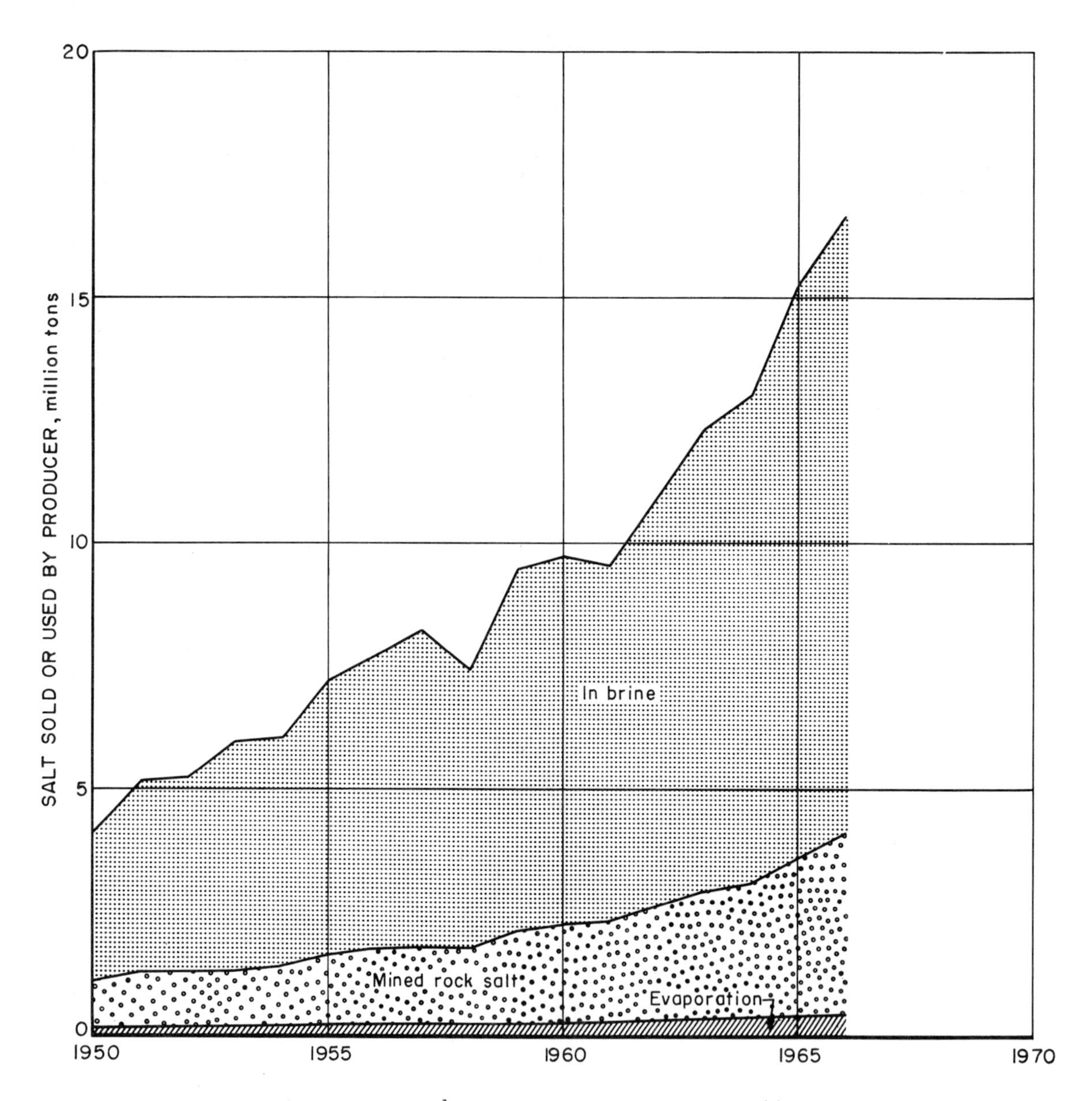

FIG. 4. Gulf Coast salt production, 1950-66.

End-use consumption pattern in the Gulf Coast differed somewhat from the U. S. pattern. Almost 83 percent of the Gulf Coast consumption was used for chemical feedstock. About 65 percent was used in the manufacture of chlorine, 13 percent for soda ash manufacturing, and 5 percent for making other chemicals and soap. Less than 5 percent of the Gulf Coast region salt end use was for snow and ice removal and roadbed stabilization, as compared with 11.5 percent of the U. S. total for the same purposes. Industrial use, including water-softener manufacturers and service companies, consumed about 3 percent of the Gulf Coast salt production, whereas food processing, agriculture, and home uses consumed about 2 percent each.

ECONOMIC IMPORTANCE OF SALT DEPOSITS

During the 106 years since discovery of the Avery Island Dome (1862), more than 330 salt domes have been proved in the Gulf Coast region of Texas, Louisiana, Mississippi, Alabama, and adjacent tidelands. These salt domes and related geologic structures have been of great economic importance, having provided a significant part of this Nation's salt, elemental sulfur, and petroleum. For example, in 1966 production from these domes amounted to about 45 percent of the salt, 85 percent of the elemental sulfur, and about 12 percent of the total crude oil produced in the United States. The economic importance of salt deposits is further emphasized in that about 57 percent of the underground liquid hydrocarbon storage capacity is in Gulf Coast salt domes.

Although the importance of rock salt deposits in the Gulf Coast region has been recognized for many years, substantial growth in salt production has occurred in comparatively recent years. Louisiana and Texas replaced Michigan and New York as the leading salt-producing States in 1959. Salt produced in the Gulf Coast region from 1957 to 1966 is listed in table 2. During the decade ending January 1, 1967, annual salt production in the Gulf Coast region increased from 8.2 million tons having an estimated value of $36.5 million in 1957 to 16.6 million tons having an estimated value of $78.7 million in 1966. Gulf Coast salt production was 34.3 percent of the U. S. production of 23.9 million tons in 1957 and 45.7 percent of the U. S. production of 36.5 million tons in 1966. For 1957 and 1966 the world production of salt was respectively 77.4 and 122.8 million tons. The portion of world salt produced in the Gulf Coast region increased from 10.6 percent in 1957 to 13.5 percent in 1966.

The market value of 1966 salt production in Louisiana was second only to sulfur in nonfuel mineral production, and in Texas the market value of produced salt ranked fifth in nonfuel mineral production. Because of increasing demands for brine as a feedstock for the chemical industry, principally the manufacture of chlorine gas and the coproduct caustic soda, the recovery of salt by solution-mining methods in 1966 was almost double the 1957 amount. The economic importance of salt production is expected to increase as technological advances in chemical derivatives are developed.

Table 2. Salt produced in the Gulf Coast region, thousand short tons.

Year	Evaporated	Mined rock salt	In brine	Total	Value, millions
1950	161	1,032	2,937	4,130	$ 9.8
1951	207	1,168	3,764	5,139	11.7
1952	210	1,125	3,881	5,216	12.2
1953	233	1,094	4,652	5,979	14.3
1954	233	1,231	4,566	6,030	20.5
1955	227	1,455	5,541	7,223	28.5
1956	235	1,552	5,960	7,747	32.3
1957	255	1,556	6,406	8,217	36.5
1958	248	1,557	5,636	7,441	34.6
1959	272	1,852	7,366	9,490	39.0
1960	294	2,002	7,419	9,715	40.8
1961	298	2,036	7,236	9,570	41.3
1962	338	2,267	8,342	10,947	47.2
1963	356	2,589	9,381	12,326	53.3
1964	358	2,799	9,833	12,990	65.4
1965	365	3,289	11,620	15,274	73.3
1966	387	3,768	12,493	16,648	78.7

Industrial growth in the Gulf Coast region has been accompanied by increasing demands for salt. Because of the wide variety of usages, abundance of supply, and intense competition of producers, prices for salt have remained relatively stable. The average price of salt produced in the Gulf Coast region was $4.44 per ton in 1957 as compared with $4.73 per ton in 1966. Recent prices per ton reported to the Bureau of Mines ranged from $23.77 to $28.67 for evaporated salt, from $5.62 to $13.72 for rock salt, and the average price for brine was $3.65 per ton. Because of the bulkiness of rock salt and the fact that it is necessary to transport about 3 tons of water for 1 ton of salt in brine, the costs of transportation and distribution become significant in determining the extent of marketing areas.

Salt recovery operations (Fig. 3) are principally in the areas of nearby industrial complexes and in areas accessible to water transportation

Table 3. Summary of LPG storage operations on Gulf Coast salt domes, 1967.

Number	Dome	County or parish	Company	Address	Capacity (M bbl)	Materials stored
			TEXAS			
1	Barbers Hill.....	Chambers....	Humble Pipe Line Co................	Box 2180, Houston, Tex.........	1,400	Ethylene.
	...do...........	...do......	...do..............................	...do..........................	900	Isobutane.
	...do...........	...do......	Sinclair Petrochemical Co..........	Petroleum Bldg., Houston, Tex...	2,500	NGL.
	...do...........	...do......	...do..............................	...do..........................	650	Ethylene.
	...do...........	...do......	Tenneco Chemical Co................	Tennessee Bldg., Houston, Tex...	4,171	Butane-propane.
	...do...........	...do......	Texas Eastern Transmission Corp.....	Box 1189, Houston, Tex.........	9,137	Do.
	...do...........	...do......	Warren Petroleum Corp..............	Box 14121, Houston, Tex........	12,344	NGL.
2	Big Hill........	Jefferson...	Union Oil Co. of California........	Box 2107, Fort Worth, Tex......	314	Butane.
	...do...........	...do......	...do..............................	...do..........................	326	Propane.
3	Blue Ridge.......	Fort Bend...	Tuloma Gas Products Co. (Subsidiary of Standard Oil Co. of Indiana).	Americana Bldg., Houston, Tex...	300	Do.
4	Butler..........	Freestone...	Northern Propane Gas Co. (United Transport and Storage Div.).	Box 600, Corsicana, Tex.........	400	Do.
	...do...........	...do......	...do..............................	...do..........................	175	Butane-propane mix.
5	Clemens.........	Brazoria....	Phillips Petroleum Co..............	Bartlesville, Okla.............	7,400	Light hydrocarbons.
6	Day.............	Madison.....	Union Oil Co. of California........	Box 2107, Fort Worth, Tex......	150	Propane[1].
7	Fannett.........	Jefferson...	Gulf Oil Corp......................	Box 14121, Houston, Tex........	650	LPG.
8	Hainesville......	Wood........	Enterprise Petroleum Gas Co........	Hainesville, Tex...............	600	Propane.
	...do...........	...do......	...do..............................	...do..........................	400	Butane.
9	Hull............	Liberty.....	Mobil Oil Co. (Div. of Socony Mobil Oil Co.).	Box 177, Houston, Tex..........	759	Propane.
	...do...........	...do......	...do..............................	...do..........................	1,071	Butane.
	...do...........	...do......	...do..............................	...do..........................	743	Isobutane.
	...do...........	...do......	...do..............................	...do..........................	310	LPG.
	...do...........	...do......	...do..............................	...do..........................	238	Propylene.
	...do...........	...do......	...do..............................	...do..........................	754	Ethylene.
10	Markham.........	Matagorda...	Seadrift Pipeline Corp. (Union Carbide Co.).	Bank of Southwest Bldg., Houston, Tex.	500	LPG.
	...do...........	...do......	...do..............................	...do..........................	1,500	Do.
	...do...........	...do......	...do..............................	...do..........................	700	Ethylene.
11	Pierce Junction..	Harris......	Wanda Petroleum, Inc...............	Melrose Bldg., Houston, Tex.....	1,000	Propane.
	...do...........	...do......	...do..............................	...do..........................	3,000	Butane.
12	Sour Lake........	Hardin......	Texaco, Inc........................	Box 2332, Houston, Tex.........	833	Do.
13	Stratton Ridge...	Brazoria....	Dow Chemical Co....................	Drawer K, Freeport, Tex........	1,000	Olefins, LPG.
14	Tyler, East......	Smith.......	Texas Eastman Co...................	Longview, Tex..................	2,400	Ethylene, propane, propylene.
			LOUISIANA			
15	Anse la Butte....	St. Martin..	Union Texas Petroleum Co. (Div. of Allied Chemical Corp.).	Saratoga Bldg., New Orleans, La.	1,000	Butane.
	...do...........	...do......	...do..............................	...do..........................	1,000	Propane.
	...do...........	...do......	...do..............................	...do..........................	100	Gasoline.
	...do...........	...do......	Wanda Petroleum, Inc...............	Melrose Bldg., Houston, Tex.....	800	Propane.
	...do...........	...do......	...do..............................	...do..........................	500	Butane.
	...do...........	...do......	...do..............................	...do..........................	800	Isobutane.
16	Arcadia.........	Bienville...	Tuloma Gas Products Co. (Subsidiary of Standard Oil Co. of Indiana).	Highway 80, Arcadia, La.........	500	Butane-propane.
	...do...........	...do......	Warren Petroleum Corp..............	Box 1590, New Orleans, La.......	450	LPG.
17	Bayou Choctaw....	Iberville...	Dow Chemical Co....................	Midland, Mich. (Home office)....	600	Ethylene-LPG.
18	Gibsland........	Bienville...	Hercules Petroleum Corp............	Gibsland, La...................	480	Propane.
19	Hackberry, East..	Cameron.....	Cities Service Oil Co..............	Cities Service Bldg., Bartlesville, Okla.	690	Butane.
	...do...........	...do......	...do..............................	...do..........................	680	Isobutane.
	...do...........	...do......	...do..............................	...do..........................	920	Propane.
20	Pine Prairie.....	Evangeline..	Continental Oil Co.................	Box 2197, Houston, Tex.........	930	Do.
21	Port Barre.......	St. Landry..	Warren Petroleum Corp..............	Box 14121, Houston, Tex........	153	LPG[1].
22	Sorrento.........	Ascension...	Humble Oil and Refining Co.........	Box 988, Baton Rouge, La........	1,100	Butane.
	...do...........	...do......	...do..............................	...do..........................	450	Isobutane.
	...do...........	...do......	...do..............................	...do..........................	870	Propane.
	...do...........	...do......	...do..............................	...do..........................	1,000	Ethylene.
	...do...........	...do......	Shell Oil Co.......................	Box 127, Metairie, La..........	550	Butane.
	...do...........	...do......	Texaco, Inc........................	P. O. Box 2332, Houston, Tex....	400	Propane.
	...do...........	...do......	...do..............................	...do..........................	350	Butane.
	...do...........	...do......	...do..............................	...do..........................	200	Refinery intermediates.
23	Sulphur Mines....	Calcasieu...	Runnels Gas Products Corp..........	Basile, La.....................	1,150	Ethylene.
24	Venice..........	Plaquemines.	Getty Oil Co.......................	Box 1404, Houston, Tex.........	228	Butane-propane.
	...do...........	...do......	Gulf Oil Corp......................	Box 14121, Houston, Tex........	2,500	NGL.
			MISSISSIPPI			
25	Petal...........	Forrest.....	Mobil Oil Co. (Div. of Socony Mobil Oil Co.).	Box 177, Houston, Tex..........	1,012	Propane.
	...do...........	...do......	Shell Oil Co.......................	Box 127, Metairie, La..........	1,800	Do.
	...do...........	...do......	Union Texas Petroleum Co. (Div. of Allied Chemical Corp.).	811 Rusk Ave., Houston, Tex.....	500	Do.
	...do...........	...do......	...do..............................	...do..........................	600	Butane.
	...do...........	...do......	Warren Petroleum Corp..............	Box 1269, Petroleum Bldg., Jackson, Miss.	2,000	LPG.

1/ Storage operations are inactive.

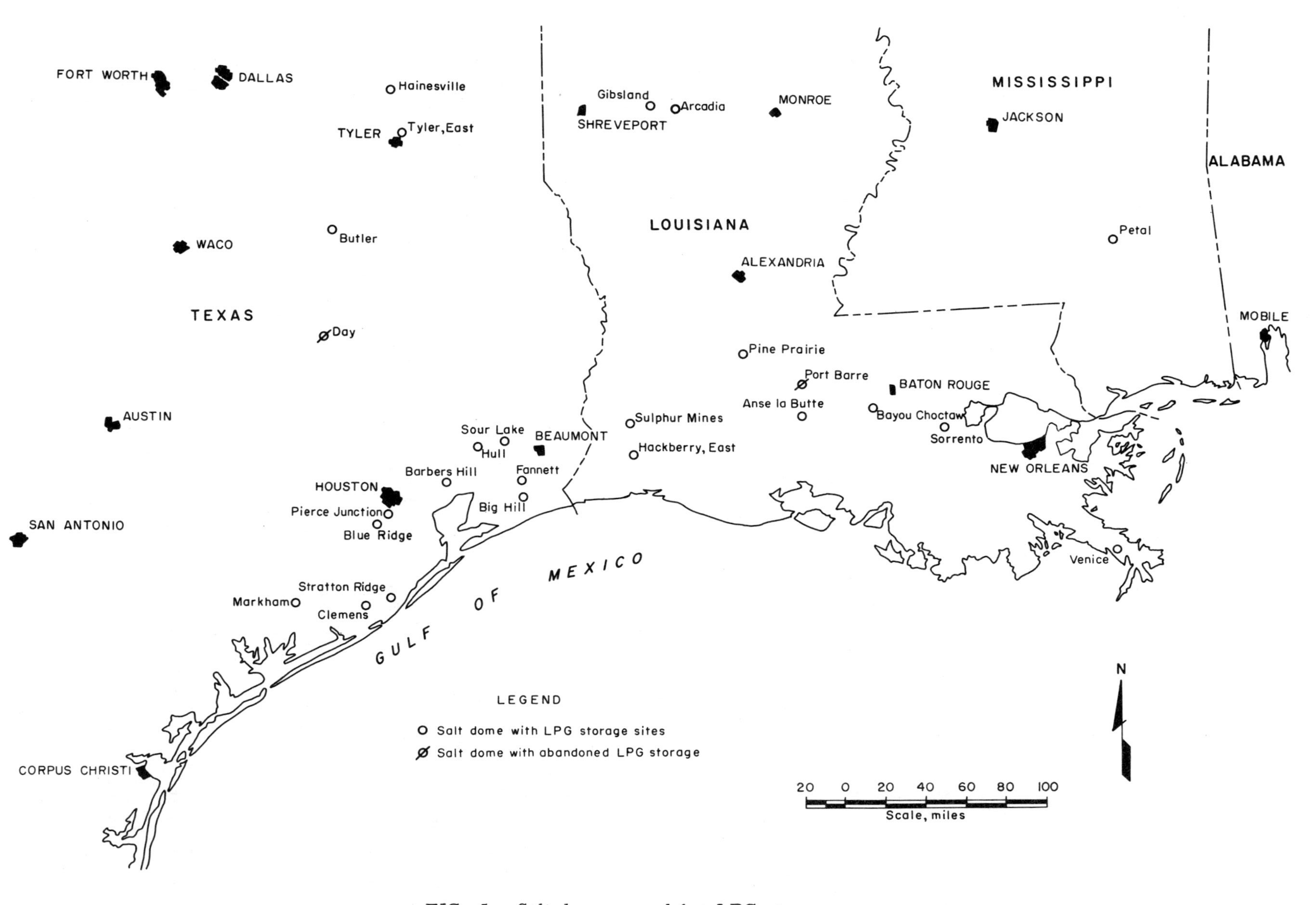

FIG. 5. Salt domes used for LPG storage.

facilities. Normally, transportation by water is less expensive than by rail or truck; therefore, larger distribution areas can be serviced competitively.

For Federal tax purposes there is a 10 percent depletion allowance permitted for salt. The Tariff Classification Act that became effective on August 31, 1963, set the duty on U. S. salt imports as follows: Brine, 10 percent ad valorem; bulk, $0.017 per 100 pounds; and other (packaged), $0. 035 per 100 pounds. As a result of the tariff rates established by the Kennedy Round Trade Agreement, tariffs have been lowered on salt imports. Effective January 1, 1968, tariff rates became: Brine, 9 percent ad valorem; bulk, $0. 015 per 100 pounds; and other (packaged), $0.025 per 100 pounds. Largest salt imports are from Canada ($1. 3 million tons in 1966). About half of the import from Canada is brine received through the Michigan customs district.

The use of cavities created in salt domes for underground LPG storage has increased substantially in the number of projects and the volume of storage. From inception of underground storage operations in 1951, storage capacity increased to about 2 million barrels in 1954 and to approximately 98 million barrels in 1964 (Bizal, 1963). By the end of 1967 the U. S. underground storage capacity had increased to 141 million barrels, about two-thirds of which was in the Gulf Coast States. Texas with 73. 9 million barrels ranked first, and Louisiana with 18. 8 million barrels ranked third (Kennedy, 1967). About 80 million barrels, or approximately 57 percent, of the U. S. underground storage capacity was in the Gulf Coast salt domes. It is estimated that above-ground steel storage capacity of 80 million barrels could have cost as much as $2 billion, as compared with a cost of less than $25 million for the present underground storage capacity. Figure 5 shows the location of domes having LPG storage operations. Table 3 is a listing of domes, operators, and underground storage capacity in the Gulf Coast region.

Two of the largest underground storage operations are being conducted at the Barbers Hill salt dome in Chambers County, Texas. Capacity of these two storage complexes is equivalent to one-seventh of total underground liquid hydrocarbon storage in the United States.

Changes in production and consumption rates have alleviated the need for underground LPG storage in some areas. As a consequence, storage facilities for the Day Dome in Texas and the Port Barre Dome in Louisiana are not being used.

Underground storage of liquid hydrocarbons in salt layers has also been satisfactory. In the Panhandle area and West Texas there are 60 active storage operations in underground salt layers in 19 counties. Storage capacity in salt-layer operations at the end of 1967 was about 17.3 million barrels, or almost one-fourth of the entire underground LPG storage capacity in the State. Because the storage cavities in salt layers are usually smaller than in domes, the average cost per barrel of storage in layers is considerably higher than in domes.

OUTLOOK

The growth rate of salt production in the United States from 23.9 million short tons in 1957 to 36.5 million tons in 1966 averaged 4.8 percent per year and was slightly higher than the growth rate of the gross national product. For the last half of the decade (1962-66) the growth rate was 6.2 percent which, projected to 1980, indicates U. S. production in that year of approximately 75 million tons. Comparable salt production in the Gulf Coast States should amount to 50 million tons.

Before 1955 the manufacture of soda ash (sodium carbonate, Na_2CO_3) by the Solvay process was the principal use for salt (Kerns, 1965). Recent increases in the demand for chlorine (8.6 million tons in 1957 to 15. 4 million tons in 1966) has increased chlorine-caustic manufacture 6. 8 percent per year. The largest percentage growth in salt use (11. 8 percent per year) was for snow and ice removal.

REFERENCES

Bizal, R. B. (1963) New caverns boost liquefied-gas storage capacity 10-1/2%: Oil and Gas Jour., vol. 61, no. 42, Oct. 21, pp. 104-115.

Fisher, W. L. (1965) Rock and mineral resources of East Texas: Univ. Texas, Bur. Econ. Geology Rept. Inv. 54, pp. 307-320.

Hawkins, M. E., and Jirik, C. J. (1966) Salt domes in Texas, Louisiana, Mississippi, Alabama, and offshore tidelands: A survey: U. S. Bur. Mines Inf. Circ. 8313, 78 pp.

Kennedy, J. L. (1967) LP-gas storage capacity sets record year around the globe: Oil and Gas Jour., vol. 65, no. 42, Oct. 16, pp. 119-124, 126.

Kerns, W. H. (1965) Sodium, in Mineral facts and problems, 1965: U. S. Bur. Mines Bull. 630, pp. 823-846.

Lang, W. B. (1957) Annotated bibliography and index map of salt deposits in the United States: U. S. Geol. Survey Bull. 1019-J, pl. 4.

Watt, R. S. (1962) A method of analyzing demand for mineral commodities, a case study of salt: U. S. Bur. Mines Inf. Circ. 8057, 35 pp.

PHOSPHATE IN THE ATLANTIC AND GULF COASTAL PLAINS*

James B. Cathcart
U. S. Geological Survey
Denver, Colorado

ABSTRACT

Phosphate pellets are widespread in marine sedimentary rocks of Cretaceous to Holocene age on the Coastal Plains of the Eastern and Southern United States. Economic deposits of phosphorite are confined to the Atlantic Coastal Plain, and are known only in rocks of middle Miocene age or in younger rocks that derived much or all of their phosphate from middle Miocene rocks.

In the Atlantic Coastal Plain, phosphate pellets are particularly widespread from North Carolina to the southern tip of Florida in rocks of middle Miocene age, although some pellets are found in rocks of Cretaceous to Holocene age. In the Gulf Coastal Plain, phosphate is widespread only in rocks of Cretaceous and Paleocene age. Younger rocks contain phosphate pellets only at a few localities where they were probably reworked from older rocks.

Economic phosphorite deposits are in part structurally controlled. All are in basins on the flanks of positive areas that were rising at the time of phosphate deposition; all are on the north or east sides of the positive areas, except for those of the land-pebble district of south Florida; all are in positions that suggest that phosphorus could have been supplied by cool, southward-moving near-shore ocean currents. Phosphate was precipitated in the basins when cool water, diverted by the positive areas, was turbulently mixed with warm waters of the Florida Current and the Gulf Stream. The scarcity of phosphate in Tertiary rocks of the Gulf Coast is probably due to the position of the Floridian Plateau, which diverted currents away from the Caribbean and the Gulf Coast.

Economic deposits of North Carolina, South Carolina, South Georgia--North Florida, and the central peninsula of Florida are similar in gross features. They are weakly consolidated, sedimentary rocks consisting of quartz and phosphate sand, clay, and in minor amounts, limestone and dolomite. The primary phosphate mineral in all of the deposits is a carbonate fluorapatite. Differences in the economic deposits include variation in chemical composition, size distribution of phosphate particles, clay mineralogy, the amount and intensity of leaching, and the formation of aluminum phosphate minerals. Differences seem to vary systematically from north to south.

Reserves are measured in billions of tons of recoverable phosphate particles containing a minimum of 30 percent P_2O_5. Resources of phosphate not minable under present conditions (particularly in the phosphatic limestone of the Hawthorn Formation) have not been measured, but they are likely to amount to scores of billions of tons.

INTRODUCTION

Phosphate pellets are widespread in rocks of all geologic periods from Cretaceous to Quaternary in the Atlantic and Gulf Coastal Plains of the Eastern and Southern United States, but economic or potentially economic phosphate deposits are known only in the Atlantic Coastal Plain in rocks of middle Miocene age or in younger rocks in which much or all of the phosphate was derived from middle Miocene rocks.

The Atlantic Coastal Plain consists of land between the Atlantic Ocean and the crystalline rocks of the Appalachians and extends from Long Island, New York, to the southern tip of Florida. For the purposes of this report, all of Peninsular Florida is considered to be a part of the Atlantic Coastal Plain. The Gulf Coastal Plain extends from the panhandle of Florida to Mexico (Fig. 1), and is generally much wider than the Atlantic Coastal Plain.

GENERAL GEOLOGY OF THE COASTAL PLAINS

The Atlantic and Gulf Coastal Plains are underlain by sedimentary rocks of Cretaceous to Holocene age that generally dip very gently seaward. Cretaceous rocks crop out at the landward edge of the Coastal Plains except where they are

*Publication authorized by the Director, U. S. Geological Survey.

covered by Quaternary deposits in the Mississippi Embayment. Rocks of Cretaceous age are covered seaward by successive sequences of rocks of Tertiary and Quaternary age. In general, rocks of each age group contain coarse clastic sedimentary rocks at the landward edge that in turn give way to finer grained rocks and finally to chemical sediments of carbonate and evaporite in their seaward extension. The Tertiary and Quaternary sediments of the Central Gulf Coastal Plain form a seaward-thickening sedimentary sequence composed largely of deltaic deposits in the Gulf Coast geosyncline (Murray, 1947).

Sedimentary rocks of much of the Coastal Plain are unconsolidated or poorly consolidated, outcrops are poor, and a cover of Pleistocene and Holocene deposits obscures many of the older geologic features. The gentle seaward dips are interrupted by broad, gentle anticlinal or synclinal folds that have northeast or northwest trends (Fig. 1).

Table 1. Known occurrence of phosphate, Gulf and Atlantic Coastal Plains.

Age	State	Stratigraphic unit	Lithology	P_2O_5 content (percent)	References
Holocene and Pleistocene	Florida, Georgia, South Carolina	Bars, floodplains along modern rivers	Sand of quartz grains and phosphate pellets[1/]	Phosphate pellets 25	Mansfield (1942) Rogers (1914)
Pleistocene	South Carolina	Ladson Formation	Clay, quartz sand,[1/] phosphate pellets	Phosphate pellets 21-29	Malde (1959)
Pliocene	Florida	Bone Valley Formation	Clay, sand, phosphate pellets and nodules[1/]	Phosphate grains 30-38	Cathcart (1963a)
	North Florida South Georgia	Unnamed	do.[1/]	Phosphate pellets 20-35	Olson (1966)
Miocene	Florida, Georgia, South Carolina	Hawthorn Formation	Clay, sand, calcite, dolomite, phosphate	Phosphate pellets in weathered rock, 25-35; in fresh rock, 15-25	Olson (1966) Sever et al. (1967) Heron and Johnson (1966) Cathcart (1963b)
	North Carolina	Yorktown Formation	Clay, sand, limestone, some phosphate pellets and granules[1/]	Phosphate pellets 30±	Brown (1958) Kimrey (1965) Gibson (1967) Rooney and Kerr (1967)
	North Carolina	Pungo River Formation	Phosphate pellets, sand, dolomite, clay[1/]	Phosphate pellets 31	do.
	Florida Georgia	Tampa Formation	Limestone, clay, sand, some phosphate pellets	Phosphate pellets 10-20 Whole rock 2-4	Sever et al. (1967) Carr and Alverson (1959)
Oligocene	South Carolina	Cooper Marl	Marl, minor phosphate pellets	Phosphate pellets 10-20 Whole rock 2-9	Malde (1959)
Eocene	North Carolina	Castle Hayne Limestone	Limestone, some phosphate pellets and granules	Phosphate pellets 30 Whole rock, 6-7	Dabney (1885)
	Georgia	Tallahatta Formation	Quartz sand, some phosphate grains[1/]	No data	Owen (1963)
	Alabama	Claiborne Group	Phosphatic shell marl	Phosphate pellets 20	Smith (1892)
Paleocene	Texas	Midway Group	Phosphate pellets in greensands and marls	Phosphate pellets 20 Matrix, 1-2	Weaver et al. (1963)
	Maryland	Brightseat Formation	Phosphate pellets in calcareous sand[1/]	No data	Adams et al. (1961)
Cretaceous	Alabama Mississippi Georgia	Prairie Bluff Formation of Selma Group and equivalents	Phosphate pellets in chalk	Phosphate pellets 30 Matrix, 2-4	Smith (1892) McCallie (1896) Monroe (1941) Stephenson and Monroe (1940)
	Texas	Eagle Ford Formation	Phosphate conglomerate at base of shale[1/]	Phosphate, 24 Whole rock, 5-10	Weaver et al. (1963)

[1/] Reworked, at least in part, from older rocks.

Stratigraphy

A complete discussion of the stratigraphy of the Coastal Plains is beyond the scope of this paper, but summary data on all known occurrences of phosphate pellets in sedimentary rocks of the Coastal Plains are given in table 1, which lists the formations, by age and State, in which phosphate occurs. Formations that contain phosphate are either quartzose or calcareous. The quartzose phosphate deposits are, at least in part, reworked, but the calcareous ones may be primary.

Phosphatic sedimentary rocks in the Gulf Coast are confined to the Paleocene and the Cretaceous, and generally are associated with marls or glauconitic sediments, although some are lag gravels at unconformities (Stephenson, 1929; Weaver and others, 1963). No commercial deposits of phosphate are known in rocks of Tertiary age on the Gulf Coast; known occurrences of phosphate are restricted to scarce pellets found at a few outcrops. A survey of the literature indicates that there is little or no phosphate in rocks of Eocene to Pleistocene age in the Gulf Coast.

North Carolina

The phosphate deposits of North Carolina are in Beaufort County, where phosphorite (table 2) is found only in the subsurface in the Pungo River Formation (Kimrey, 1965). Brown (1958) tentatively assigned a middle Miocene age to the phosphorite on the basis of fossils found in the upper part of the formation, but pointed out that the lower part, which does not contain recognizable fossils, might be older than Miocene, and perhaps as old as Oligocene. Gibson (1967), in a more detailed study of the fossils in the formation, con-

Table 2. Generalized stratigraphy of phosphorite deposits of Tertiary and Quaternary age, Atlantic Coastal Plain.

		NORTH CAROLINA	SOUTH CAROLINA Charleston Area	SOUTH CAROLINA Beaufort County	SOUTH GEORGIA NORTH FLORIDA	SOUTH FLORIDA
Holocene and Pleistocene		Surficial sand	Surficial sand and Ladson Formation (phosphorite)	Surficial sand	Surficial sand	Surficial sand
Pliocene					Unnamed unit (phosphorite)	Bone Valley Formation (phosphorite)
Miocene	Upper	Yorktown Formation (trace phosphorite)				
	Middle	Pungo River Formation (phosphorite)		Hawthorn Formation (phosphorite)	Hawthorn Formation (phosphorite)	Hawthorn Formation (phosphate pellets abundant)
	Lower				Tampa Formation (trace phosphate)	Tampa Formation (trace phosphate)
Oligocene			Cooper Marl (trace phosphate)		Suwanee Limestone	Suwanee Limestone
Eocene	Upper		Jackson, Claiborne, and Wilcox Groups		Ocala Limestone	Ocala Limestone
	Middle	Castle Hayne Limestone (trace phosphate)		Santee Limestone	Avon Park Limestone	Rocks of Claiborne age
References		Gibson (1967)	Malde (1959)	Heron and Johnson (1966)	Olson (1966) Sever, Cathcart, and Patterson (1967)	Cathcart (1963a, b)

firmed the middle Miocene age of the top of the formation, and pointed out that the lower part contained only molluscan molds and phosphatized foraminifera that have greater affinities to Oligocene than to Miocene species.

The Pungo River Formation consists of sand, silt, clay, diatomaceous clay, dolomite, and in the top part of the formation, limestone and shell "hash" beds. Although phosphate particles occur in all the lithologic units, they are abundant only in the more sandy beds. The amount of recoverable phosphate ranges from about 2 percent in some of the clay and dolomite beds to 60 percent in some of the sandy beds.

The formation, which thickens eastward, ranges in thickness from 0 to about 200 feet. The thickening is due to the intercalation in a seaward direction of beds of dolomite and clay. Thus, in a drill hole close to the western margin of the basin, the Pungo River Formation is about 25 feet thick, and consists of a single bed of phosphate and quartz sand, while seaward at the eastern edge of the basin the formation is about 175 feet thick and consists of 5 beds of phosphorite sand, separated by 6 beds of relatively barren clay and dolomite. Individual beds cannot be traced between drill holes that are spaced 2 to 5 miles apart, but the formation can be divided into an upper unit consisting of limestone and shell hash in a distinctively chartreuse clay that Gibson (1967) calls the bioclastic carbonate bed, a middle unit that contains abundant phosphate, and a lower unit that contains only minor phosphate.

The Pungo River Formation unconformably overlies the Castle Hayne Limestone of Eocene age and is unconformably overlain by the Yorktown Formation of late Miocene age.

The surface is mantled by deposits of Pleistocene age (Gibson, 1967).

South Carolina

The phosphate deposit of the Charleston area, South Carolina, is at the base of the Ladson Formation of Pleistocene age (Malde, 1959), which rests unconformably on the Cooper Marl of Oligocene age (table 2). This is the so-called land rock. The river rock (phosphate concentrated as bars along the modern streams) is Holocene (Malde, 1959). The phosphate member of the Ladson Formation consists of gravel, sand and clay. The gravel is composed of irregular pieces of phosphate rock, rounded pebbles of phosphate, internal phosphatized fossil molds, partly phosphatized shell material, bones and fish teeth, and quartz grains. According to Malde (1959), the phosphate is reworked and phosphatized Cooper Marl. Malde did not propose any source for the solutions that phosphatized the Cooper Marl. The Hawthorn Formation of early and middle Miocene age contains abundant phosphate, and is present south of the Charleston area (Heron and Johnson, 1966). Malde said (1959, p. 28) "...it appears that the Hawthorn Formation in the Charleston area has been removed by erosion." The source of the P_2O_5 that phosphatized the upper part of the Cooper Marl probably was phosphate pellets in the Hawthorn Formation.

The phosphate deposit of Beaufort County, South Carolina, is in the Hawthorn Formation (table 2), which is underlain by the Santee Limestone of Eocene age and overlain by surficial materials of Pleistocene age (Heron and Johnson, 1966). The phosphate occurs in sand or clayey sand and is in two beds, an upper bed that averages about 5 feet in thickness, and a lower bed that averages about 15 feet in thickness. The lower phosphate bed, at the base of the formation, rests directly on the Santee Limestone and is separated from the upper bed by 25 to 60 feet of green clayey sand, sandy clay, and thin beds of limestone. The upper bed is overlain by as much as 35 feet of overburden which consists of nonphosphatic beds of the Hawthorn Formation and surficial material of Pleistocene age.

North Florida--South Georgia

Phosphate deposits of the north Florida--south Georgia field are in the Hawthorn Formation and in rocks of Pliocene age (table 2). According to Olson (1966, p. 79), the phosphate deposit at the mine of the Occidental Corporation of Florida is Pliocene in age, but the phosphate probably was reworked from beds of Hawthorn age. Much of the phosphate in this area, however, is in the Hawthorn Formation (Sever, Cathcart, and Patterson, 1967).

Land-pebble district, Florida

The phosphate deposit of the land-pebble district of Florida is in the Bone Valley and Hawthorn Formations (Cathcart, 1963a, b; Altschuler, Cathcart, and Young, 1964).

The Hawthorn Formation of early and middle Miocene age consists of interbedded and lenticular limestone and dolomite, sand, clayey sand, and clay, all containing phosphate pellets. The Hawthorn exposed in the mining pits is a soft, buff to yellow, sandy phosphate-bearing dolomite and dolomitic limestone.

The Bone Valley Formation of Pliocene age consists of two units. A lower unit 0 - 80 feet thick is composed of interbedded and lenticular sand, clay, and clayey sand, all of which contain abundant phosphate particles. An upper unit 0 - 70 feet thick is composed of clayey sand and sandy clay that contains only a few phosphate pellets, mostly concentrated in the basal part of the unit.

The surface is blanketed by loose sand that

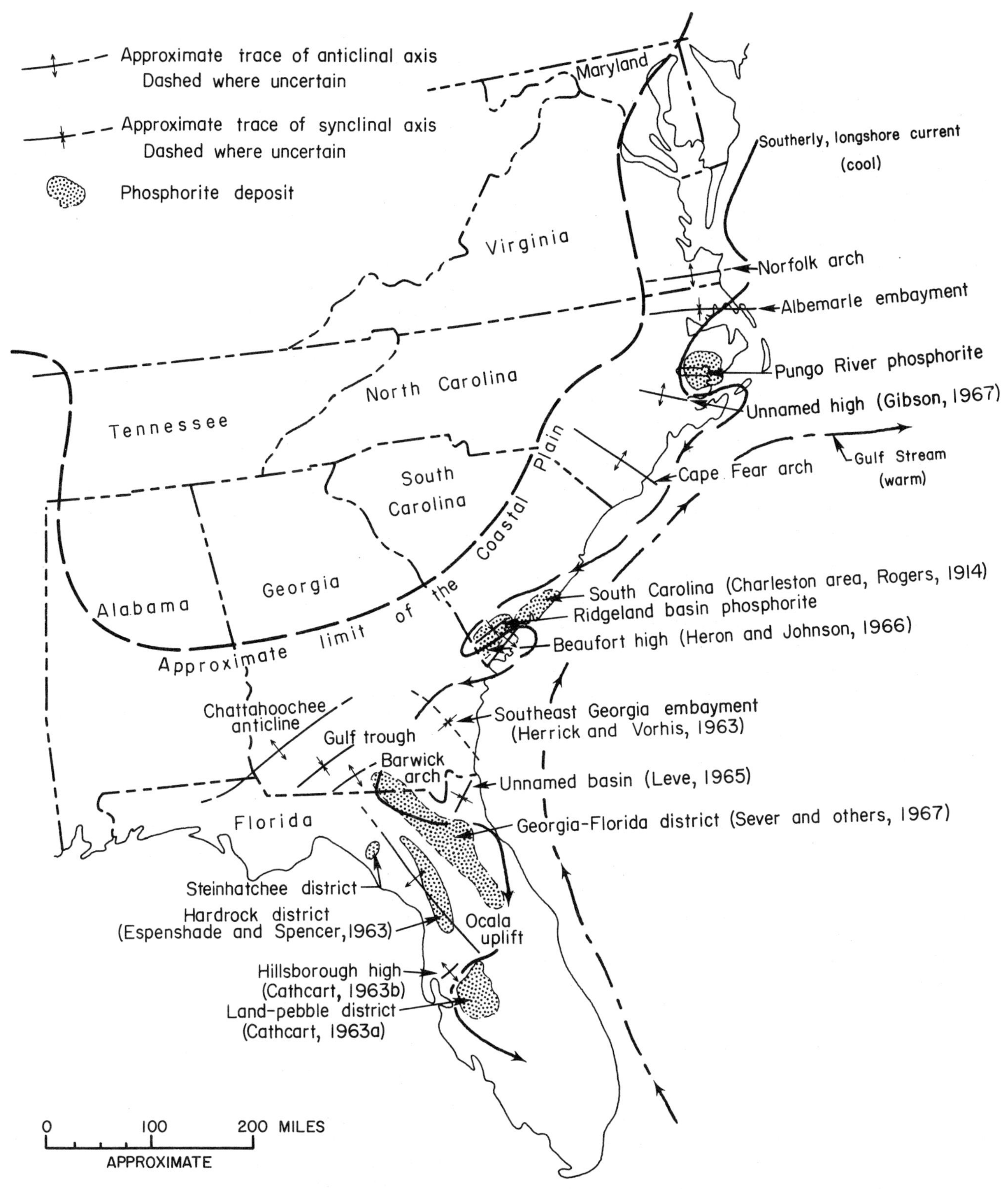

FIG. 1. Sketch map of southeastern United States, showing structure of the Coastal Plain, phosphorite deposits of Tertiary and Quaternary age, and paleocurrents.

contains a few vertebrate fossils of Pleistocene age, and wind-blown sand and swamp deposits of Holocene age.

Structure

Structurally, the Coastal Plain is characterized by beds tilted seaward at such low angles that the dips are expressed in feet per mile. The general seaward dip of the rocks on the Atlantic Coastal Plain is interrupted by a series of very gentle anticlines and synclines that strike northwest or northeast, except in the northern part of the area where the strike gradually becomes more nearly east (Fig. 1).

Phosphate deposits are related to structures in the Coastal Plain (Cathcart and Osterwald, 1954). Each deposit is in a basin on the flank of an anticline that was rising at the time of phosphate deposition. The Pungo River Formation of North Carolina was deposited in a broad shallow bay or basin on the north flank of a rising positive area that is just to the north of the Cape Fear arch. The basin was open to the sea on the east or northeast. The phosphate deposit in Beaufort County, South Carolina, is in the Ridgeland Basin, which probably was open to the sea on the northeast, and is on the flank of the Beaufort high (Heron and Johnson, 1966). Phosphate deposits of the south Georgia--north Florida field are in basins on the east flank of the Barwick arch and the north flank of the Ocala uplift. The basins were open to the sea to the north and east. The phosphate deposits of the land-pebble district of Florida are on the flanks of the Ocala uplift and the Hillsborough high (Cathcart, 1963b).

PHOSPHORITE

Phosphorite deposits consist of an unconsolidated mixture of phosphate pellets, quartz grains, some clay, and minor amounts of limestone and dolomite.

The phosphate pellets are similar in all of the deposits even though there are many forms and sizes of the particles. The forms are phosphatized rock fragments, mostly limestone; internal phosphatized molds of gastropods and pelecypods; phosphatized sharks' teeth; phosphatized fish bones and teeth; other phosphatized bone material, such as the phosphatized manatee bones so abundant in the land-pebble deposits of Florida; rod-shaped grains that may be fecal pellets; phosphatized shell material, foraminiferal in some areas; and rounded ovoid pellets that appear to have formed by precipitation around a nucleus. A few of these particles are oolitic; most, however, have only an outer rim of a color that contrasts with that of the dense, structureless interior. Flattened but rounded ovoid pellets ("button shaped") are common in deposits of the land-pebble district.

The phosphate particles are black, brown, tan, red, green, gray, and white, and all intermediate shades. Some characteristic color differences are noted from north to south. Phosphate in North Carolina generally is dark brown to black; in South Carolina it is brown; in south Georgia it is light brown, gray, and white. Phosphate from the land-pebble district of Florida generally is varicolored. The phosphate from mines on the ridge tends to be dark, but that from the mines of the lower flatwoods and adjacent to the rivers tends to be light. White phosphate pellets are much more common in the southern than the northern areas of the Coastal Plain.

The dark phosphate particles so characteristic of the North Carolina deposit contain abundant organic matter. Dark particles from deposits in other States do not contain substantial amounts of organic matter; their dark color is caused by disseminated fine particles of pyrite or marcasite.

Many of the phosphate pellets are reworked; the pellets are broken and rerounded and have overgrowth rims of later phosphate. Compound pebbles showing several generations of phosphate are especially common in the land-pebble deposits of Florida (Altschuler, Cathcart, and Young, 1964).

Most of the particles are highly polished and well rounded, although normally not spherical except in the deposits of the Pungo River Formation in North Carolina. Particles with dull luster generally are in the weathered zones near the surface of the deposits and have been bleached and altered by solution. White color or dull luster alone, however, is not evidence of weathering and alteration.

Phosphate pellets generally range in diameter from less than 0.1 mm to several centimeters; some phosphate boulders as large as 30 cm in diameter have been found in Florida and South Carolina.

Mineralogy

The phosphate mineral of all of the deposits is a carbonate fluorapatite that has the formula $Ca_{10}(PO_4, CO_3)_6F_{2-3}$ (Altschuler and others, 1958). Although the phosphate mineral is the same in all of the deposits, X-ray diffractometer patterns of the phosphate pellets from the different deposits indicate a slight, but definite change from north to south. The phosphate from North Carolina contains the most carbonate, and that from Florida the least. This change is apparently due to a variation in the content of CO_3 in the apatite molecule. Samples heated to 850-1000°C for 30 minutes and then X-rayed were identical in their patterns, and the pattern was that of a fluorapatite.

The gangue minerals occurring with the phosphate are carbonate, quartz, and clay.

Detrital quartz grains of varying size are common everywhere, but are most abundant in the reworked phosphate deposits. Quartz and phosphate are the only minerals in some reworked deposits. Deposits in North Carolina contain diatoms.

Carbonate minerals are major constituents in all of the primary phosphorites in the Atlantic Coastal Plain. Dolomite is the principal carbonate mineral in the deposits of North Carolina, but calcite is the principal carbonate mineral in deposits elsewhere. Dolomite of possible secondary origin does occur, however, in the Hawthorn Formation of Florida.

Unweathered phosphate deposits contain one or more of the clay minerals illite, montmorillonite, attapulgite, and sepiolite. Kaolinite is a predominant clay mineral where the rocks are weathered.

The clay minerals are different in the various deposits. Illite and montmorillonite are the common clay minerals of the North Carolina phosphorite (Rooney and Kerr, 1967); abundant sepiolite and montmorillonite and minor illite characterize the Hawthorn phosphorite of South Carolina (Heron and Johnson, 1966); montmorillonite and attapulgite are the predominant clay minerals of the south Georgia--north Florida deposits (Olson, 1966); montmorillonite is the principal clay of the phosphate in the land-pebble district of Florida, but some attapulgite also occurs in the Hawthorn Formation in this area (Altschular, Cathcart, and Young, 1964).

Small amounts of iron minerals, mainly pyrite, hematite, goethite, and limonite, are in all of the phosphate deposits. The iron phosphate minerals--vivianite and dufrenite--occur in weathered material of the deposits of Florida.

Glauconite occurs in trace amounts in many of the deposits, and is found almost entirely within the phosphate pellets.

The heavy minerals which occur in trace amounts in the deposits are chiefly ilmenite, zircon, rutile, staurolite, sillimanite, tourmaline, and garnet.

Clinoptilolite, a zeolite mineral that possibly formed as an alteration of volcanic ash (Rooney and Kerr, 1967), occurs in small amounts in the Pungo River Formation of North Carolina, and has been reported in one sample from South Carolina (Heron and Johnson, 1966).

Aluminum phosphate minerals, wavellite, crandallite, and millisite, are important secondary minerals in the Florida deposits.

Trace Elements

Many metallic and nonmetallic elements are present in phosphate rock in trace amounts, and some of these may have use as byproducts. For example, vanadium is being recovered as a byproduct of the chemical treatment of western phosphate rock; uranium was recovered from phosphoric acid made from Florida rock; fluorine is being recovered as a byproduct of phosphoric acid manufacture in Florida; and rare-earths are present in the Florida phosphate rock in amounts large enough that they are a potential byproduct (Altschuler, Berman, and Cuttita, 1967).

Data on the contents of trace elements in the phosphate rock of the eastern United States are not abundant. Certain elements needed in plant nutrition in trace amounts include boron, copper, iron, manganese, molybdenum, and zinc (Mortvedt, 1967). Semiquantitative spectrographic analyses indicate that all of these elements are present in trace amounts in samples of phosphorite from the land-pebble field and from the south Georgia--north Florida field.

Uranium is present in all the marine phosphate deposits in sufficient quantity that the deposits can be delineated by the use of gamma-ray detection devices. Uranium in phosphate has been thoroughly investigated only in the deposits of Florida and the western United States, but some work has been done on the other deposits of the Atlantic Coastal Plain. In general, in south Florida, the uranium content varies with particle size (Cathcart, 1956); coarser particles contain more uranium than the finer particles, and Altschuler, Clarke, and Young (1958) pointed out that those coarser particles that are compound, and show evidence of several reworkings in a marine environment, contain much more uranium than simple pellets that were not reworked. The relation of uranium to P_2O_5 content is not simple but, in general, is inverse--that is, coarse particles contain high uranium and low P_2O_5, whereas fine particles contain low uranium and high P_2O_5. When sized particles are compared, however, the relation is direct, but weak. Uranium came from sea water, and coarser particles, exposed for longer periods of time or those particles that were reworked, and exposed to sea water more than once, contain more uranium than the simple particles that were never reworked. For example, particles in the Hawthorn Formation that were not reworked, and that have not been weathered, contain much less uranium than particles of the same size that were reworked into the Bone Valley Formation.

Uranium content of the aluminum phosphate zone is very high because of secondary enrichment from solutions which have leached apatite and uranium from above.

Uranium content of the phosphate deposits in North Carolina is uniform, and relatively low--about the same as the uranium content of the phosphate pellets from the Hawthorn Formation and much less than the uranium in the phosphate of the Bone Valley Formation.

Uranium is present in phosphate pellets of the other deposits, but there are too few analyses to allow any generalizations.

ORIGIN

Phosphate was precipitated in marine basins on the flanks of anticlines that were rising from a relatively stable shelf. The position of the deposits suggests that phosphorus was derived from currents that moved southward along the Atlantic Coast. These cool, temperate waters (Gibson, 1967), diverted by the rising anticlinal structures of the Atlantic Coastal Plain (Fig. 1), were turbulently mixed with the warmer waters of the Gulf Stream, causing phosphate to precipitate. Most of the phosphate was precipitated during middle Miocene time. Younger deposits were derived from the reworking of the phosphate particles in the middle Miocene rocks. Phosphate pellets in the surficial materials under the Atlantic Ocean today are thought to have been derived from Miocene rocks (Pilkey and Luternauer, 1967).

Phosphate deposits are not present in the Gulf Coast. The deposits of the Central Gulf Coastal Plain are deltaic (Murray, 1947), an unfavorable environment for phosphate. Other sedimentary deposits, particularly in the Western Gulf Coastal Plain, are similar to sediments of the Atlantic Coastal Plain, but contain no phosphate because cooler waters containing phosphorus were diverted to the south and east by the Floridian Plateau.

Pevear (1966) has suggested that phosphorites of the eastern United States were deposited in an estuarine environment. The nutrient-laden waters typical of estuaries may have been the loci for the start of precipitation of phosphate. Certainly a part of the land-pebble deposit of Florida was deposited in an estuary. The estuarine environment, however, cannot account for the vast, widespread phosphorite of the middle Miocene throughout the Atlantic Coastal Plain, and conversely, cannot account for the lack of phosphate in the rocks of the Gulf Coast, some of which must have been deposited in estuarine environments.

Both Gibson (1967) and Rooney and Kerr (1967) attach significance in the origin of the phosphorite of North Carolina to clinoptilolite derived from volcanic ash. Rooney and Kerr (1967, p. 747) suggest that: "Widespread ash falls of long duration killed large numbers of marine organisms whose subsequent decay contributed phosphate." Neither report, however, gives any data regarding the amount of clinoptilolite. Recent work by the writer shows that clinoptilolite occurs in small amounts and only in the fine fraction (-200 mesh) of the phosphorite. The -200 mesh fraction averages 16 percent, by weight, of the total rock, and the amount of clinoptilolite in the -200 mesh fraction (estimated from X-ray diffraction studies) ranges from a trace to about 30 percent. Thus, the amount of clinoptilolite in the whole rock ranges from a trace to about 5 percent.

Elsewhere in the Atlantic Coastal Plain, clinoptilolite has been reported only from one sample from South Carolina (Heron and Johnson, 1966). It has not been reported from offshore deposits in the Atlantic, nor has it been observed in any X-ray patterns made by the writer from other deposits in the Atlantic Coastal Plain.

It seems likely that the very small amount of clinoptilolite in the North Carolina phosphorite deposit is merely coincidental and has no relation to the origin of the phosphate deposit.

ECONOMIC GEOLOGY

Beneficiation

The phosphorite deposits, as mined, require beneficiation to make a commercial product. The mined material, called matrix, is disaggregated and screened, yielding three size fractions. The -0.1 mm fraction is a waste product called slime. The -1 mm + 0.1 mm fraction is the flotation feed consisting of phosphate pellets and quartz grains. The quartz and phosphate particles are separated by flotation into an economically valuable phosphate concentrate and a quartz sand tailing. The +1 mm fraction, called pebble, may be an economic product, as in the land-pebble district of Florida where the P_2O_5 content of the fraction meets specifications of commercial grade.

Pebble Fraction

The pebble fraction varies in amount and composition, and the composition of this fraction is important in the economic geology of the deposits. The pebble fraction of the Pungo River Formation of North Carolina consists of phosphate and quartz grains and calcite or dolomite rock fragments. The pebble fraction ranges from 7 to about 30 percent P_2O_5, and averages about 23 percent. Both acid insoluble and CaO contents are high in the pebble fraction, and added together have an inverse relation with P_2O_5 content.

The pebble fraction of the phosphate deposits

of the Charleston area, South Carolina, consists of large, irregular, rounded phosphatized marl fragments. The P_2O_5 content of the material produced ranged from 21 to about 29 percent, and averaged about 25 percent. Analytical data (Malde, 1959) indicate that silica is an important diluent, and that there are some dolomite and calcite.

The pebble fraction of the phosphate deposit of Beaufort County, South Carolina, consists of phosphate particles, quartz grains, and fragments of clay, limestone, and dolomite. Clay fragments could be eliminated by more thorough washing and disaggregation, and the pebble fraction would then contain quartz as the principal diluent. Phosphate pellets from the pebble fraction have not been analyzed, but X-ray diffraction traces were made of the hand-sorted phosphate grains. The pebbles are largely apatite, but contain some quartz and a trace amount of clay mineral. The P_2O_5 content of the phosphate grains is probably high--perhaps greater than 30 percent--but the phosphate content of the pebble fraction must vary greatly because of the different amounts of diluents present.

The pebble fraction of the south Georgia--north Florida district consists of phosphate particles, quartz grains, clay, and fragments of carbonate rock. The fraction ranges from 6.3 to 30.8 percent P_2O_5, and averages about 27 percent. Acid insoluble content ranges from 3.1 to 65.7 percent, and averages about 21 percent, and the relation of acid insoluble content to P_2O_5 content is inverse. Carbonate is not an important diluent in the pebble fraction.

The pebble fraction of the land-pebble phosphate district of south Florida is an economic product because it consists almost entirely of phosphate particles. The pebble fraction ranges from 25 to about 37 percent P_2O_5. The principal diluent is quartz, most of which is within the phosphate pellet. X-ray patterns of the separated nodules show that quartz always occurs in trace to minor amounts, and that trace amounts of calcite occur in some samples, but dolomite is not recognizable. To the south of the main land-pebble district, in Manatee, Hardee, and DeSoto Counties, the pebble fraction contains phosphate particles, fragments of sandy dolomite, clay, and minor quartz. The separated phosphate grains are composed of apatite, minor quartz, and some calcite and dolomite. Dolomite is more abundant and occurs in more of the samples in DeSoto County. Thus, the amount of dolomite in the phosphate particles increases southward.

Concentrate Fraction

Phosphate particles in the concentrate fraction of the North Carolina deposit are remarkably uniform in P_2O_5 content. Samples taken from drill holes over the entire basin of deposition ranged from 25-31 percent P_2O_5, and no single sample contained as much as 32 percent P_2O_5. Only a few samples contained less than 27 percent P_2O_5, and all of these samples contained abundant crystalline rhombs of dolomite--accounting for the low phosphate content.

Sample data are too sparse in the deposits of Beaufort County, South Carolina, to permit generalization regarding grade, but a few samples analyzed by X-ray diffraction indicate the concentrate fraction to be composed of apatite, trace amounts of quartz, and a very slight trace of clay. At least some samples of phosphate concentrate are high in P_2O_5 content--probably as much as 35 percent.

Concentrate from potentially economic deposits in south Georgia--north Florida ranges from about 27 to about 35 percent P_2O_5, and averages about 32 percent. According to Olson (1966), light-colored particles are characteristic of sandy, more permeable zones, and are possibly enriched in P_2O_5, as well as leached by ground water. There is some evidence that light-colored particles are higher in P_2O_5 content than darker colored particles; close to the ground surface, light-colored particles have been bleached and altered by weathering.

Complete chemical analyses are not available for the material from the south Georgia--north Florida deposits, but the few samples that were analyzed by X-ray indicate that quartz is the principal diluent in the particles. Iron and alumina, reported as total Fe_2O_3 plus Al_2O_3, together average about 2.6 percent, and their combined content more or less uniform, so that there is no relation between these elements and P_2O_5 content. Silica is the principal diluent in the pellets, although carbonate may be present in small amounts.

The concentrate fraction of the land-pebble deposits of south Florida ranges from 25 to about 37 percent P_2O_5, but the phosphate content varies with the stratigraphic position (Cathcart, 1964). The phosphate content of pellets in the Hawthorn Formation is generally lower than the content in the Bone Valley Formation, because of a greater amount of diluent in the pellets. In the Hawthorn pellets, calcite is an important diluent, and dolomite is present. Carbonate material is virtually absent in the Bone Valley pellets, where quartz is the principal diluent. Clay and iron oxide minerals are also important diluents in some of the pellets.

The concentrate fraction is distinctly higher in P_2O_5 content than the pebble fraction in the south Florida deposits, and this generalization is true for individual drill holes as well as for averages for entire mine areas.

Amount of Phosphate

The amount of recoverable phosphate particles, measured as tons per acre, varies greatly in the different deposits.

In North Carolina, phosphate in the Pungo River Formation ranges from 20,000-30,000 tons per acre, even though the thickness of the formation ranges from 20 to 200 feet.

Not many data are available from the deposits in South Carolina, but the thickness of the phosphate beds ranges from 0 to about 20 feet and the tonnage presumably varies accordingly. In the main part of the basin, however, the thicknesses are more uniform, and in one measured sample, the tonnage of phosphate in the upper bed was about 3,000 tons per acre, whereas that of the lower bed (about twice as thick) was only 2,000 tons per acre. Elsewhere, the bottom bed contains a greater percentage of recoverable phosphate, so the total tonnage of both beds probably ranges from 1,000 to about 15,000 tons per acre.

In the south Georgia--north Florida field, tonnages range from less than 1,000 tons per acre to about 20,000 tons per acre.

Tonnages in the land-pebble district range from less than 1,000 tons per acre to 50,000 tons per acre. The very high tonnages are in locally overthickened areas, such as sinkhole fillings.

The ratio of pebble tonnage to concentrate tonnage in the land-pebble district varies from 0 (no pebble) to infinity (no concentrate). Dominant pebble areas are on subsurface ridges; concentrate is on the flanks of the ridges and in the valleys between (Cathcart and Davidson, 1952; Davidson, 1952). Pebble deposits on the ridges are thought to be due to submarine winnowing--removal of fines from the ridges, and their deposition in lower areas.

Reserves

Reserves in all of the deposits are measured in terms of recoverable tonnage of phosphate particles--either concentrate or pebble or both.

Total reserves in the land-pebble district of Florida are hundreds of millions of tons of high-grade ore and billions of tons of low and medium-grade ore, not necessarily minable under present conditions.

Measured reserves of phosphate in the south Georgia--north Florida field are hundreds of millions of tons, but total resources are probably billions of tons of low and medium grade.

Reserves of phosphate in the Hawthorn Formation of South Carolina are probably scores, perhaps hundreds of millions, of tons, of at least medium-grade material.

Reserves in the Charleston area of South Carolina are scores of millions of tons, but the grade is low, and individual deposits are small.

Reserves in North Carolina are measured in billions of tons of low- to medium-grade phosphate concentrate.

Total resources of recoverable phosphate particles in the Atlantic Coastal Plain, mostly in the concentrate fraction, are measured in scores of billions of tons. Offshore deposits, known to be present from North Carolina south to Florida, are not included in this total, nor are the vast reserves in carbonate rock of the Hawthorn Formation.

Mining and Production

Large capacity draglines are used to mine phosphate from open pits in the Atlantic Coastal Plain. The largest draglines used in Florida have a capacity in excess of 40 cubic yards; they swing from 200-foot booms and have a digging depth of about 65 feet. These draglines can complete a digging cycle in about 1 minute and, thus, can move rock at the rate of about 1 ton per second. The dragline at the mine of the Texas Gulf Sulphur Company in North Carolina has a 70-yard capacity bucket.

In 1966, production of phosphate rock in the United States reached slightly more than 36 million tons--28 million (78%) from Florida and North Carolina, 3 million tons (8%) from Tennessee, and about 5 million tons (14%) from the western United States (Lewis, 1967). The rock sold had an average value of slightly less than $7.00 per ton (Lewis, 1967).

Mining began in 1867, in South Carolina, and in the first full year of operation (June 1867--June 1868) production was about 12,000 tons that had an average value of about $6.00 per ton. Total production of phosphate rock in the eastern United States in past 100 years amounted to about 400 million tons. Production generally has doubled in each decade since mining began.

REFERENCES

Adams, J. K., Groot, J. J., and Hiller, N. W., Jr. (1961) Phosphatic pebbles from the Brightseat Formation of Maryland: Jour. Sed. Petrology, vol. 31, pp. 546-552.

Altschuler, Z. S., Berman, Sol, and Cuttitta, Frank (1967) Rare earths in phosphorites--geochemistry and potential recovery, in Geological Survey Research, 1967: U. S. Geol. Survey Prof. Paper 575-B, pp. B1-B9.

__________, Cathcart, J. B., and Young, E. J. (1964) The geology and geochemistry of the Bone Valley Formation and its phosphate deposits, west-central Florida: Guidebook field trip no. 6, Geol. Soc. America, ann.

meeting, Miami Beach, 68 pp.
__________, Clarke, R. S., Jr., and Young, E. J. (1958) Geochemistry of uranium in apatite and phosphorite: U. S. Geol. Survey Prof. Paper 314-D, 90 pp.

Brown, P. M. (1958) The relation of phosphorites to ground water in Beaufort County, North Carolina: Econ. Geology, vol. 53, pp. 85-101.

Carr, W. J., and Alverson, D. C. (1959) Stratigraphy of middle Tertiary rocks in part of west-central Florida: U. S. Geol. Survey Bull. 1092, 111 pp.

Cathcart, J. B. (1956) Distribution and occurrence of uranium in the calcium phosphate zone of the land-pebble phosphate district of Florida, in Page, L. R., Stocking, H. E., and Smith, H. B., compilers, Contributions to the geology of uranium and thorium by the United States Geological Survey and Atomic Energy Commission for the United Nations International Conference on peaceful uses of atomic energy, Geneva, Switzerland, 1955: U. S. Geol. Survey Prof. Paper 300, pp. 489-494.

__________(1963a) Economic geology of the Keysville quadrangle, Florida: U. S. Geol. Survey Bull. 1128, 82 pp.

__________(1963b) Economic geology of the Plant City quadrangle, Florida: U. S. Geol. Survey Bull. 1142-D, pp. D1-D56.

__________(1964) Economic geology of the Lakeland quadrangle, Florida: U. S. Geol. Survey Bull. 1162-G, 128 pp.

__________ and Davidson, D. F. (1952) Distribution and origin of phosphate in the land-pebble district of Florida: U. S. Geol. Survey TEI-212, issued by U. S. Atomic Energy Comm. Tech. Inf. Service, Oak Ridge, Tenn., 12 pp.

__________ and Osterwald, F. O. (1954) Relation of structure to phosphate deposition (abs.): Amer. Inst. Mining and Metall. Engrs., Ind. Min. Div. Meeting, Tampa, Florida, 1954.

Dabney, C. W., Jr. (1885) The phosphate exploration: North Carolina Agric. Exp. Sta. Ann. Rept. 1884.

Davidson, D. F. (1952) Relation of the "topography" of the Hawthorn Formation to size of phosphate particles in the deposits, and to topography, in the northern part of the land-pebble phosphate field, Florida: U. S. Geol. Survey TEM-337, issued by U. S. Atomic Energy Comm. Tech. Inf. Service, Oak Ridge, Tenn., 17 pp.

Espenshade, G. H., and Spencer, C. W. (1963) Geology of phosphate deposits of northern peninsular Florida: U. S. Geol. Survey Bull. 1118, 115 pp.

Gibson, T. G. (1967) Stratigraphy and paleoenvironment of the phosphatic Miocene strata of North Carolina: Bull. Geol. Soc. America, vol. 78, pp. 631-649.

Heron, S. D., Jr., and Johnson, H. S., Jr. (1966) Clay mineralogy, stratigraphy, and structural setting of the Hawthorn Formation, Coosawatchie district, South Carolina: Southeastern Geology, vol. 7, pp. 51-63.

Herrick, S. M., and Vorhis, R. C. (1963) Subsurface geology of the Georgia Coastal Plain: Georgia Dept. Mines, Mining, and Geol. Inf. Circ. 25, 67 pp.

Kimrey, J. O. (1965) Description of the Pungo River Formation in Beaufort County, North Carolina: North Carolina Div. of Min. Resources Bull. 79, 131 pp.

Leve, G. W. (1965) Ground water in Duval and Nassau counties, Florida: U. S. Geol. Survey open-file report, 102 pp.

Lewis, R. W. (1967) Phosphate rock, calendar year 1966: U. S. Bur. Mines, Mineral Industry Survey, 4 pp.

Malde, H. E. (1959) Geology of the Charleston phosphate area, South Carolina: U. S. Geol. Survey Bull. 1079, 105 pp.

Mansfield, G. R. (1942) Phosphate resources of Florida: U. S. Geol. Survey Bull. 934, 82 pp.

McCallie, S. W. (1896) A preliminary report on a part of the phosphates and marls of Georgia: Georgia Geol. Survey Bull. 5A, 101 pp.

Monroe, W. H. (1941) Notes on deposits of Selma and Ripley age in Alabama: Alabama Geol. Survey Bull. 48, 150 pp.

Mortvedt, J. J. (1967) The secondary and micronutrients: Agronomic aspects: Seminar for Latin American Fertilizer Executives. Tennessee Valley Authority, National Fertilizer Development Center, pp. 57-58.

Murray, G. E. (1947) Cenozoic deposits of central Gulf Coastal Plain: Bull. Amer. Assoc. Petr. Geol., vol. 31, pp. 1825-1850.

Olson, N. K., ed. (1966) Geology of the Miocene and Pliocene Series in the north Florida--south Georgia area: Guidebook, Southeastern Geol. Soc. 12th Ann. Field Conf., 94 pp.

Owen, Vaux, Jr. (1963) Geology and ground-water resources of Lee and Sumter counties, southwest Georgia: U. S. Geol. Survey Water-Supply Paper 1666, 70 pp.

Pevear, D. R. (1966) The estuarine formation of United States Atlantic Coastal Plain phosphorites: Econ. Geology, vol. 61, pp. 251-256.

Pilkey, O. H., and Luternauer, J. L. (1967) A North Carolina shelf phosphate deposit of possible commercial interest: Southeastern Geology, vol. 8, pp. 33-51.

Rogers, G. S. (1914) The phosphate deposits of South Carolina: U. S. Geol. Survey Bull. 580, pt. 1, pp. 183-220.

Rooney, T. P., and Kerr, P. F. (1967) Mineralogic nature and origin of phosphorite, Beaufort County, North Carolina: Bull. Geol.

Soc. America, vol. 78, pp. 731-748.

Sever, C. W., Cathcart, J. B., and Patterson, S. H. (1967) Phosphate deposits of south-central Georgia and north-central peninsular Florida: Georgia State Div. Conservation, South Georgia Minerals Program Project Rept. 7, 62 pp.

Smith, E. A. (1892) On the phosphates and marls of Alabama: Alabama Geol. Survey Bull. 2, 82 pp.

Stephenson, L. W. (1929) Unconformities in the Upper Cretaceous series of Texas: Bull. Amer. Assoc. Petr. Geol., vol. 13, pp. 1323-1334.

__________ and Monroe, W. H. (1940) The Upper Cretaceous deposits: Mississippi State Geol. Survey Bull. 40, 296 pp.

Weaver, H., Campbell, D., and Northcutt, N. (1963) Phosphate exploration in Texas, A progress report: 41 pp. Deposited with Bureau of Economic Geology, Univ. Texas, on open-file status.

CHEMICALS FROM THE SEA

W. F. McIlhenny
The Dow Chemical Company
Freeport, Texas

ABSTRACT

The oceans of the world cover nearly three-quarters of the surface of the earth to an average depth of over 12,000 feet. The presence of this vast quantity of water gives our planet a special position in the solar system. No other is believed to have substantial quantities of liquid water.

The water of the oceans is quite probably the most complex solution that will ever be encountered. It was in the oceans that life first appeared; even now, our biological chemistry and the composition of our body fluids are closely related to the original saline environment.

More than 77 elements have been found dissolved in sea water. It is quite probable that all naturally occurring elements occur in sea water, and detection depends only upon our analytic ability. Despite the complexity of the solution, the technical chemistry of sea water is determined by the nine major components, and it is from these nine that industrial products are now recovered. The cationic elements are sodium, magnesium, calcium, potassium, and strontium. Anions are chloride, sulfate, bicarbonate, and bromide. All other constituents of sea water total less than one percent of the weight of the dissolved materials.

The economic usage of the oceans is quite impressive. It has been estimated that the value of ocean activities exceeds 8 billion dollars per year, not including transportation, conventional recreation, and most naval operations, and that the annual growth rate of ocean-connected markets is between 8 and 15 percent.

Of this total, more than 700 million dollars is the present annual value of chemically related materials taken from the ocean, and nearly 250 million dollars of this total is recovered from the dissolved state in sea water.

From sea water, common salt, magnesium metal and magnesium compounds, bromine, and water are now produced. Potassium compounds have been produced from residual solar salt bitterns. In addition, authigenic minerals, peculiar to the marine environment, are formed from the dissolved content of the ocean by precipitation, such as manganese nodules, phosphorites, glauconites, and oolitic aragonite. Both calcium carbonate and silicon dioxide are removed from sea water by biological action to serve as skeletons or protective mantles for marine organisms. Iodine has been recovered from sea plants after concentration from sea water.

The types of minerals capable of being recovered from sea water are those which are widely distributed and produced. They are heavy chemicals, basic to a chemical industry. They are sold at low prices, and the shipping costs are relatively important.

The largest sea water processing plant in the world is that of The Dow Chemical Company at Freeport, Texas. In this complex, magnesium, magnesium compounds, bromine, and fresh water are produced.

INTRODUCTION

From its beginning, the chemical industry has been closely associated with the oceans. Sodium chloride, sodium carbonate, bromine, magnesium salts, and potassium salts were first recovered in industrial quantities from sea water. Sea plants were for a period the primary source of iodine, bromine, and potash. The recovery of salt, Glauber's salt, and Epsom salt from the sea in New England by John Sears was the first chemical manufacturing operation in the western hemisphere, and the first patent issued in America was for making sea salt.

Nearly 400 million dollars worth of chemicals or chemically related materials are recovered throughout the world (Fig. 1) from the dissolved state in sea water in chemical plants or by chemical engineering processes. In order of value, salt, magnesium metal, water, bromine, and magnesium compounds (table 1) are now being industrially recovered. In addition, small amounts of potassium and calcium compounds are being produced from solar silt bitterns.

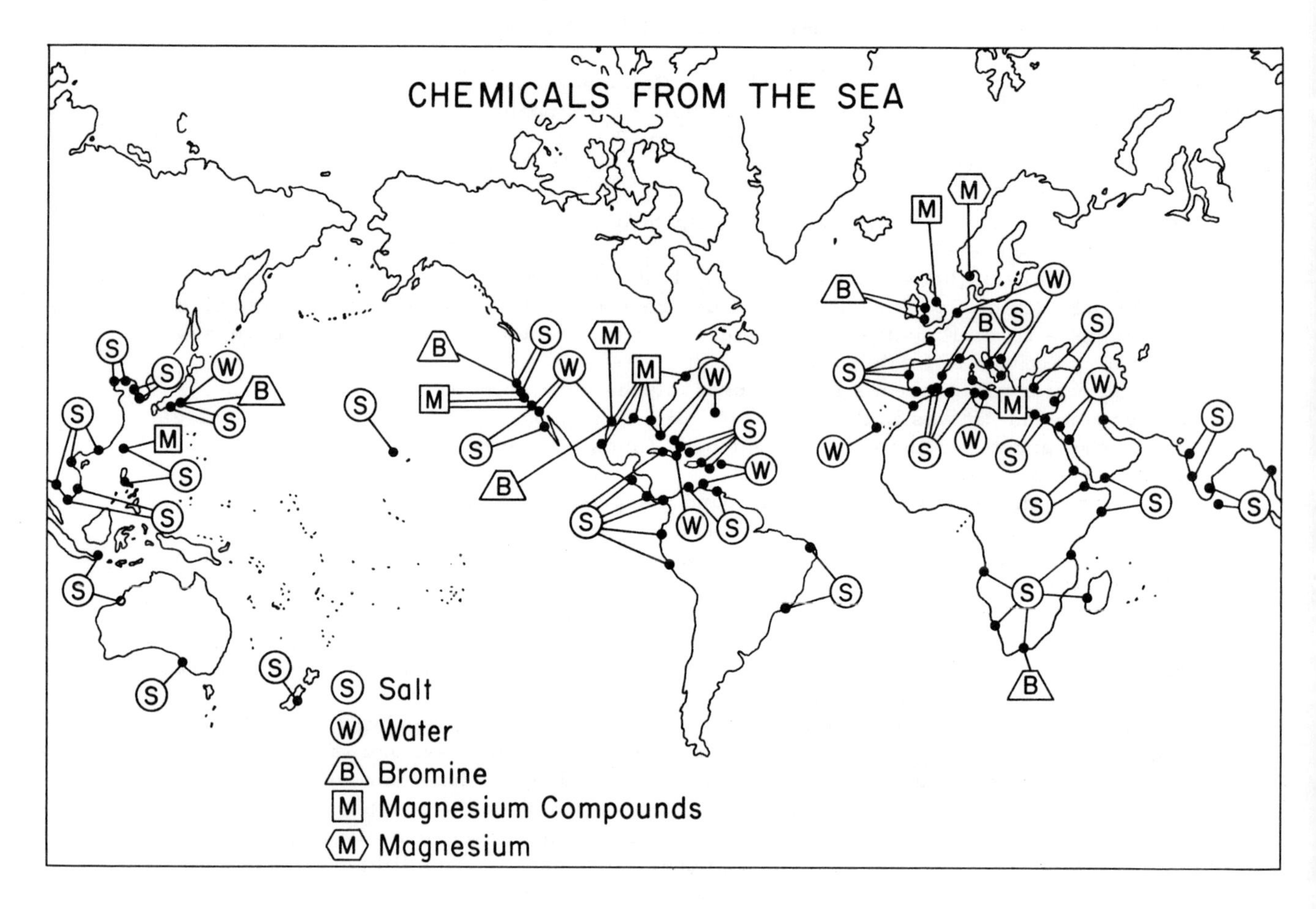

FIG. 1. Chemicals from the sea.

Table 1. World production of chemicals from sea water.

Commodity	Total Annual World Production (tons)	Annual Production From Sea Water (tons)	% of Total	Annual Value of Sea Water Production ($)
Salt (NaCl)	118,600,000	34,600,000[1]	29.2	173,000,000
Magnesium (Mg)	174,000	106,300[2]	61.1	74,700,000
Desalinated water	241,000,000	142,000,000	58.9	51,000,000
Bromine (Br)	146,000	101,600[3]	69.5	44,700,000
Magnesium compounds	11,400,000[1]	690,000[1,2]	6.1	41,000,000
Total annual value of sea water-based production				384,400,000

1/ Estimated. Figures not available.

2/ Includes magnesium from dolomitic lime.

3/ Includes sea salt bittern.

In the United States, including Puerto Rico, about $136,000,000 worth of chemicals is produced from sea water, including solar salt in California, Hawaii, and Puerto Rico; magnesium metal in Texas; magnesium compounds in California, Texas, Alabama, Florida, and New Jersey; bromine in California and Texas; and desalinated water at a number of seaside locations. Four of the five major sea water-based commodities are produced at Freeport, on the Texas Gulf Coast.

Chemical raw materials, in addition to those recovered from the dissolved state, are produced from marine beaches, from the floors of bays and estuaries, from the adjacent continental shelf, and from marine plants and animals.

The chemical industry uses the ocean as a major source of raw materials, as an artery of commerce, as a heat sink, and as a repository for industrial waste products. The total economic impact of the ocean on the chemical production industry certainly exceeds a billion dollars each year.

SEA WATER

Sea water is quite probably the most complex solution (table 2) that technology is ever likely to encounter. Some 77 elements, including the atmospheric gases, have been detected in sea water. It is quite likely that all naturally occurring elements exist in the ocean. The lack of detection of those existing in trace components is due to our analytical limitations. Only a few of the dissolved elements occur as simple ions. The great majority are dissolved as complexes, ion pairs, or highly oxygenated ions. In addition, sea water contains a wide variety of organic compounds, both in true solution and in the bodies of marine plants and animals.

This bewildering array of materials is in

Table 2. Chemical composition of sea water (after Goldberg).

Element	Concentration in Sea Water (Mg/Liter)	Principal Species in Solution	Element	Concentration in Sea Water (Mg/Liter)	Principal Species in Solution
Chlorine	19,000	Cl^-	Phosphorus	0.07	$H_xPO_4^{(-)}$
Sodium	10,500	Na^+	Iodine	0.06	IO_3^-, I^-
Magnesium	1,350	Mg^{+2}, $MgSO_4$	Barium	0.03	Ba^{+2}, $BaSO_4$
Sulfur	885	SO_4^{-2}	Aluminum	0.01	
Calcium	400	Ca^{+2}	Iron	0.01	$Fe(OH)_3$
Potassium	380	K^+	Zinc	0.01	Zn^{+2}, $ZnSO_4$
Bromine	65	Br^-	Molybdenum	0.01	MoO_4^{-2}
Carbon	28	HCO_3^-	Selenium	0.004	SeO_4^{-2}
Strontium	8	Sr^{+2}, $SrSO_4$	Copper	0.003	Cu^{+2}, $CuSO_4$
Boron	4.6	$B(OH)_3$	Arsenic	0.003	$H_xAsO_4^{(-)}$
Silicon	3	$Si(OH)_4$	Tin	0.003	
Fluorine	1.3	F^-	Uranium	0.003	$Uo_2(CO_3)_3^{-4}$
Nitrogen	0.5	NO_3^-, NO_2^-, NH_3^+	Vanadium	0.002	$VO_2(OH)_3$
Lithium	0.17	Li^+	Silver	0.0003	$AgCl_x^{(-)}$
Rubidium	0.12	Rb^+	Gold	0.000004	$AuCl_4$

dynamic equilibrium as a result of a very large number of chemical and biological reactions with reaction rate constants varying from extremely rapid to some of the slowest known to science. It is believed that the composition of the ocean has changed very little for more than 500 million years.

The chemistry of the ocean, however, is determined by the nine major dissolved components (and the water itself), and it is from these that industrial products are being recovered today. The major cations are sodium, magnesium, calcium, potassium, and strontium. Anions include chloride, sulfate, bicarbonate, and bromide. All other dissolved constituents total less than 1 percent of the total weight of the dissolved materials.

Processes have been developed for the extraction of all the major components and some of the minor constituents. Whether a material is being recovered or not depends primarily upon the relative attractiveness of the marine material as an economic source rather than upon the development of recovery and processing technology.

Salt

Sodium chloride is, except for water, the most abundant chemical existing in its primary form on the surface of the earth. It is one of the simplest of molecules, yet it is essential for the existence of life and is indispensable to the chemical industry. Common salt is the single chemical with the longest technological history. It has been connected by mankind with the ocean since prerecorded history.

Salt was the world's first article of commerce, and a large part of the earliest salt was recovered from the ocean. Trade for salt established the location of the earliest trade routes and the location of many centers of population.

Although salt is a dietary necessity, only a small part is actually used in food. Sodium compounds play a part indirectly or directly in everything man uses. The chemical usages for sodium compounds are so extensive that salt is one of the primary foundations upon which the chemical industry rests. About two-thirds of the salt consumed in the United States is by the chemical industry. About half of this is used for the electrolytic production of chlorine and caustic soda and about one-third in the production of soda ash.

Almost certainly the first widely used salt was obtained from the ocean by solar evaporation. Shore-dwelling tribes quite probably noticed the natural deposition from evaporating tidal pools and then deliberately impounded sea water to produce similar evaporation and crystallization.

In all solar sea salt processes (Fig. 2), sea water is spread in a series of shallow basins where it evaporates, gradually increasing in concentration until the solution becomes saturated with sodium chloride and salt begins to crystallize. Contamination with other precipitating salts, tur-

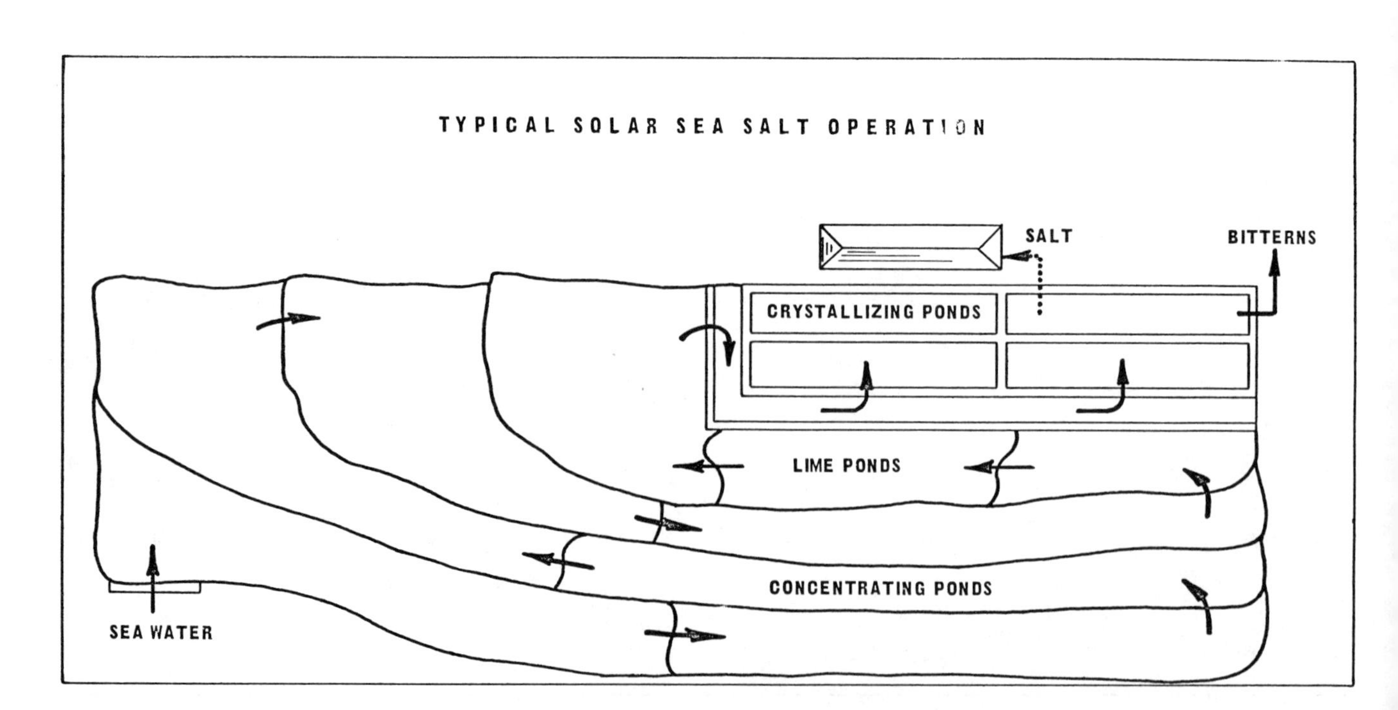

FIG. 2. Typical solar salt operation.

bidity from the initial sea water, and organic matter from biological growth occur to a greater or lesser degree. The purity of the salt produced is dependent upon the technical sophistication and desires of the salt manufacturer.

There are generally three sets of evaporating areas in solar sea salt operations: concentrating ponds, pickle ponds, and crystallizing ponds. Most of the water is evaporated in the evaporation ponds, and they occupy the largest area. Individual ponds are often 500 to 1,000 acres in size.

The operation is designed to obtain maximum vaporization of water, minimum leakage of concentrated solutions, and minimum contamination of the salt product by other materials. Each set of basins has a purpose and is designed differently.

In the concentrating ponds, the incoming sea water is held until the specific gravity of the solution is increased to 1.16. Ferric hydroxide and calcium carbonate precipitate in the initial ponds, which can be uneven in depth and which generally follow the ground contours. Most of the water is evaporated in these ponds which occupy the largest area. The concentrated sea water is further evaporated in a series of lime ponds in which gypsum ($CaSO_4 \cdot 2H_2O$) precipitates. The brine is held here until a specific gravity of 1.21 is reached. The lime (or pickle) ponds can also be uneven and follow land contours. Dyes are sometimes added to increase absorption of the solar radiation.

The brine is moved to a series of harvesting ponds and held until a density of 1.25 to 1.27 is reached. Salt precipitates as large crystals and settles to the pond floor. The concentration of the final liquor is controlled to prevent deposition of magnesium salts. The residual liquor, termed bittern, is often further processed to produce magnesium salts or bromine. The evaporation cycle is continued until a sufficient depth of salt has accumulated. The pond is drained and the salt harvested by large, especially designed harvesters and conveyed to a stockpile. The raw salt is washed and sold in bulk, dried and bagged, or further processed by recrystallization to purer grades.

The harvesting or crystallizing ponds are quite carefully designed. They are normally rectangular with flat or gently sloping floors with properly shaped containing levees. Harvesting ponds are commonly located near storage areas.

A solar salt operation must be located near a market or close to favorable, inexpensive transportation to a more distant market. The necessity for location on the sea coast makes deep draft shipping relatively accessible, and modern plants are usually located to take advantage of ocean transportation. The terrain must be suitable, preferably a low-lying tidal area with an impervious clay soil. The number of suitable sites is quite limited.

The evaporation rate or, more properly, the net excess of water evaporated over rainfall must be sufficiently high. Rates range from 150 inches per year in arid, tropical countries to a useful minimum of 20 inches. A value of 40 to 50 inches per year is normally considered to be the lower limit for general consideration.

A substantial amount of salt is produced in Japan for table use by the direct evaporation of sea water or sea-water concentrates in coal or electrically heated units, and a three-stage, forced-circulation evaporator and salt crystallizer has been operated for a few years in Kuwait, producing 20 tons of salt per day, using as feed the brine effluent stream from the large multiple-stage flash evaporators installed in that country.

There was no substantial sea salt production in the United States until 1853 when the production of salt along the southern shores of San Francisco Bay was begun. Today, solar operations are carried on by two companies in Alameda, San Mateo, and Santa Clara counties and processed and refined in plants at Newark and Mount Eden. Leslie Salt Company supplies crude salt to the adjacent Morton Salt Company refinery and sells the bitterns to FMC Corporation which recovers magnesium hydroxide and bromine in a nearby plant. Crude salt is also harvested from solar ponds along San Pablo Bay in Napa County on the northern arm of San Francisco Bay.

Crude salt is produced by solar-evaporation at Moss Landing on Monterey Bay and sold to local ice companies, water softening firms, and food processers. Solar salt is harvested from ponds on the South Bay at San Diego and is processed in a Chula Vista plant for sale to local customers. Bitterns are sold to a nearby plant for the production of magnesium chloride.

Substantial quantities of solar salt are recovered from salt ponds in the Mayaguez District of Puerto Rico, and commercial salt is produced in Hawaii on the island of Oahu.

Figures are not directly available for solar sea salt production in the United States and Puerto Rico, but it is estimated to be about 4 percent of the United States' total consumption or nearly 1,400,000 tons in 1965.

Salt from the ocean is produced in commercial quantities in 60 countries. An additional 30 or so nations produce small amounts for local or individual consumption. The total production of salt from the ocean is estimated to be about 34,600,000 tons or more than 29 percent of the total world production of salt.

Large new solar sea salt industries have been established in the last decade at Long Island in the Bahamas, Bonaire in the Netherlands Antilles, Port Hedlund in western Australia, and at Guerrero Negro on the west coast of the Mexican State of Baja California.

Magnesium Compounds

Magnesium is the third most abundant dissolved component of sea water, and substantial amounts of both magnesium metal and magnesium compounds are produced from the ocean. Magnesia (magnesium oxide) is the principal product of the magnesium compounds industry. It is widely used as a basic refractory for metallurgical furnaces and as an electrical resistor, an absorbent, a catalyst, and as a component of magnesium oxysulfate cements. Other magnesium compounds are produced by neutralization of magnesia or magnesium hydroxide with various acids.

All sea water magnesium, whether destined for magnesium metal or for magnesium compounds, is precipitated from sea water by an alkali. Calcined dolomite, a widely available double carbonate of magnesium and calcium, is often used as the source of alkalinity. In this case roughly half of the magnesium in the hydroxide product originates from the dolomite and half from the sea water. Lime, produced by calcining shell or limestone, and caustic soda are also used as commercial precipitants. Oyster shell is the principal source of lime for magnesia production on the Texas, Louisiana, and Alabama coasts.

In a typical sea-water magnesia plant (Fig. 3), the incoming sea water is screened and treated to prevent calcium contamination of the magnesium hydroxide product. The calcined limestone or dolomite is added in a water slurry to the sea water in a reactor. Magnesium hydroxide is precipitated and the resultant slurry settled in a series of thickeners. The spent sea water is discharged and the concentrated magnesium hydroxide washed with a countercurrent stream of water and filtered.

Three types of magnesia are normally produced: a dense periclase for refractory use; a chemically active, finely divided magnesium for chemical and physical uses; and a dense, very high purity, chemically inactive magnesia grain. Modern practice seems to be to produce the active oxide by calcination at a relatively low temperature (~600-900°F) in a multiple hearth or rotary furnace. A portion of this oxide is commonly milled, mixed with water, briquetted, and recalcined in a rotary kiln at temperatures of 3200°F or higher to produce the high purity grain.

Part of the thickened, washed hydroxide is mixed in a pug mill with selective additives to lower the sintering temperature and then calcined, usually in rotary, gas-fired kilns at temperatures up to 3300°F. The product is a dense periclase grain with a purity of 84 to 97 percent magnesium oxide.

There are now nine manufacturers of magnesium compounds in the United States who depend upon sea water as a raw material. Four plants in other countries produce magnesia from sea water. These include the world's largest facility at Hartlespool on the northeast coast of England which has been operating since 1937. In recent years, modern magnesia plants have been constructed at Porsgrunn in Norway, at Cagliari in Sardinia, and near Tampico in Mexico. A facility is being constructed at Agua Thuna in Newfoundland and plans have been announced for the construction of sea water magnesium plants in Greece and Saudi Arabia, as well as an additional facility in Mexico.

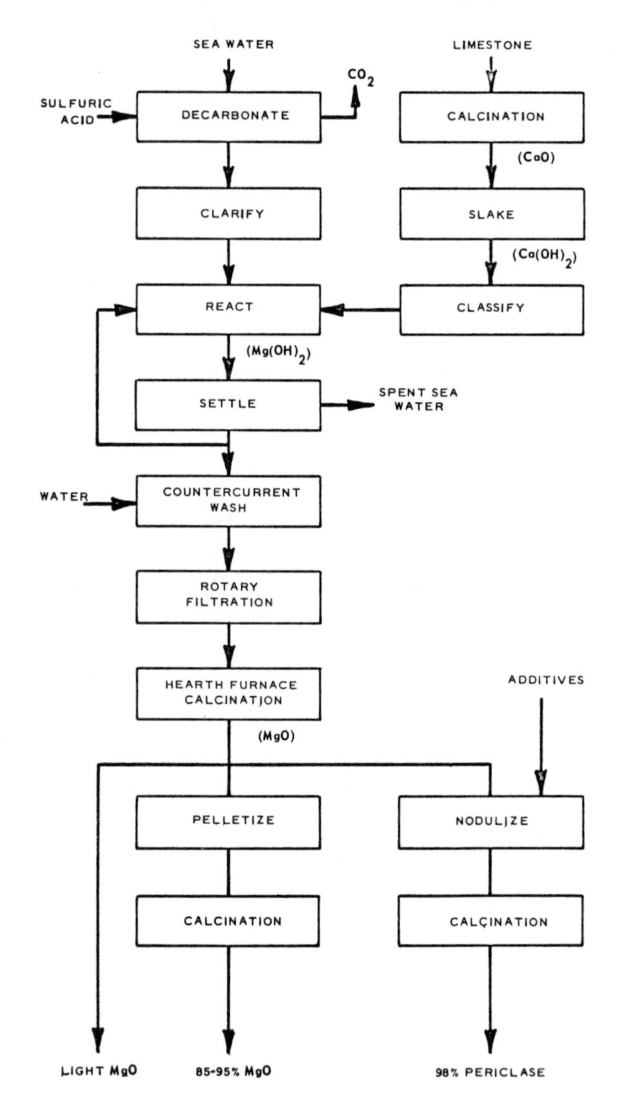

FIG. 3. Magnesia from sea water. Qumica del Mar S. A., Tampico, Mexico.

Magnesium Metal

Sea water is the principal source of magnesium metal. Some 61 percent of the world's production comes from the only two plants which process sea water. These are the Texas Division of The Dow Chemical Company at Freeport, Texas, and the facilities of Norsk Hydro-Elektrisk at Heroya in Norway.

The electrolytic magnesium production processes used in Texas and in Norway are considerably different, but both depend upon an initial precipitation from sea water as magnesium hydroxide. Oyster shell is used in Texas and dolomitic limestone in Norway. The process used in Norway was developed before World War II in Germany. In this process, the precipitated hydroxide is calcined to magnesium oxide which is chlorinated with returned chlorine from the electrolytic cell to produce an essentially anhydrous $MgCl_2$ which is electrolyzed in a refractory-lined cell to produce magnesium metal.

Freeport is located at the mouth of the Brazos River in an advantageous hydrological position in the center of abundant, low-cost raw materials. At this location, magnesium metal, magnesium compounds, bromine, and fresh water are produced from sea water. More than two million gallons per minute of sea water passes through the Dow complex, an amount equal to that of all other process users of sea water in the world.

Sea water is taken in from the bottom of the harbor (to obtain dense, cool, concentrated water) through one of six intakes, is screened, chlorinated, and pumped into one of several central plant flumes. It is pumped from the flume into a concrete flocculator and mixed with a slurry of lime produced by the slaking of calcined oyster shells. The resultant magnesium hydroxide precipitate is settled in large, 500-foot-diameter Dorr thickeners and separated from the spent sea water.

The settled hydroxide is filtered, washed, and neutralized with byproduct hydrochloric acid to produce a concentrated solution of magnesium chloride. The magnesium chloride solution is purified and dried in fluo-solid driers to produce a free-flowing granular solid of the approximate composition: $MgCl_2 \cdot 1.5\ H_2O$.

This granular cell feed is added to a fused salt bath in large bathtub-shaped steel pots. High amperage electric current is passed through the bath using anodes of compacted graphite. During electrolysis, elemental magnesium and chlorine are generated. The magnesium is lighter and rises to the top of the bath where it is collected and cast into ingots.

The two present plants--Dow, with a capacity of 120,000 tons per year, and Norsk Hydro-Elektrisk, with a capacity of 38,000 tons per year--produce the majority of magnesium outside of the Communist bloc nations. The Dow plant at Freeport produces about 92 percent of the total United States production. Both plants have been modernized and expanded in the last few years.

Bromine

A reddish-brown, volatile halogen, bromine, was first discovered in the bitterns remaining after French sea salt production. Bromine is the most marine of all elements; the majority of the world's available bromine is found in the ocean. It exists as bromide ion at a concentration of about 70 parts per million, making sea water as used for bromine production one of the least concentrated raw materials used by any chemical production plant.

Two separate processes (Fig. 4) have developed for the production of bromine, and all commercial bromine production uses one of these two. In the older process, developed in Germany, bromine is stripped from a laden brine by steam distillation after it is chemically oxidized to the free element. This process seems to be more economical when the raw material is a concentrated brine such as sea salt bittern.

The newer process uses a countercurrent air stripping of the oxidized elemental solution followed by absorption. The blowing-out process was developed for use on underground brines by Dr. H. H. Dow at Midland, Michigan, but has found its greatest utility in the recovery of bromine from lean solutions such as sea water, and the blowing-out process is used for all sea water-fed operations.

In the blowing-out process, incoming sea water is filtered and treated with acid to reach a pH of about 3.5. Chlorine is added to oxidize the bromide to bromine which is stripped from the sea water by a countercurrent stream of air in a large packed tower. The bromine-laden vapor is led into a baffled mixing chamber where sulfur dioxide is added, and the reaction products are absorbed in an aqueous acid solution. The acid solution is rechlorinated and steam-stripped to produce a high quality bromine which can be redistilled to elemental bromine or reacted with ethylene to produce ethylene dibromide.

In the steaming-out process, the bromine-laden liquor is preheated, chlorinated, and stripped by steam in specially designed towers. The vapors are condensed, and the halogen phase rectified in a steam-heated reboiler to produce a satisfactory bromine purity.

Bromine has been produced from sea salt bitterns in the United States since 1925 at Chula Vista, and at Newark on San Francisco Bay using the steaming-out process.

There are now six bromine production plants using sea water as a raw material feed. The

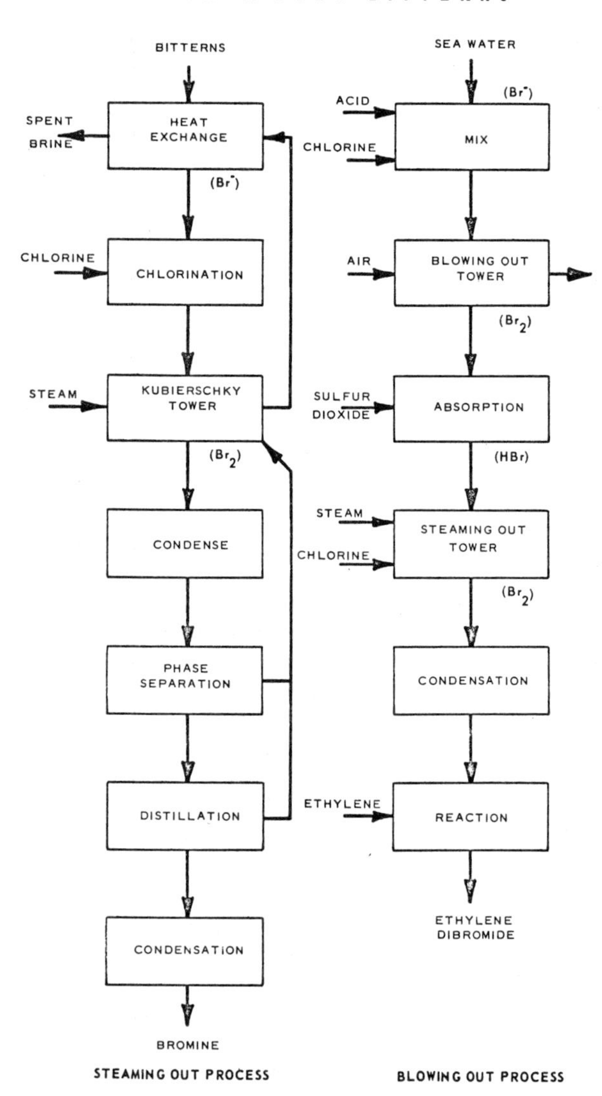

FIG. 4. Bromine from sea water or solar salt bitterns.

Freeport, Texas, operation (the only one in the United States) is the largest. Other large production units are located at Port de Bouc in southern France, in Cornwall and Wales in Great Britain, and in Sicily. Bittern-based production is generally smaller and located adjacent to substantial sea salt operations. Bromine is produced from bitterns in India, Japan, Spain, and Italy. Several of these operations have been recently expanded.

About three-fourths of the 158,000 tons per year of bromine produced in the United States is used in the manufacture of ethylene dibromide, and the percentage of world production so directed is even larger. This dependence on a single use determines the markets and the economic production patterns.

Potassium

Potassium salts are essential to plant growth and are a necessary component of a balanced agricultural fertilizer. For this use, potassium has no substitute.

Because of the agricultural necessity, and because potassium is a principal constituent of the bitterns remaining after solar salt production, and because potassium salts are differentially concentrated by certain marine algae, potassium salts of marine origin have been recovered by a number of processes at various locations. Some potassium chloride is being recovered from Chinese and Japanese bitterns, and about 2,000 tons per year of potassium sulfate are being recovered in India from salt bitterns. No potassium salts are now recovered from any United States sea salt production facilities, and no potassium salts are produced directly from sea water anywhere in the world.

From the 17th century on, seaweed ashes, or kelp as it was known, was a principal source of potash, particularly in Scotland and France, although the industry flourished primarily for the production of soda ash and, later, iodine. California seaweeds were the source of some 25,000 tons of potash during the First World War.

Several bittern processes have been developed for the production of potassium salts. These have been fractional crystalline processes using rather complicated flow schemes and solar evaporation. Potassium can be recovered by arranging the conditions so that carnallite ($KCl \cdot MgCl_2 \cdot 6H_2O$) is able to precipitate. Carnallite is then disproportioned by water leaching to produce KCl. A double sulfate salt ($K_2SO_4 \cdot MgSO_4 \cdot 6H_2O$) is produced by solar evaporation of bitterns from which potassium chlorine can be crystallized after the magnesium and calcium are removed by fractional crystallization.

Both of these processes have been used industrially. The carnallite process has been used to produce potassium salts in France and in California. The Nicolli process was operated in Italy and in Eritrea, and is being used in India.

Water

A surprisingly large amount of sea water is desalinated daily to produce fresh water for municipal, industrial, and agricultural use. At the end of 1966, there were 153 sea water desalination plants with daily capacities of more than 25,000 gallons per day in operation or under construction. The total daily production is 93,046,000 gallons. The largest single unit is being built at Terneuzen, Holland, which will produce 7.65 million gallons, although the total distillation capacity of the multiple units in Kuwait is 24 million gallons per day.

A chemical engineering manipulation must be performed to produce potable water by either separating the salt or water components from a saline feed solution. In practice, all sea water desalination plants use a modification of distillation. Neither electrodialysis nor reverse osmosis are suitable for sea water, and no truly commercial plants using a freezing process are in operation.

Almost all large desalination units now in operation or under construction use multiple-stage flash distillation. In this type of distillation, incoming sea water and recirculated brine are preheated by the condensation of product water in a series of stages at consecutively higher temperatures. The heated sea water is elevated to the maximum operating temperature by condensing steam from an external source. The hot sea water is allowed to flash to vapor in a successive series of flash chambers, each at a lower temperature and pressure. The flashed vapor is condensed by the circulating sea water and is collected as the water product. A final condenser, using additional cooling sea water, maintains the final vacuum.

Flash distillation is relatively new, having been developed in the past fifteen years as a technically useful method. The quality of the water product is excellent, the process is thermally efficient, and sufficient design and operating experience has accumulated so that warranted operation and off-the-shelf designs are now available.

Vapor compression distillation is used in a number of smaller, commonly portable units. These are powered by diesel or gasoline engines driving a compressor. The energy input is mechanical. A few larger vapor compression distillation units are in operation using sea water as a feed.

Multiple-effect, falling-film distillation has been successfully used for the production of water from sea water in the Office of Saline Water Test Facility at Freeport, Texas. Sea water is evaporated from a thin film on the interior periphery of an evaporator in a series of effects. Heat released by the condensation of vapor from a previous effect is used to vaporize the water and the condensed vapor is collected as the plant product. A large distillation plant now under construction in the Virgin Islands will use multiple-effect, falling-film distillation.

The rate of growth of the amount of sea water desalinated has been about 30 percent per year for the last decade. It can be predicted with considerable confidence that the installed capacity will increase to 1,000 million gallons per day by 1978. There now is an investment of about $200 million in equipment, and the annual value of the water produced is estimated to exceed $50 million. Most plants are located in specialized locations where water is necessary, and the cost of water is not an important criteria. The cost of desalinated water will be gradually lowered as the installed equipment increases.

Other Minerals

Calcium precipitates as gypsum during the preconcentration of sea water for salt production. Precipitated gypsum has been used as a supplementary raw material for cement production in Indonesia. The amount is small, and because of the cheapness and wide availability of calcium compounds, sea water is unlikely to be a major source.

Gold occurs in sea water at a very low concentration, primarily as chloride complexes. There have been many attempts to recover gold by a bewildering variety of schemes, none of which have been economically successful.

Fritz Haber, the most famous chemist of his time, undertook to relieve the German war debt by extracting gold from the ocean. He developed careful analytical techniques, and in a series of voyages across the Atlantic in the late 1920's, Haber succeeded in identifying gold in sea water but not in economic amounts.

Uranium is present in sea water as the carbonate complex at a concentration about 1,000 times as high as gold and ten times as high as silver. Processes have been developed for the extraction of uranium from sea water but are not able to compete with alternative sources.

Selective extraction methods are available for almost all of the dissolved components of sea water. In all but a few of these cases, the value of the material recovered is insufficient to cover the cost of pumping, equipment amortization, and labor.

REFERENCES

Dow Chemical Company (1967) A feasibility study on the utilization of waste brines from desalination plants, Part I: Office of Saline Water Res. & Dev., Rept. No. 245.

Goldberg, E. D. (1963) in M. N. Hill, ed., The sea, vol. 2: Interscience Publishers, John Wiley & Sons, Inc., New York.

Gross, W. H. (1967) Magnesium and magnesium alloys: Encyclopedia of Chem. Tech., 2d ed., vol. 12, pp. 661-707, Interscience.

Jolles, Z. E. (ed.) (1966) Bromine and its com-

pounds: Ernest Benn, Ltd., London.
Kaufman, D. W. (ed.) (1960) Sodium chloride--the production and properties of salt and brine: ACS Monograph No. 145, Reinhold, New York.
McIlhenny, W. F. (1966) Problems and potentials of concentrated brine, in Water production using nuclear engineering: Univ. Arizona Press, Tucson.
Mero, J. L. (1965) The mineral resources of the sea: Elsevier Publ. Co., New York.
Shigley, C. M. (1951) Minerals from the sea: Jour. Metals (January).
VerPlanck, W. E. (1958) Salt in California: Bull. 175, California Div. of Mines, San Francisco.

INDUSTRIAL CARBONATES OF THE TEXAS GULF COASTAL PLAIN

G. K. Eifler, Jr.
Bureau of Economic Geology
The University of Texas at Austin
Austin, Texas

ABSTRACT

Industrial carbonates of the Texas Gulf Coastal Plain are limestone, chalk, dolomite, and oyster shell. Carbonate rocks are contained in Cretaceous formations, which crop out along the western part of the Coastal Plain; oyster shell composes Recent reefs in bays along the coast between Sabine Lake and Corpus Christi Bay. In the extensive area between Cretaceous outcrop and bays are minor occurrences of thin, impure Tertiary limestone, local limestone associated with salt domes, and Quaternary caliche; all of these contribute less than 5 percent to the total carbonate industry of the Coastal Plain. The value of lime and cement utilizing Cretaceous carbonates and oyster shell as source material amounts to more than $100, 000, 000 annually. No data are available for the combined value of crushed limestone, crushed shell, and associated products.

Chief source of crushed stone is the Edwards Formation (Lower Cretaceous), which is composed of pure, durable limestone and minor dolomite. Crushed stone and allied products are utilized mainly as aggregate, ballast, base material, fluxstone, and agricultural limestone.

Owing to high purity, Edwards limestone is suitable for calcination to lime. Six plants use Edwards rock as source material and one plant uses Walnut (Lower Cretaceous) limestone. In 1966 these plants produced and sold 738, 000 tons of lime valued at $8, 500, 000. This lime was consumed mainly in soil stabilization and in chemical and associated industries. At one plant dolomite is deadburned for use as a refractory.

Chalk of the Austin Group (Upper Cretaceous) is utilized in the manufacture of cement by eight plants located on the Cretaceous outcrop; one plant uses argillaceous limestone of the Duck Creek Formation (Lower Cretaceous). In 1966 these plants produced approximately 15, 700, 000 barrels of cement valued at $48, 750, 000.

Along the coast crushed oyster shell is used principally as aggregate and base material in road construction, generally no farther inland than 75 miles. Ground shell is used extensively in feed mixes for chickens and cattle.

Oyster shell is the chief source material calcined by plants along the coast. In 1966 seven coastal plants produced 735, 000 tons of lime valued at $10, 000,000. This lime was used mainly by the producer for chemical and other industrial purposes. Six plants utilizing oyster shell produced 11, 500, 000 barrels of cement valued at $37, 000, 000. Coastal as well as inland cement plants are located in metropolitan areas for proximity to local markets.

Reserves of Cretaceous limestone and chalk are infinitely great for the foreseeable future. Although reserves of oyster shell are not known, it has been variously estimated that they will be exhausted in about 15 to 30 years.

INTRODUCTION

Limestone and related carbonates including oyster shell constitute a major resource among the nonfuel minerals of Texas. Production of these raw materials in 1966 amounted to more than $40, 000, 000. In addition, 10 million tons of carbonates were calcined into lime and cement valued at $120, 000, 000. Of this $160, 000, 000 total value, production in the Texas Gulf Coastal Plain accounted for 80 percent or $128, 000, 000.

The Texas Gulf Coastal Plain (Fig. 1) lies gulfward of a line which runs from Red River at the Montague-Cooke County boundary south-southwestward to the Brazos River, down the river to Waco, and thence through Austin, San Antonio, and Del Rio (Fenneman, 1928). The western part of the Coastal Plain is formed by the outcrop of Cretaceous formations which dip southeastward beneath Tertiary strata. Between Austin and San Antonio faults of the Balcones System sharply demarcate the Coastal Plain from the Edwards Plateau; elsewhere the western boundary is less distinct. Cretaceous limestones cropping

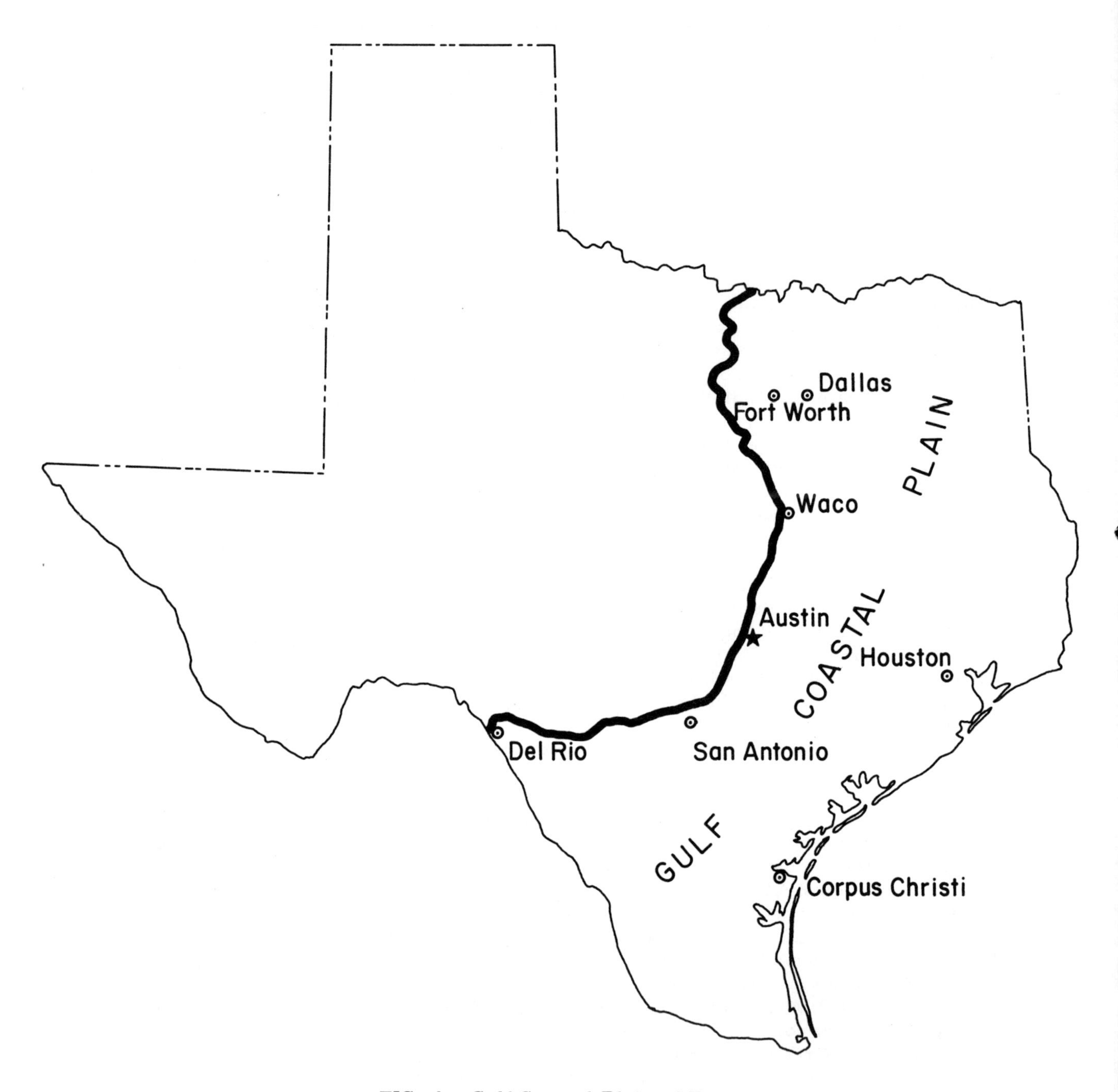

FIG. 1. Gulf Coastal Plain of Texas.

out just west of the inland boundary of the Texas Gulf Coastal Plain are considered in this paper to be deposits within the Coastal Plain.

Industrial carbonates of the Texas Gulf Coastal Plain are limestone, chalk, dolomite, and oyster shell. Carbonate rocks are contained in Cretaceous formations and oyster shell occurs in Recent reefs of bays along the coast. Between the Cretaceous outcrop and the coast is an extensive area generally lacking in carbonates, with the minor exception of limestone associated with a few salt domes, thin impure Tertiary limestone, and Quaternary caliche. The contribution of these minor occurrences amounts to only 3 or 4 percent of the total Gulf Coast carbonate industry.

CARBONATES OF THE CRETACEOUS ROCKS

In Central Texas the Cretaceous System is approximately 2,500 feet thick, and it thickens along the outcrop both northward to Red River and southwestward to the Rio Grande. About half of the section is composed of limestone and chalk

(Adkins, 1933). In the Cretaceous section two units furnish practically all of the industrial carbonates. These are the Edwards Formation (Lower Cretaceous), which is composed of limestone, and the Austin Group (Upper Cretaceous), which comprises mainly chalk.

Edwards Formation

Lithology. --The Edwards Formation is predominantly limestone, but contains dolomite and chert. Although several Lower Cretaceous formations contain potential industrial carbonates, only the Edwards Formation is being extensively exploited at the present time. This is due to the fact that Edwards limestone is superior stone, is exceptionally pure, and much of it is accessible to railroad and highway.

The Edwards Formation was deposited on an extensive, shallow clear-water shelf, the Comanche Platform (Fisher and Rodda, 1967, p. 53), which was bounded on the north, east, and south by basins of deeper water. From Central Texas the formation grades by change of facies northward into the Comanche Peak and Goodland formations (Rodda et al., 1966, p. 5) and southwestward into the Devils River and associated formations (Lozo and Smith, 1964). In thickness the formation increases from 20 feet in northern Johnson County to 300 feet at Austin and to 500 feet north of San Antonio.

North of the Austin area the formation is composed largely of bioherms and biostromes with associated shell debris (Nelson, 1959). Bioherms range in thickness from 10 to 50 feet and are subcircular to elliptical in plan. In some areas bioherms are abundant and locally they coalesce. Biohermal framework is composed mainly of rudistids and caprinids with minor corals and the matrix is a mixture of foraminiferal shells and microgranular calcite. Coarse-grained calcite occurs in shell cavities, in veins, and as cement between shells.

Poorly defined beds with dips up to 25 degrees flank the reefs. These beds are made up of angular shell fragments, minor whole shells, and fine-grained calcite. Tongues of reef core extend locally into reef flanks and interreef deposits.

Interreef deposits are evenly bedded layers up to 3 feet thick and are composed of fine- to coarse-grained limestone containing moderately rounded shell debris.

Biostromes range up to 8 and 10 feet in thickness. They are regularly bedded and are similar in composition to the bioherms.

Southwest of the Austin area bioherms largely disappear and the Edwards is composed mainly of limestones resembling interreef deposits to the north. Biostromes, miliolid limestones, and fine-grained limestones of probable chemical origin occur in varying proportions. The biostromes are up to 10 feet thick and maintain constant thickness and lithology for considerable distances. They are composed chiefly of whole and fragmental shells of rudistids and other pelecypods. Limestones of the Edwards Formation are notably pure. Forty nondolomitic samples from 16 localities between Johnson City and the San Antonio area have an average calcium carbonate content of 97.8 percent (Rodda et al., 1966, pp. 43-277).

Along the Edwards outcrop there are 12 quarries operating on a large scale. However, more than 95 percent of the tonnage is produced in the belt of outcrop between Georgetown and San Antonio, a distance of about 100 miles. In this belt the thickness ranges from 175 feet in the north to 500 feet in the south. In addition to considerable thickness, other favorable quarrying features are good drainage and gravity haul situations.

Utilization. --Since limestones of the Edwards Formation are compact, tough, durable, and pure, they are quarried extensively for a multiplicity of industrial uses. In 1966 approximately 10,000,000 tons of rock were quarried from the Edwards Formation. Of this amount more than 8,750,000 tons of rock were crushed stone, which in small part was a by-product of quarrying high-calcium limestone.

More than 7,000,000 tons of this crushed limestone were used in aggregate and road material and in the production of alkalis. Fluxstone and agricultural limestone each accounted for approximately 275,000 tons. Minor uses included asphalt filler, stone sand, roofing material, and mineral foods.

The lime industry along the inner margin of the Texas Gulf Coastal Plain is tied closely to the Edwards Formation. In Texas the lime industry started more than a century ago in the Austin area. Lime was used in the construction of the State Capitol Building, which was completed in 1888. Texas manufactures 8 to 9 percent of the lime produced in the United States, ranking 4th among lime-producing states.

The great bulk of the Edwards limestone is suitable for calcination into lime, since the rock is typically 96 to 97 percent calcium carbonate and the minimum requirement is 96 percent. Six of the seven lime plants along the inner margin of the Gulf Coastal Plain use Edwards limestone as source material. The seventh plant uses a limestone from the Walnut Formation, which is stratigraphically below the Edwards.

Unlike the cement plants of the Coastal Plain, most lime plants are not located in metropolitan areas, probably because of a broader market (Fig. 2).

In 1966 these seven lime plants produced and sold 738,000 short tons of lime valued at

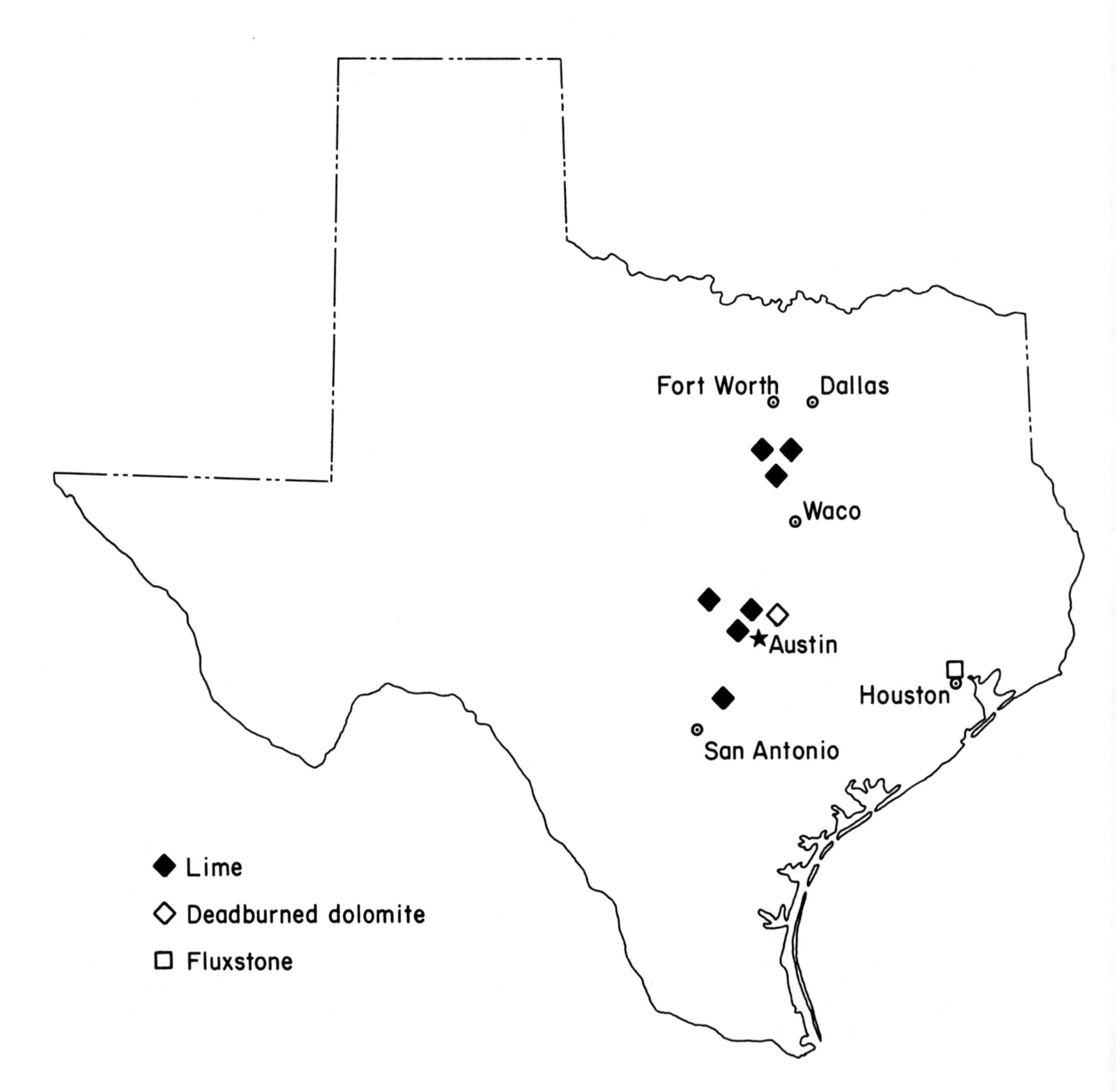

FIG. 2. Lime plants utilizing Cretaceous rocks.

$8,470,000, or an average of approximately $11.70 per ton. This production consumed about 1,200,000 tons of raw limestone. A new lime plant will be completed this year at San Antonio; the plant will use Edwards limestone and have a capacity of 220 tons a day.

Approximately 48 percent of the lime output was sold for soil stabilization, mostly for road beds; 38 percent was sold for chemical and other industrial purposes; and the remaining 14 percent was used in construction, agriculture, and miscellaneous uses of minor importance. Most lime used in road stabilization is fed into delivery trucks equipped with spreaders in order to eliminate second handling. Considerable lime was used in purification of water in suburban and other areas and in water used in the Frasch process of sulfur recovery.

Dolomite from the Edwards Formation is deadburned for use as refractory material in construction of open hearth and blast furnaces. The rock must contain at least 18 percent magnesium and less than 1 percent each of alumina, silica, and iron oxide.

In the Edwards Formation, dolomite occurs in two ways: (1) in individual layers or stratal dolomite, and (2) in thick masses or massive dolomite (Fisher and Rodda, 1967, p. 58). Stratal dolomite occurs as laterally persistent beds up to 3 feet thick; the rock is fine grained and probably formed in the prelithification stage. Massive dolomite occurs in irregular bodies 10 to 15 feet thick, but locally is much thicker. The rock is commonly coarse grained and formed in the post-lithification stage. Massive dolomite is locally associated with outer parts of bioherms and flanking deposits.

Dolomite is locally abundant in the Edwards Formation near Round Rock where the rock is deadburned at one plant. This is the only plant calcining dolomite and production figures are not available.

Edwards limestone mixed with clay and iron ore would be suitable for manufacture of cement; however, none is so used along the Cretaceous outcrop of the Coastal Plain. The reason for this is that the Edwards limestone is locally dolomitic and is hard to grind. A softer, yet suitable material is abundant and accessible. This material is chalk from the Austin Group of the Cretaceous Gulf Series.

Austin Group

Lithology.--The Austin Group in Central Texas is divided into six formations with a combined thickness of 300 to 500 feet. These are dominantly thick, white, fairly pure chalk and minor, thin marl (Adkins, 1933, p. 446). The formations are sometimes referred to collectively as the Austin Chalk. The rock is composed typically of fine-grained calcite, foraminiferal and molluscan shells, and debris of finely comminuted shells and other organic matter. Calcium carbonate content in general ranges between 80 and 85 percent. Northward from Austin the Austin Chalk generally thickens and crops out in an almost straight belt to Dallas. Southwestward from Austin the outcrop extends through San Antonio and Uvalde, but it is locally cut by faults of the Balcones System.

The Austin Chalk is underlain by shale of the Eagle Ford Formation and overlain by marl of the Taylor Group. Westward from San Antonio the basal Taylor marl is replaced by hard, argillaceous limestone of the Anacacho Formation.

Utilization.--Chalk of the Austin Group is utilized in the manufacture of cement by eight of the nine plants located along the Cretaceous outcrop (Fig. 3). The ninth plant uses argillaceous limestone of the Duck Creek Formation of the Comanche Series.

The argillaceous component for the raw mix with chalk is obtained from either the underlying shale of the Eagle Ford Formation, as done at Dallas, Midlothian, and Waco, or the overlying argillaceous limestone of the Anacacho Formation as done at San Antonio.

In 1966, these plants produced 15,700,000 barrels of cement valued at $48,700,000. This production represents 49 percent of the total Texas output and 57 percent of the total production of the Texas Gulf Coastal Plain.

OYSTER SHELL INDUSTRY

As previously stated, the other major source of industrial carbonates besides the Cretaceous formations is oyster shell along the coast.

History

Commercial dredging of oyster shell along the Texas coast is said to have started at Galveston in 1880 with the sale of a cargo of shell obtained by shovel and wheelbarrow from an exposed reef in Galveston Bay. Improved methods of dredging and a steady increase in shell demand on the coast have made the present industry a multi-million dollar business with a $22,000,000 investment in dredges, barges, tugs, and auxiliary equipment and with an annual payroll of $7,500,000. Shell is the second largest cargo item of Port Houston; petroleum products rank first. Texas is the largest producer of oyster shell in the United States.

The average annual production of dredged shell for the past 10 years is 11,500,000 cubic yards and accumulated production for the past 50 years is approximately 275,000,000 cubic yards. An average cubic yard of shell weighs between 1,600 and 1,700 pounds. Oyster shell occurring in bays along the coast is the property of the State of Texas. In the last 10 years the State has collected more than $10,000,000 in royalties from the dredging industry. The current royalty is 15¢ per cubic yard for coarse shell and 13¢ per cubic yard for fine shell, which is approximately 25 percent of the total production.

In the early years of this century crushed shell was used in quantity only for road material. At that time it was probably supposed that not many roads would be built and that the supply of shell was almost without limit. Impetus to the shell industry occurred in 1916 when shell was first used in the manufacture of cement at a Houston plant. Demand was increased in 1929 when it was discovered that shell could be made into lime and was increased again in 1941 when Dow Chemical Company at Freeport began to use lime made from shell in the extraction of magnesium from sea water. Production has leveled off in the last few years, probably because gravel was, to some extent, substituted for shell as

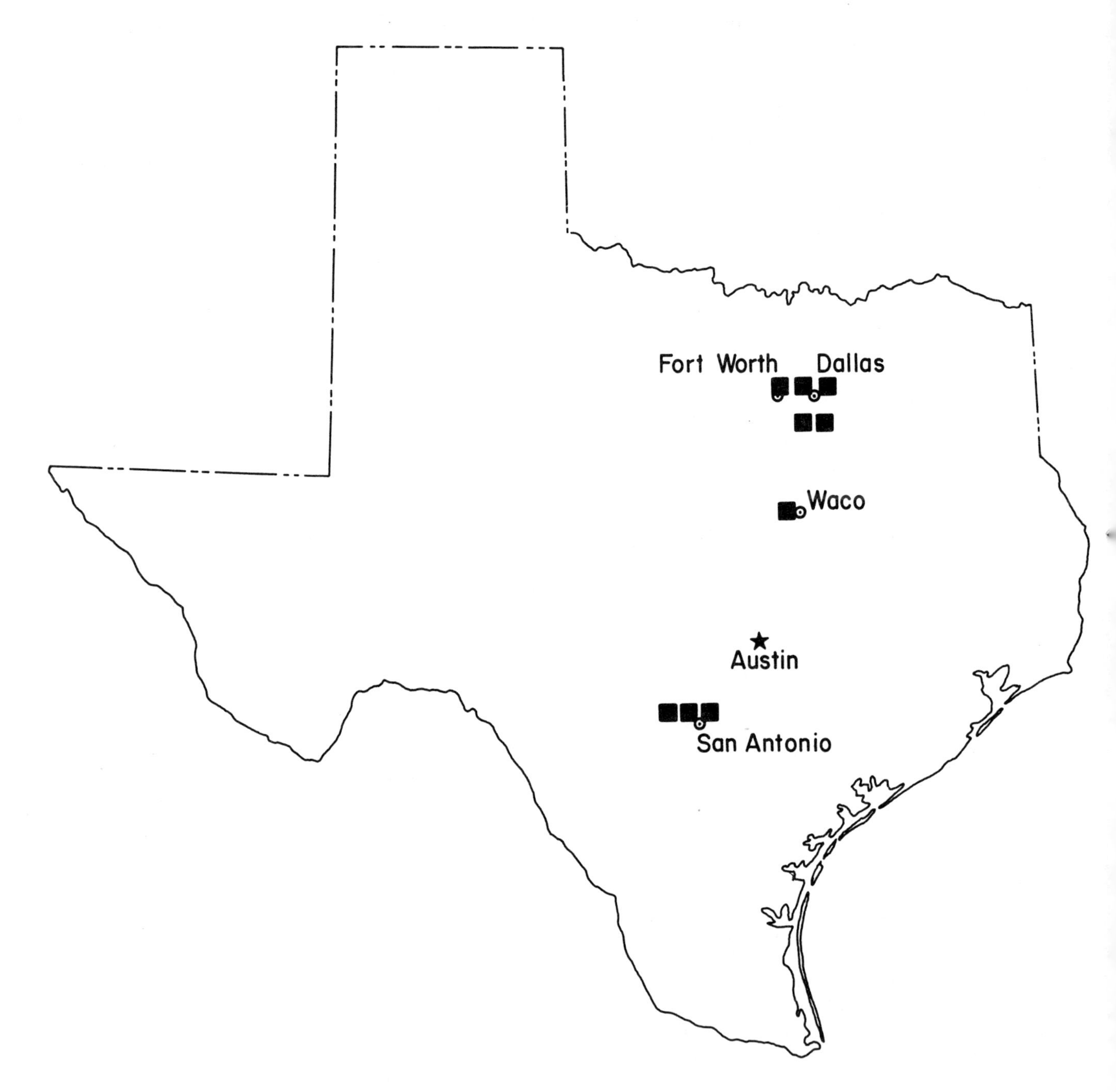

FIG. 3. Cement plants utilizing Cretaceous rocks.

road material (Fig. 4).

Distribution of Reefs

Oyster reefs occur in bays from Sabine Lake to Corpus Christi Bay. Shell is currently being produced from five bays: Sabine Lake, Galveston-Trinity Bay, Lavaca Bay, San Antonio Bay, and Nueces Bay (Fig. 5). In the past, Galveston-Trinity Bay has accounted for about 75 to 80 percent of shell, but last year it produced only 63 percent of the total amount. San Antonio Bay has an increase of 50 percent over its production two years ago. No dredging has been done in Corpus Christi and Aransas Bays since the 1930's or in Matagorda Bay since 1961. Copano Bay has had only limited and intermittent production in the last few decades.

Dredging Procedure

In 1953 dredgers were prevented by the State of Texas from operating within 1,500 feet of live reefs. In 1963 dredgers, having difficulty in

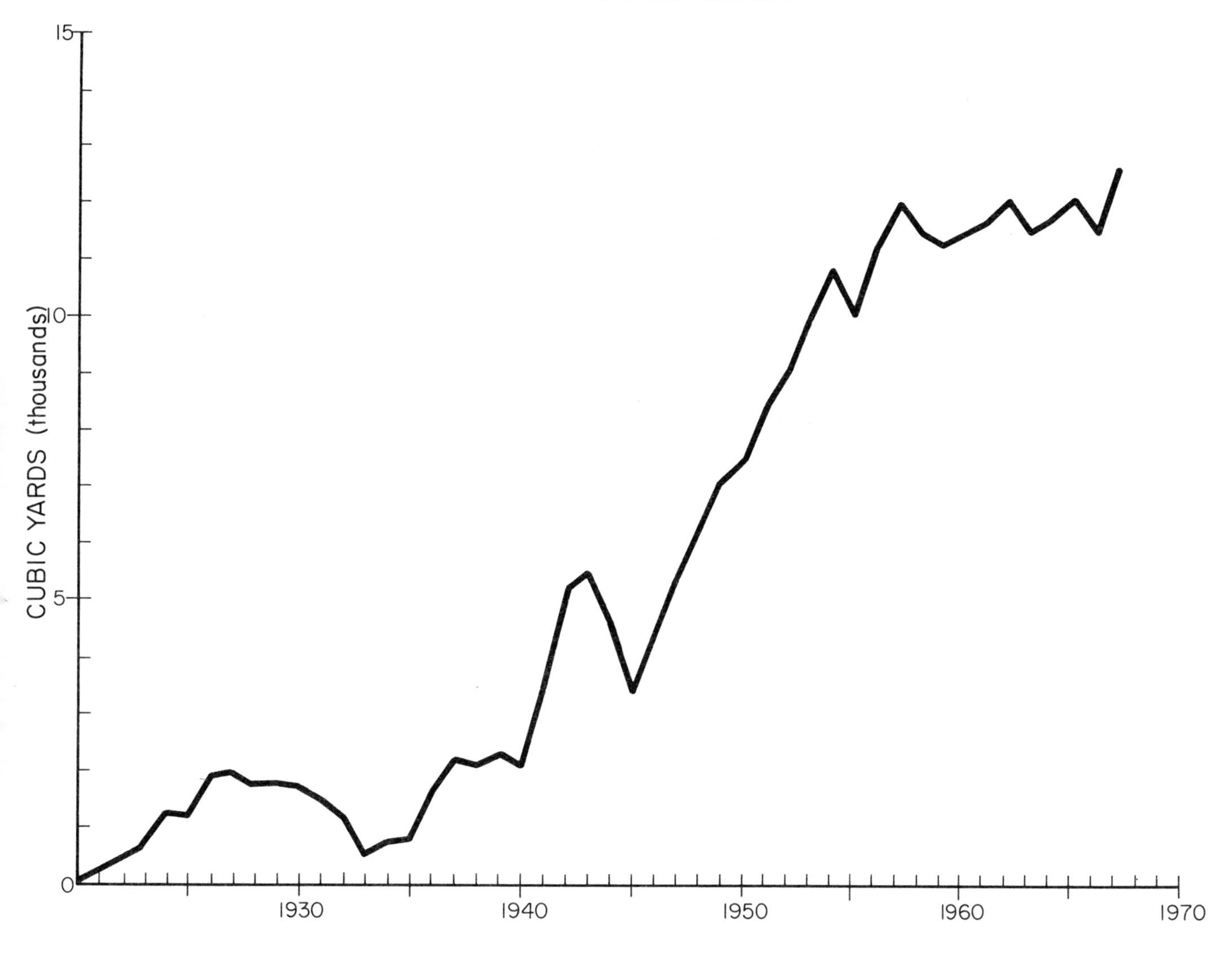

FIG. 4. Oyster shell production in Texas bays, 1923-67.

getting sufficient shell, requested permission to take all the shell in Galveston-Trinity Bay. The Texas Wildlife Commission ordered that the 1,500-foot limit be reduced to 300 feet. After 4 months, difficulties arose and it was then ordered that all dredging within the established limit be done under the surveillance of the Commission. This is the current procedure.

Modern dredgers operating in 5 to 10 feet of water in reef areas are effective to a depth of 30 to 35 feet below water level. It is generally thought that very little shell exists below this level. The dredge cutting tool loosens the shell, which is pulled by suction through a tube onto the dredge where mud is eliminated by rotary washers. The yield is generally 95 percent shell and shell debris.

Composition of Shell

The calcium carbonate content of shell compares favorably with that of high-calcium limestone.

A batch of thoroughly washed shell is 95 percent calcium carbonate. Shell with more than 3 percent silica is generally rejected. A typical analysis of washed shell is given below.

	Percent
$CaCO_3$	93-97
$MgCO_3$	1.0
SiO_2	0.5-2.0
$CaSO_4$	0.3-0.4
Fe_2O_3 / Al_2O_3	0.2-4.0
Others	0.85-1.0

Utilization.--Principal use of crushed shell is in road construction, which consumes approximately 28 percent of total shell production. An excellent road base is made by the use of a mixture of two parts of shell to one part of sand, which is equal in strength to a base of 100 percent crushed limestone. Large quantities of crushed shell are used as aggregate in concrete

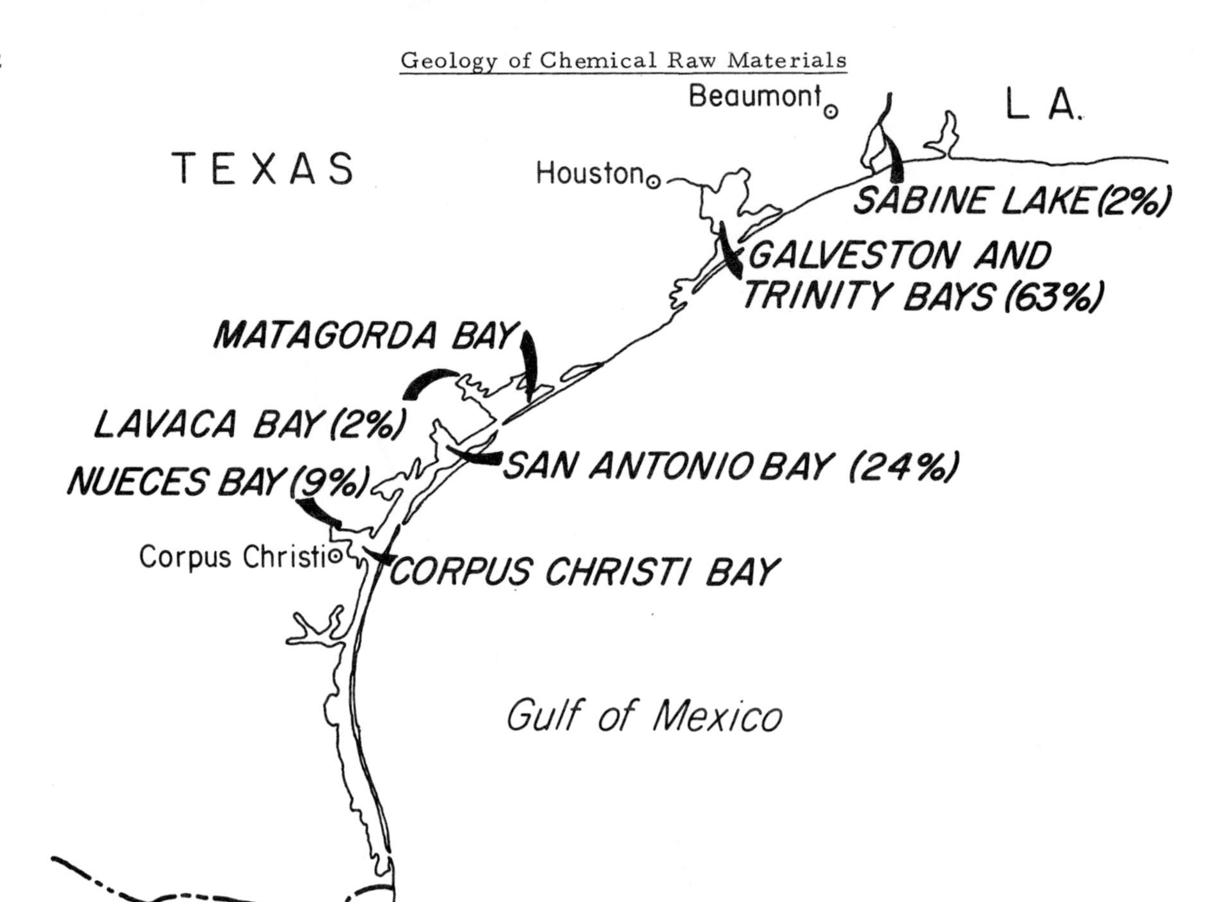

FIG. 5. Oyster shell production, 1966-67, percentage by bays.

and asphalt surfacing. In addition, 100 percent crushed shell is used on shoulders of roads and as surfacing on some county and private roads, but this usage is declining. Use of shell as road material is, in general, limited to a 75-mile-wide strip along the coast, beyond which shell runs into competition with gravel owing to transportation costs.

Oyster shell is ground and mixed with feeds for chickens and cattle. Calcium carbonate of the shell is effective in producing a thicker, stronger egg shell, which is an important factor in handling and shipping eggs. Cattle feed requires a coarser grind than chicken feed. The oyster shell ingredient serves as roughage, as a bone-builder, and as an aid in milk production. No statistics are available for the quantity of shell used in all types of feed mixes, but the industry must be considerable as chicken feeds alone have consumed more than 5,000,000 cubic yards of shell since 1931. Houston ranks second in the nation as a producer of chicken feed with admixed shell.

Shell is used in the manufacture of lime at four of the five plants along the coast (Fig. 6). The fifth uses Edwards limestone as fluxstone in the manufacture of steel. In East Texas two paper manufacturing plants regenerate lime, a process requiring small additional quantities of shell or high-calcium limestone from time to time.

In 1966 the coastal plants produced 735,000 tons of lime valued at $10,000,000. More than 96 percent of this lime was consumed by the producers for use in their respective chemical and allied industries. The remaining 4 percent was sold chiefly for use as a soil stabilizer.

Two chemical manufacturers, Dow Chemical Company and Pittsburgh Plate Glass Company, produced approximately 70 percent of the lime manufactured on the coast. At the Dow Chemical plant in Freeport, the first step in the production of magnesium is the addition of lime to seawater contained in flocculators where magnesium hydrate is then precipitated. This precipitate is treated with hydrochloric acid to produce magnesium chloride, which is dissociated electrolytically into metallic magnesium and chlorine. The plant uses approximately 1,000,000 cubic yards of shell annually in this process. Lime is a low-cost factor in the production of magnesium.

The Chemical Division of Pittsburgh Plate Glass Company uses lime directly or indirectly in production of soda ash. In 1966 this company used approximately 1,000,000 cubic yards of shell in the production of lime. Carbon dioxide and chlorine are important by-products in the manufacture of soda ash.

In the reduction of bauxite to aluminum, lime

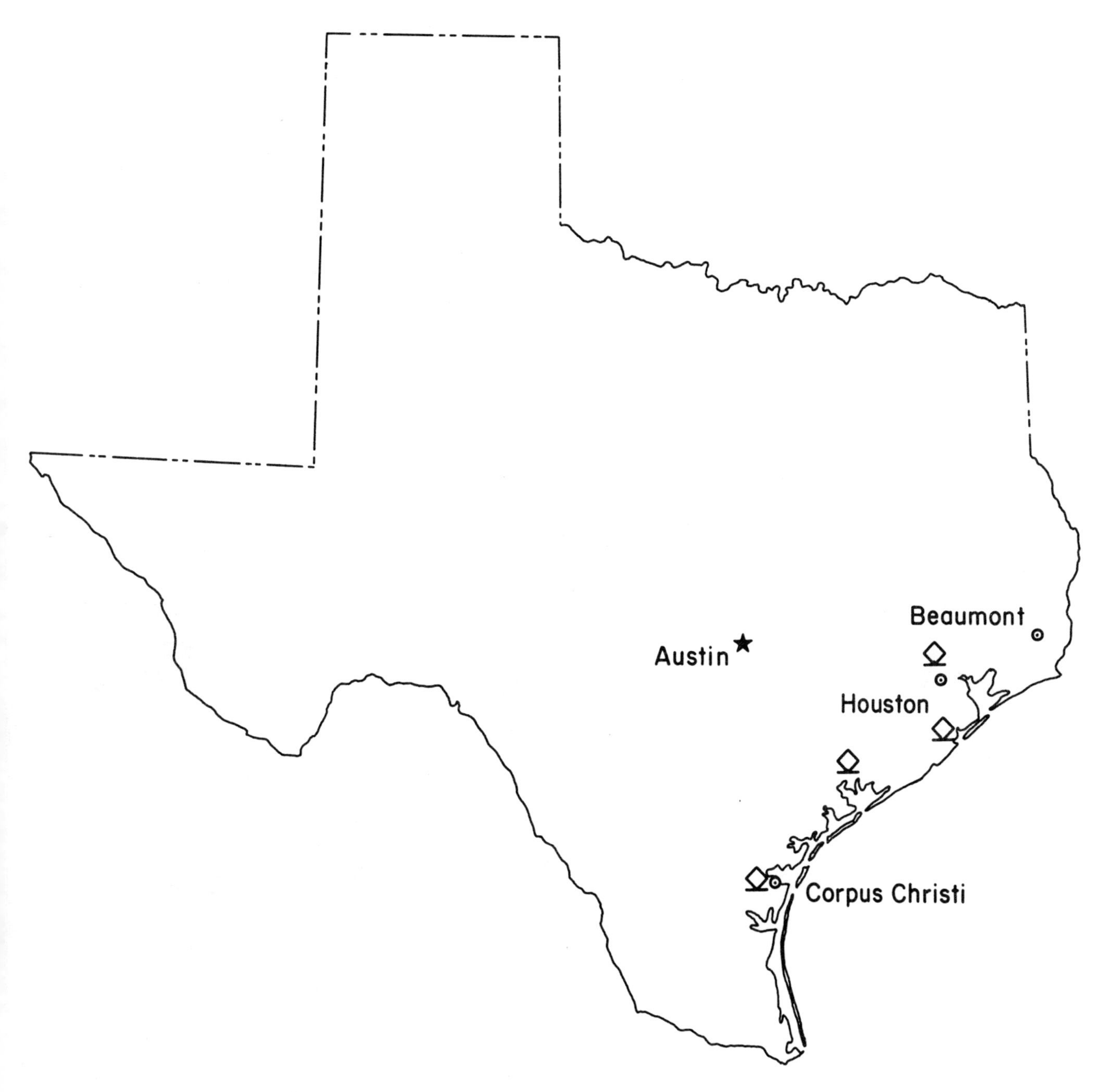

FIG. 6. Industrial plants manufacturing lime from oyster shell.

is used in the first stages to remove silica and other impurities from the bauxite ore. This use of lime is a low-cost factor in the final production of metallic aluminum. There are two aluminum plants on the Texas coast, Reynolds Aluminum Company near Corpus Christi and Aluminum Company of America on Lavaca Bay. The latter plant converts more than 100,000 cubic yards of shell to lime annually for its own use. Reynolds Aluminum, however, buys lime from Central Texas. This lime is transported by rail at a cost of approximately $4.00 per ton, or an increase of 30 percent over the average price.

In the manufacture of paper, lime is utilized in the production of calcium hypochlorite, which is the chief agent used to bleach pulp. Lime is an important ingredient of the cooking liquor used in the manufacture of paper. One large paper company at Houston uses about 40,000 cubic yards of shell annually for these purposes.

There are six cement plants on the Texas coast and all use oyster shell as the calcareous ingredient of the raw mix (Fig. 7). One plant, which is located east of Beaumont, obtains shell

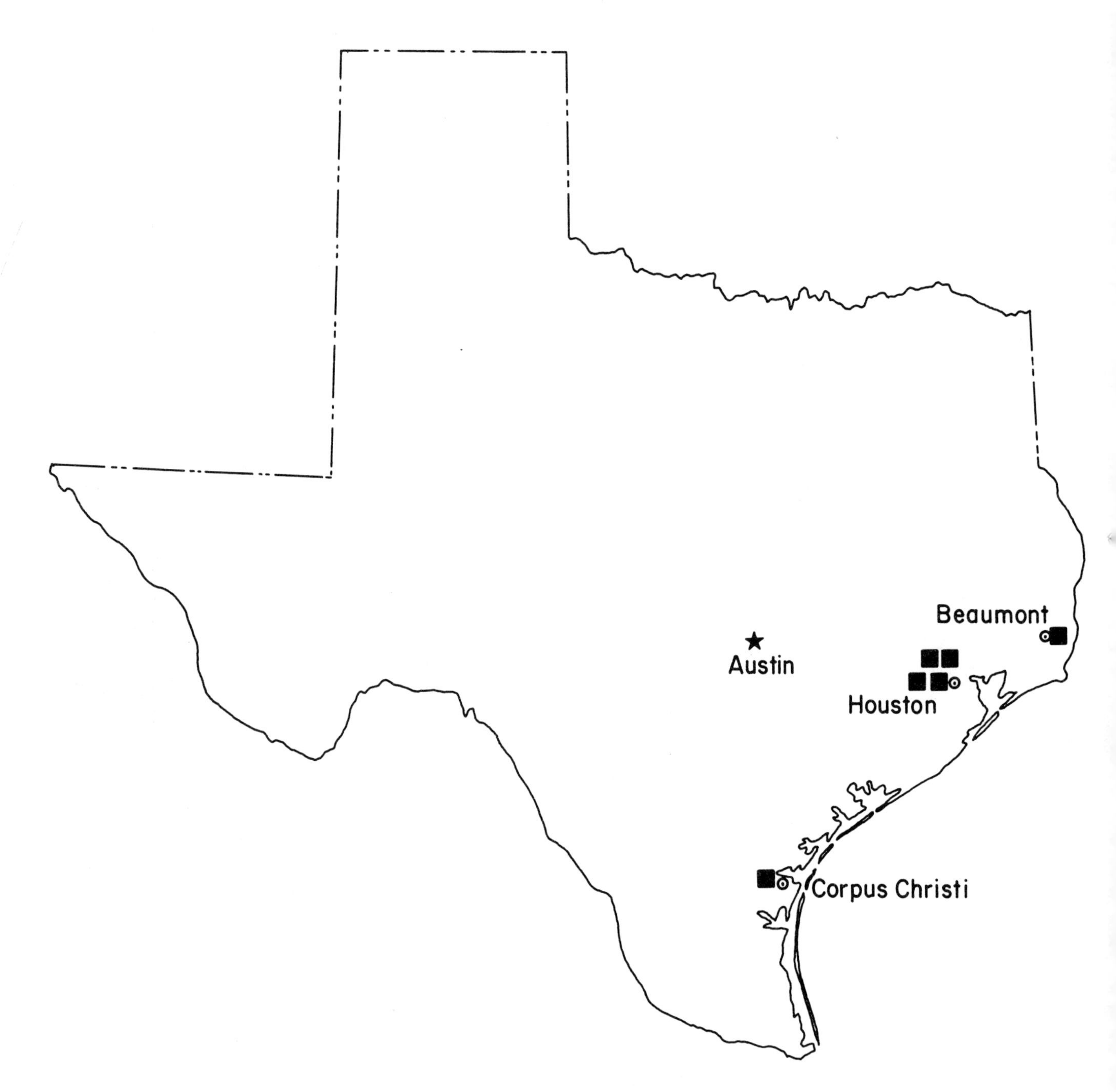

FIG. 7. Cement plants utilizing oyster shell.

from Louisiana because no dredger would make a 15-year contract to supply Texas shell. For the argillaceous component of the mix, clay is obtained from the Beaumont Formation which crops out in a broad belt contiguous to the coast. A typical mix to produce one ton of cement is:

shell		clay		iron ore		cement
2,500 lbs.	+	720 lbs.	+	110 lbs.	=	1 ton
(75%)		(22%-)		(3%+)		

In 1966 these plants produced over 11,500,000 barrels of cement valued at $37,000,000. This production is 37 percent of the total Texas output and 43 percent of the production of the Texas Gulf Coastal Plain. Local metropolitan areas constitute the major markets, but the cement is sold along the entire coast.

SUMMARY OF RELATIVE VALUES

In summary, a comparison of Cretaceous limestone production with shell production shows that crushed limestone has approximately twice the value of crushed shell and the value of cement and lime from Cretaceous rock is about 20 percent greater than the value of these products derived from shell.

RESERVES

Reserves of Cretaceous limestone, dolomite, and chalk suitable for crushed stone and for calcination into lime and cement are infinitely great. At the present rate of consumption the supply would last for a period measurable in thousands of years. On the other hand, reserves of oyster shell available under existing dredging restrictions are critically short. Although no survey of remaining shell has been made, estimates ranging between a 15- and 30-year supply have been made. The Texas Parks and Wildlife Commission will request a $285,000 appropriation at the next session of the legislature for purposes of making a survey of shell resources in the eight major bays. It should be noted that dredging shell to depletion is not an economic analogue to the mining of progressively lower grade metallic ores, since the percentage of shell recovery would be about the same up to the last barge load.

Galveston-Trinity Bay, which has produced 95 percent of all Texas shell, is now low in reserves, possibly a 10- to 12-year supply remaining under present conditions. San Antonio Bay may have appreciable reserves, but these would probably never approach the original shell reserves of Galveston-Trinity Bay. San Antonio Bay is about 150 miles from the Houston ship channel by way of the intracoastal canal, and transportation cost to the Houston area would increase the price of delivered shell by more than 20 percent.

Oyster shell has been discovered recently at depths of 65 and 80 feet, but no reef has been outlined. If large reefs were discovered at these depths they would have an overburden of mud ranging between 55 and 70 feet, so that it is doubtful that this shell could be recovered on an economic basis.

ECONOMIC CONSIDERATIONS

When shell can no longer be taken from the bays, certain economic considerations will arise. The greatest effect will be on the dredging industry, which will come to a halt along the Texas coast. Perhaps next in the line of adverse effect will be the lime and cement industries, followed by industries using lime in the manufacture of such products as magnesium, steel, aluminum, soda ash, and paper. Also involved in the depletion of shell will be road contractors and feed producers. It seems reasonable to assume that a price increase, to a greater or lesser extent, would result for all products made either directly or indirectly from shell.

Crushed shell used in road construction as base material and as aggregate in surfacing could be replaced by gravel and caliche occurring in the interior of the Coastal Plain, but at a cost increase of local variation.

The feasible sources for replacement of Texas shell used in calcination are Louisiana shell, the loose sands of calcareous oölites occurring as beach deposits in the Bahama Islands, and the Edwards Limestone of Central Texas. Transportation of Louisiana shell to Houston would result in a price increase of 35 to 40 percent, but the supply is limited so that the remedy would be temporary. Bahama lime sands might possibly be used in the manufacture of cement. However, difficulty has been encountered in pulverizing the oölites, apparently because of their shape. Freight rates vary according to classification of stone and are negotiable, but the rate for a ton of limestone transported from Central Texas to Houston would probably be between $1.80 and $1.90. An average cost of a ton of high-calcium limestone F.O.B. Central Texas quarries is approximately $1.25, or $3.05 to $3.15 delivered at Houston. A ton of Texas shell costs $1.50 to $1.60, or about one-half that of Edwards limestone delivered to the Houston area.

Two tons of shell or high-calcium limestone are required to produce one ton of lime owing to the loss of CO_2 and a small amount of calcium through minor chemical combinations. Thus a Houston plant in producing one ton of lime would use two tons of shell valued at $3.00 or two tons of Edwards limestone valued at $6.00, which would be an added cost of $3.00 per ton of lime. Present transportation cost of one ton of lime from Central Texas to Houston is a little more than $4.00. Therefore, it would be cheaper for coastal plants to produce from Edwards limestone than to buy lime from Central Texas plants.

In the absence of shell, coastal plants would probably use high-calcium Edwards limestone rather than Austin Chalk in the manufacture of cement, since clay from the Beaumont Formation is readily accessible. Added cost of the use of this limestone instead of shell would be approximately $1.75 per ton or about 30¢ per barrel. This additional cost would shorten the radius of the competitive market area of the coastal plants.

In conclusion, it seems apparent that many industries, especially lime and cement, using shell in their products, will eventually turn for source carbonates to the Cretaceous rocks, which have so well nurtured their competitors in the past.

REFERENCES

Adkins, W. S. (1933) The Mesozoic systems in Texas, in The geology of Texas, Vol. I, Stratigraphy: Univ. Texas Bull. 3232 (Aug. 22, 1932), pp. 239-518.

Fenneman, N. M. (1928) Physiography of eastern United States: McGraw-Hill Book Company, New York, N. Y., 714 pp.

Fisher, W. L., and Rodda, P. U. (1967) Stratigraphy and genesis of dolomite, Edwards Formation (Lower Cretaceous), in Proceedings of the 3d forum on geology of industrial minerals: State Geol. Survey of Kansas Spec. Dist. Pub. 34, pp. 52-73.

Lozo, F. E., and Smith, C. I. (1964) Revision of Comanche Cretaceous stratigraphic nomenclature, southern Edwards Plateau, southwest Texas: Gulf Coast Assoc. Geol. Socs. Trans., Vol. XIV, pp. 285-306.

Nelson, H. F. (1959) Deposition and alteration of the Edwards Limestone in Central Texas, in Symposium on Edwards Limestone in Central Texas: Univ. Texas Pub. 5905, pp. 21-85.

Rodda, P. U., et al. (1966) Limestone and dolomite resources, Lower Cretaceous rocks: Univ. Texas, Bur. Econ. Geology Rept. Inv. 56, 286 pp.

GULF COAST SULFUR RESOURCES

John C. Myers
Consulting Geologist
Houston, Texas

ABSTRACT

Consumption of sulfur is closely correlated with industrial progress and is sometimes used as a barometer of industrial production and a country's standard of living. In the latter part of the 18th century, the world was supplied with sulfur from sedimentary sulfur deposits on the island of Sicily. In 1894, the Frasch process of producing sulfur was first successfully applied at Sulphur Dome, Calcasieu Parish, Louisiana. Since then, out of 174 domes whose caprock lies above 3,000 feet, located on and offshore in the Gulf Coast of the United States, only 26 have been productive. In Mexico, 5 domes have produced sulfur to date, making a total of 31 domes for the Gulf Coast area of the United States and Mexico.

Sulfur is found in a transition zone in the caprock of a salt dome, below a limestone cap and above a calcium sulfate base. Thus, carbonate content increases and sulfate content decreases with an increase in sulfur.

Future prospects for production of additional sources of sulfur in the Gulf Coastal area of the United States and Mexico lie entirely with salt dome development.

GENERAL STATEMENT

The sulfur industry is an integral part of the production of nearly everything we eat, wear, and use. Sulfur is not a commodity used directly by consumers, but is used or employed in the manufacture of more than seventy products. Its greatest use is in the production of sulfuric acid and fertilizers. Sulfur production is closely correlated with industrial progress, and its rate of consumption is frequently used as a barometer of industrial production, and therefore, a country's standard of living.

HISTORY

In the latter part of the 18th century, Sicilian sedimentary deposits provided the world with sulfur, but in 1894, the Frasch process was first successfully used at Sulphur Dome, near the town of Sulphur, in Calcasieu Parish, Louisiana, and since then, Frasch producers have been in command of the supply.

Use or consumption of sulfur is closely related to its development in the Gulf Coastal area. The Union Sulphur Company operated the mine at Sulphur, Louisiana, and their judgment of the consumption rate of sulfur caused them to refuse an opportunity to operate the second mine to be put into production at Bryan Mound, Brazoria County, Texas. Freeport Sulphur Company was organized, and placed this dome into production in 1912. Freeport Sulphur, in turn, did not take advantage of the offer to operate the Gulf Hill Mine (Big Hill Gulf) in Matagorda County, Texas; thus, again a new company was organized under the name of Texas Gulf Sulphur Company to operate this mine, which began in 1919.

A total of 22 mines were operated before sulfur was found on the Isthmus of Tehuantepec, Veracruz, Mexico. The first mine to be operated in Mexico was the San Cristobal-Capoacan Dome, which commenced operations in December 1953, followed closely by the Jaltipan Dome in September of the following year.

Table I gives the name of the mine, the operating company and starting and closing dates, together with long tons of sulfur produced to the end of 1967.

Some of the mines were abandoned as uneconomical in 1954 and 1955 because of low sulfur prices, but some have since been placed back into production in secondary recovery projects due to the increased demand and price of sulfur.

At the present time, along the Gulf Coast of Texas and Louisiana, there are 10 productive and 8 depleted mines. In Mexico there are 4 productive and 1 depleted.

Figure 1 shows the relative position of mines along the Gulf Coast of the United States and Mexico. There are three plants under construction, and five secondary recovery projects working. (The Caminada mine began operations in late February 1968 and Lake Hermitage started the 4th of March. This leaves only one plant under construction at the present time.)

Since discovery of the Frasch process for mining sulfur, a total of 29 domes have been found productive in the Gulf Coastal area of the United States and Mexico. It is probable that

Table 1. History of Frasch production.

Dome	Company	Date Started	Date Closed	Accumulated Production to 1-1-68
LOUISIANA				
Bay Ste. Elaine	Freeport Sulphur Company	11-19-52	12-29-59	1,131,204
Caminada Pass	Freeport Sulphur Company	2- -68	Operating	-
Chacahoula	Freeport Sulphur Company	2-25-55	9-28-62	1,199,015
	John W. Mecom	7-28-67	Operating	6,288
Garden Island Bay	Freeport Sulphur Company	11-19-53	Operating	7,771,219
Grande Ecaille	Freeport Sulphur Company	12- 8-33	Operating	32,110,786
Grand Isle	Freeport Sulphur Company	4-17-60	Operating	5,875,190
Lake Hermitage	Jefferson Lake Sulphur Company	3- 4-68	Operating	-
Lake Pelto	Freeport Sulphur Company	11-26-60	Operating	3,110,576
Lake Peigneur (Jefferson Island)	Jefferson Lake Sulphur Company	10-20-32	6- 7-36	430,811
Starks	Jefferson Lake Sulphur Company	6-15-51	12-13-60	840,289
Sulphur Mine	Union Sulphur Company	12-27-94	12-23-24	9,412,165
	Allied Chemical Corporation	9-18-66	Operating	19,110
MEXICO				
Jaltipan	Cia. Azufrera Panamericana, S.A.	9-26-54	Operating	12,682,495
Mezquital (Salinas Mine)	Cia. de Azufre Veracruz, S.A.	5- 3-56	Operating	3,382,837
Nopalapa	Cia. Exploradora del Istmo	2- 8-57	2-17-60	322,243
	Cia. Exploradora del Istmo	10-18-67	Operating	8,080
San Cristobal-Capoacan	Cia. Azufre Mexicana, S.A.	12- 7-53	5- 5-57	152,167
Texistepec Mine*	Cia. Central Minera, S.A.	6-12-59	11-20-61	9,114
TEXAS				
Big Creek**	Union Sulphur Company	3- 6-25	2-24-26	1,450
Bryan Mound	Freeport Sulphur Company	11-12-12	9-30-35	5,001,068
	Hooker Chemical Company	6-20-67	Operating	1,620
Clemens Dome	Jefferson Lake Sulphur Company	5- 3-37	12-14-60	2,975,828
Damon Mound	Standard Sulphur Company	11-11-53	4-20-57	139,618
Fannett Dome	Texas Gulf Sulphur Company	5- 6-58	Operating	1,942,607
Gulf Hill	Texas Gulf Sulphur Company	3-19-19	8-10-36	12,349,597
	Texas Gulf Sulphur Company	10- 8-65	Operating	212,922
High Island Dome	United States Sulphur Company	3-25-60	2- 8-62	36,788
Hoskins Mound	Freeport Sulphur Company	3-31-23	5-26-55	10,895,090
Long Point	Texas Gulf Sulphur Company	3-19-30	10-19-38	402,105
	Jefferson Lake Sulphur Company	6- 7-46	Operating	4,816,204
Moss Bluff	Texas Gulf Sulphur Company	6-24-48	Operating	5,272,576
Nash Dome	Freeport Sulphur Company	2- 3-54	11-23-56	153,115
	Phelan Sulphur Company	11- 7-66	Operating	54,944
New Gulf (Boling Dome)	Union Sulphur Company	11-14-28	8-30-29	9,164
	Texas Gulf Sulphur Company	3-19-29	Operating	62,608,170
	Duval Sulphur Company	3-23-35	4-25-40	571,123
	Baker Williams	6- 2-35	12-18-35	1,435
Orchard Dome	Duval Corporation	1-29-38	Operating	5,245,345
Palangana	Duval Sulphur Company	10-27-28	3-10-35	236,662
Spindletop	Texas Gulf Sulphur Company	5-12-52	Operating	6,854,393

*Mine closed down, dome not depleted.
**Not shown on the map, as it is not considered commercial.

(Data by Pan American Sulphur Company.)

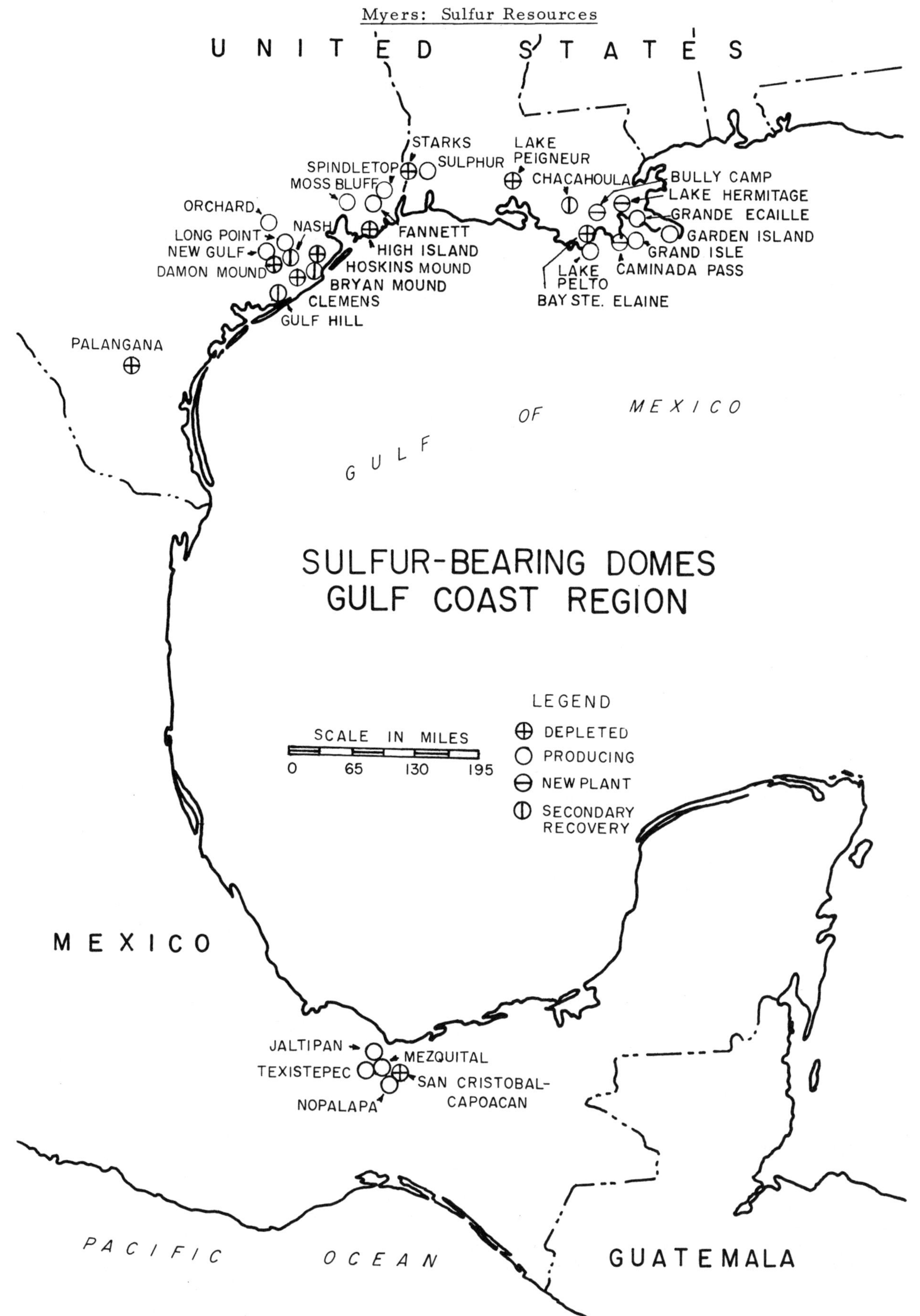

FIG. 1. Map showing the position of sulfur mines along the Gulf Coast of the United States and Mexico.

Cia. Exploradora del Istmo (a Texas Gulf Sulphur Company subsidiary) will place the Nopalapa Mine back in operation once the stockpile of sulfur has been shipped to market.

CHARACTERISTICS OF SULFUR DEPOSITS

Figure 2 illustrates the position of sulfur with relation to the shoulder of various productive salt domes. It was prepared in this manner because varying depths of the domes appear to have some relationship to the size of the deposit, as well as the size of the dome itself. Contours represent the center of the shoulder of each dome.

A salt dome has, in a general way, a gentle slope of the caprock away from a common center, and as the edge of the dome is approached, the dip increases rapidly to nearly vertical. The shoulder of a dome is defined as the area embraced by the mid-point between the gentle slope on top of the dome and the steep incline at the periphery. One observes that New Gulf (Boling) Dome has a shoulder illustrated by the 1,000-foot contour. It is of interest to note that Texistepec and Nopalapa in Veracruz, Mexico, appear to be more or less salt ridges, and it is difficult to bring them into the picture as real salt domes. However, the character of the caprock in the sulfur-producing zone is more or less identical with that of other sulfur domes of the Gulf Coast. Figures in the center of the domes show the number of acres within the contour.

It is apparent that the position of sulfur deposits with respect to the dome is not uniform, nor does the pattern repeat itself from one dome to the next. Smaller domes have sulfur completely across the top, such as at Sulphur, Calcasieu Parish, Louisiana, whereas medium-sized

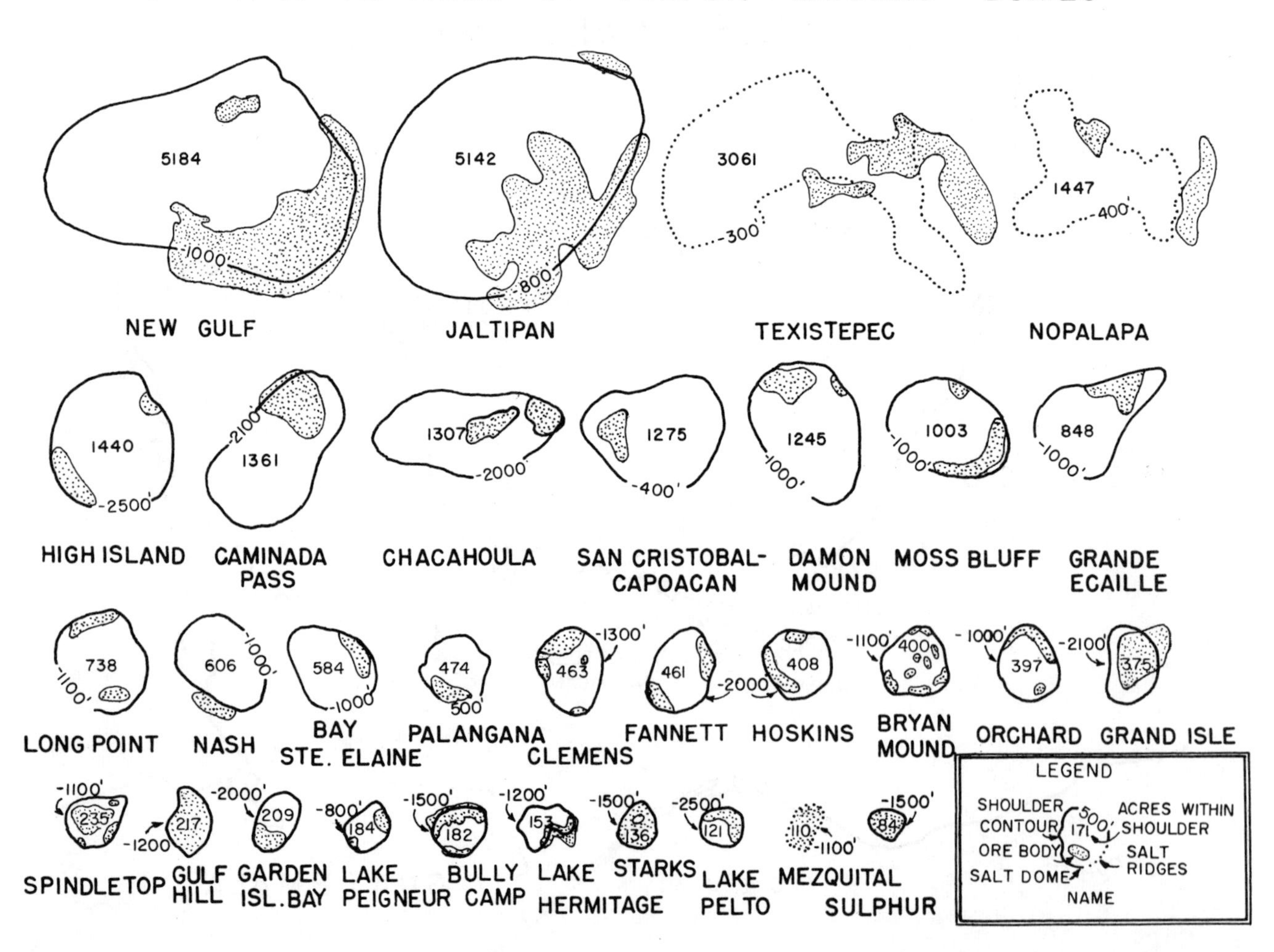

FIG. 2. The position of sulfur with respect to the shoulder of various productive salt domes.

domes have small deposits scattered about the top and periphery with the greater part near the shoulder, such as at Bryan Mound, Brazoria County, Texas. On some domes, the deposit extends from the shoulder to near the apex as at Chacahoula, LaFourche Parish, Louisiana, and at Jaltipan, Veracruz, Mexico. The observation that the deposit most commonly lies along the shoulder of the dome suggests that a common environment exists at that place. This environment reached completely across the smaller domes, and only part way to the apex on other domes.

The largest deposits occur on the larger domes such as New Gulf (Boling), Wharton County, Texas, and also at Jaltipan, Veracruz, Mexico, probably because there is a greater area of caprock exposed to a common environment.

even.

On some of the domes, alteration within the deposit is not complete from gypsum to limestone and sulfur, and the deposit is a matrix of limestone, sulfur, and gypsum in unequal quantities, both laterally and vertically. This type of mineralization appears to affect almost all of the deposit at Bryan Mound, Brazoria County, Texas. On others, this incomplete alteration occurs mostly near the edges of the deposit.

On many of the domes there is more than one horizon or mineralized zone; sometimes as many as three or four zones occur. If this condition exists between two rich deposits, it may be that only one or two zones will carry completely from one deposit to the other on the same dome. This condition exists at Chacahoula and Jaltipan.

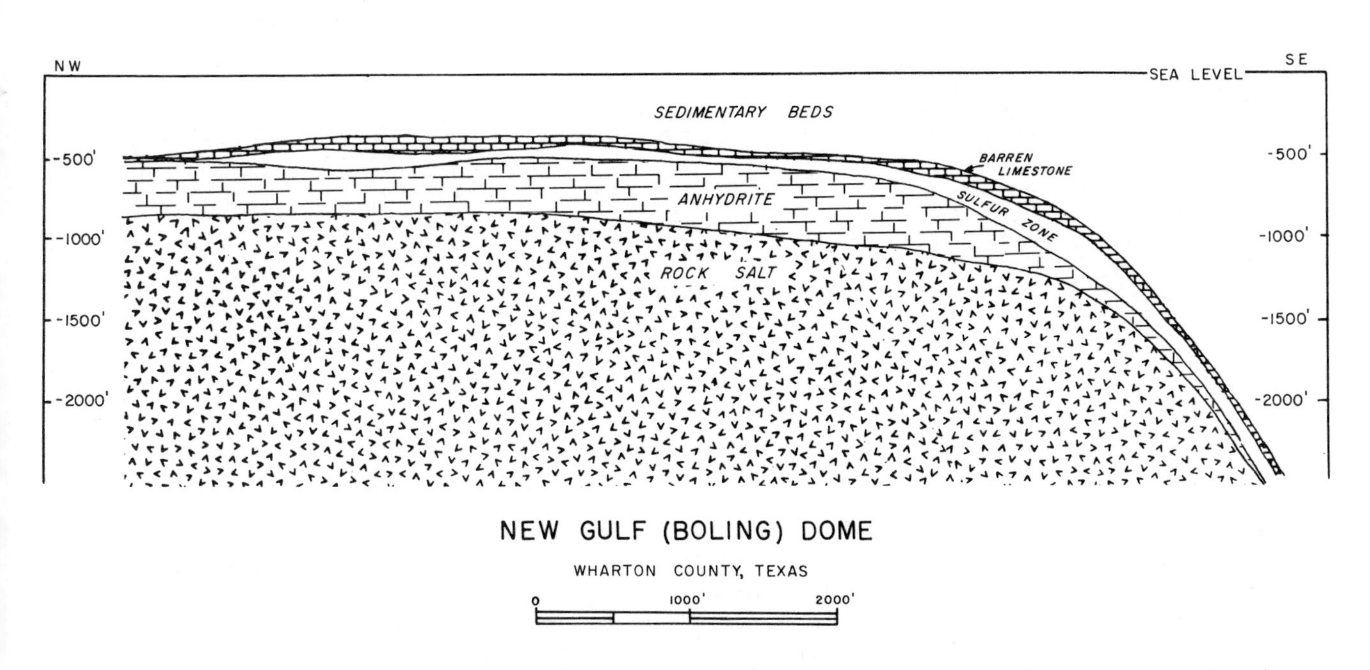

FIG. 3. Cross section of New Gulf (Boling) Dome (after Texas Gulf Sulphur Company, 1959).

At New Gulf, shown in Figure 3, the sulfur zone is thickest at the shoulder of the dome and gradually thins to a depth of about 2,500 feet below sea level. There is a low place near the interior edge of the sulfur zone which is thicker than average for the rest of the deposit on the dome. The linear distance is 7,180 feet from the -2,500-foot level to the limits of the ore deposit on higher parts of the dome. Thin sections of barren limestone occur above the sulfur zone. The thickest part of the sulfur zone, measured at right angles to the slope of the base of the zone, is about 380 feet, whereas the limestone at its maximum thickness is approximately 96 feet. The sulfur zone covers a vertical interval of 2,000 feet in depth. It is interesting to note that the top and bottom of the sulfur zone is fairly

Sometimes the caprock has been eroded over a part of the dome, suggesting that shallow water conditions existed at the time the sulfur was formed, or relatively soon thereafter. This phenomenon could also account for the absence of a sulfur body on other parts of some domes.

At Lake Peigneur (Jefferson Island) Dome in Iberia Parish, Louisiana (Fig. 4), sulfur occurs in a limestone caprock shelf which was formed at one time and then part of the dome later rose to the surface and the caprock was eroded from this later uplifted area. It is of interest that the sulfur zone of this dome has a maximum thickness of around 160 feet, whereas the limestone has a maximum thickness of about 100 feet. Beneath the sulfur zone are areas of broken caprock material, approximately 60 feet in thickness,

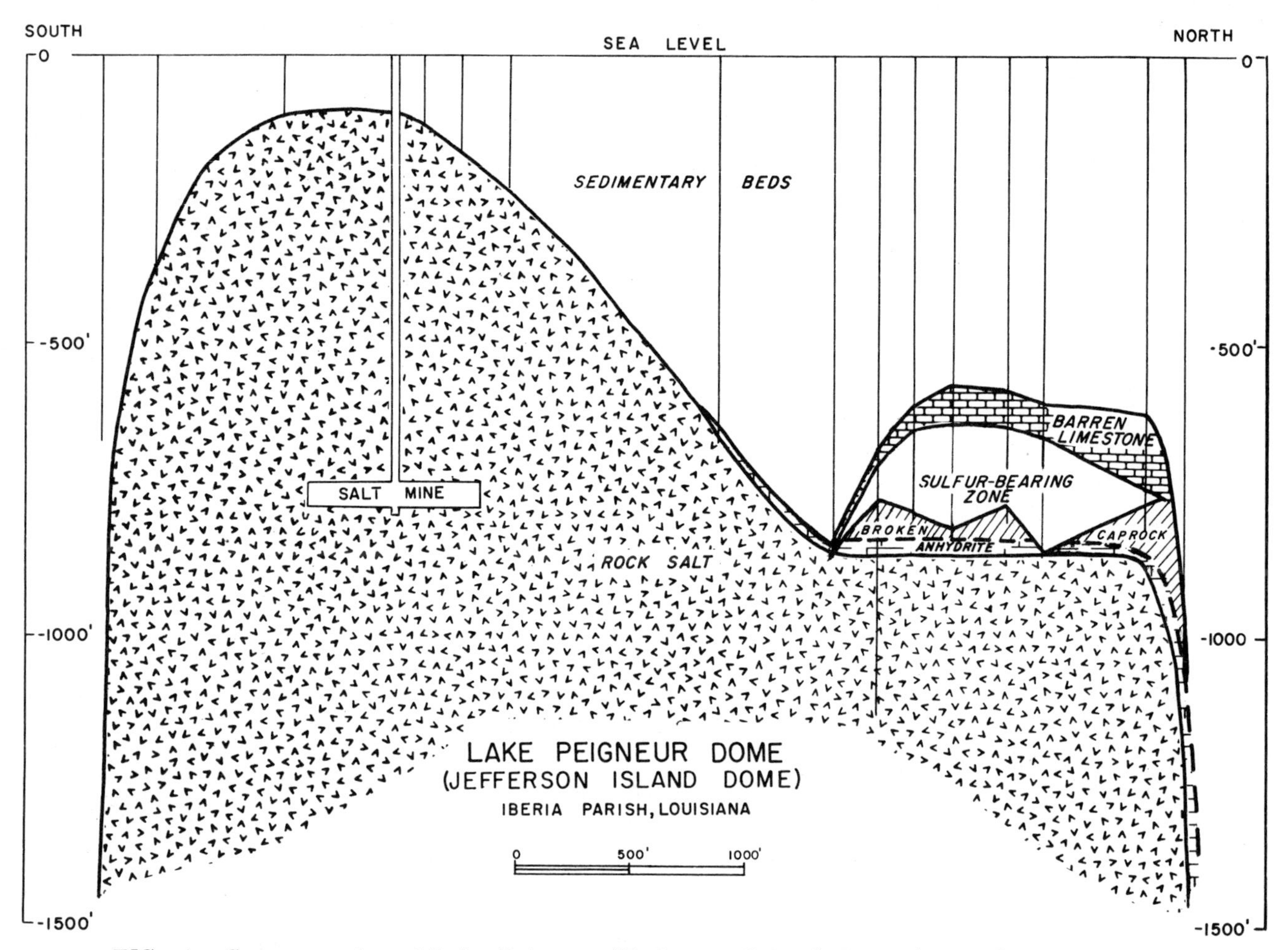

FIG. 4. Cross section of Lake Peigneur (Jefferson Island) Dome (after O'Donnell, 1935).

followed below by anhydrite not over 35 feet thick. This cross section shows the anhydrite to extend below 1,500 feet on this dome. The sulfur zone extends from a depth of 640 to 840 feet. Thus, it appears that sulfur must have formed much more rapidly than on some other domes as evidenced by the magnitude of the vertical interval through which sulfur occurs.

At Orchard Dome in Fort Bend County, Texas, the ore body repeats itself beneath an overhang of caprock. This may suggest that sulfur was forming during final movement of the dome, or that it could have formed before latest uplift, or while upward movement was still in progress. The sulfur zone occurs at approximately 350 feet on top of the dome to as low as 2,200 feet on the flanks of the dome or a vertical interval of 1,850 feet.

At Hoskins Mound, Brazoria County, Texas, shown in Figure 5, a continuous thick body of limestone, more or less barren of sulfur, extends to the transition zone of sulfur and limestone (which is the sulfur ore body), and grades immediately below into anhydrite without the occurrence of gypsum. This is in contrast to the condition at Bryan Mound, where the transition zone is composed of gypsum and sulfur with an uneven contact with the anhydrite body below. Again at Hoskins Mound, it is noted that the greatest thickness of sulfur is in a zone along the shoulder of the dome, but in some instances, it does carry to near the top of the dome, involving a vertical interval of 1,000 feet. The sulfur zone reaches a maximum thickness of about 325 feet, and has a very irregular surface at the top and bottom. Barren limestone above the sulfur zone has a maximum thickness of about 260 feet. In some places the limestone occurs both above the sulfur zone and beneath it before contact with the underlying anhydrite.

Figure 6 is a cross section of Sulphur Dome at Sulphur, Louisiana. The section extends from west to east and is typical of practically all sections drawn either north and south, east and west, or intermediate between these points. Here again there is a barren limestone body whose thickness is about 155 feet. Maximum thickness of the sulfur zone is 100 feet, with the exception of the

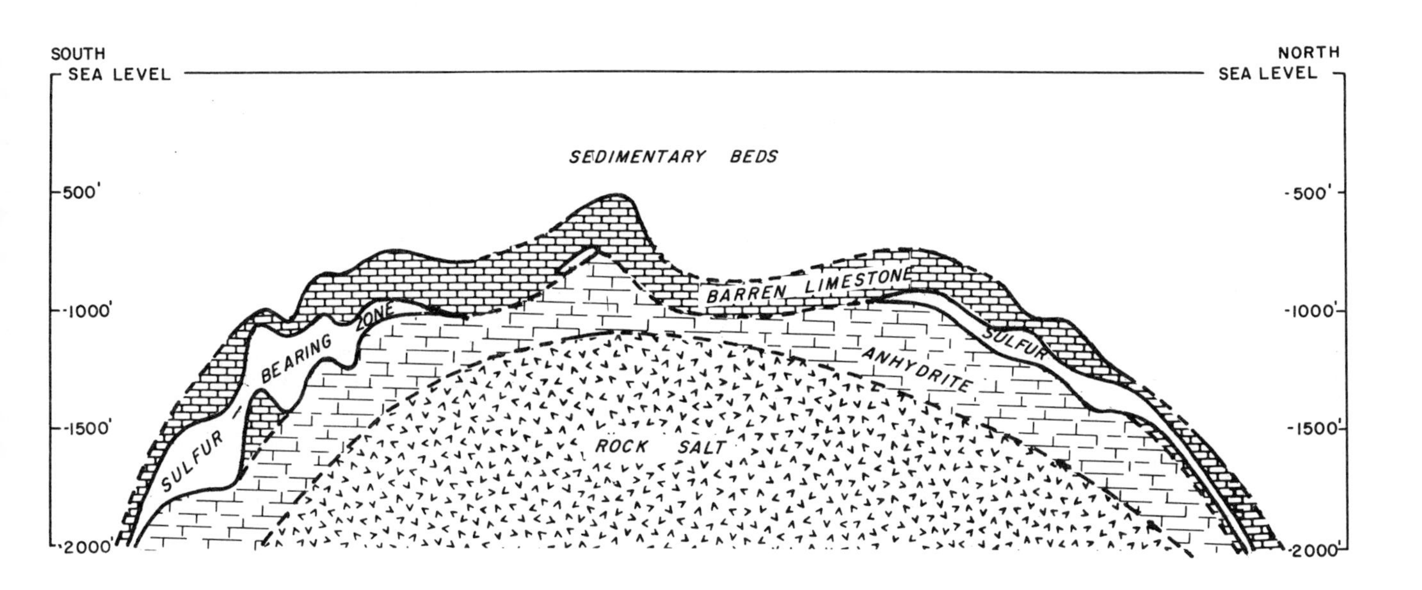

FIG. 5. Cross section of Hoskins Mound Dome (after Marks, 1936).

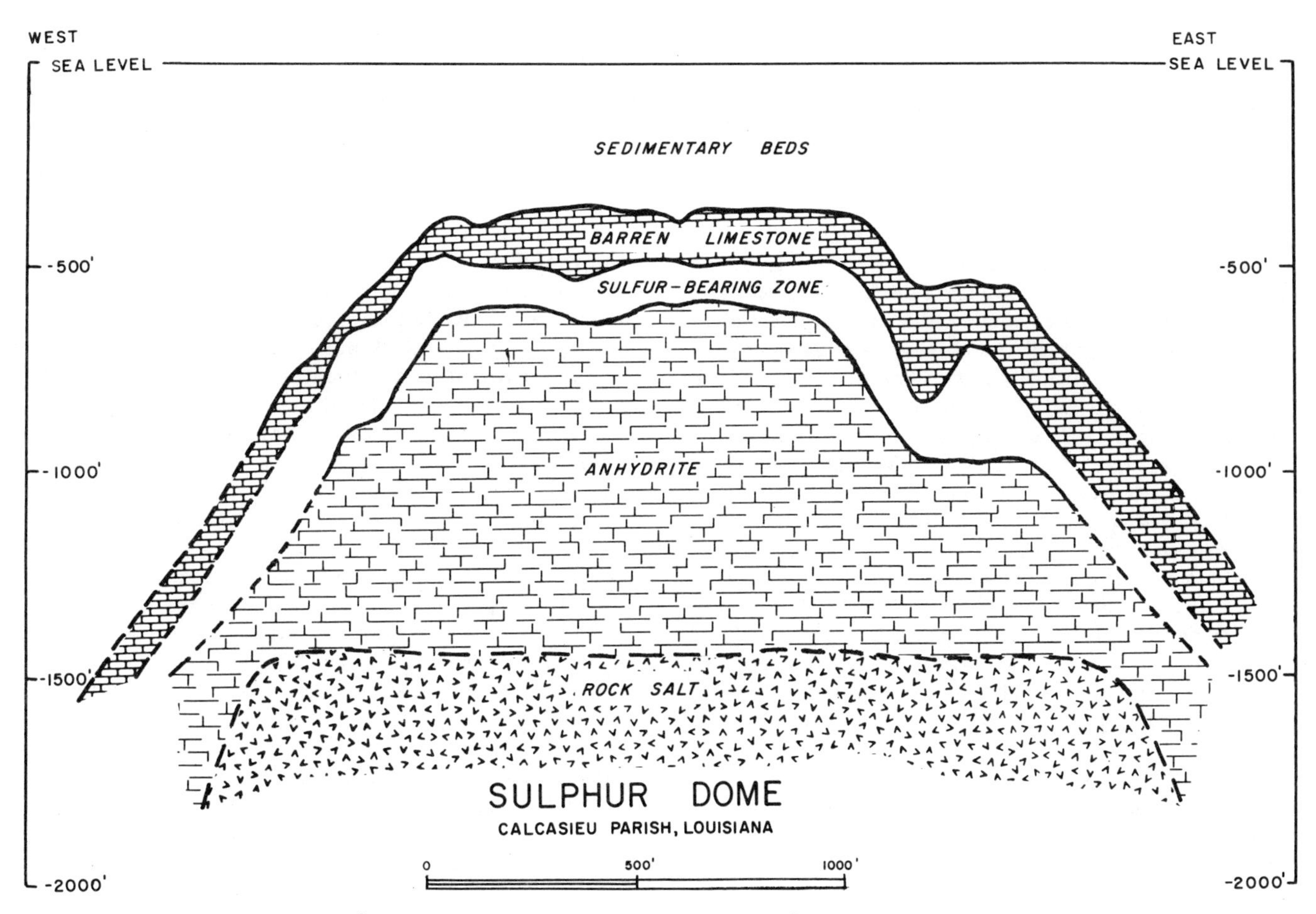

FIG. 6. Cross section of Sulphur Dome (after Kelly, 1925).

thick "hill" at the right side. It is interesting to note that the anhydrite zone has a thickness of 850 feet near the center of the dome, and that the top of the salt mass is flat. At Lake Peigneur the top of the salt was probably flat also during the formation of sulfur. The sulfur zone at Sulphur Dome covers a vertical interval of from 500 to below 1,500 feet, or through 1,000 feet of elevation. The linear distance across the dome from shoulder to shoulder is about 2,100 feet, and the top of the dome is practically circular and embraces about 84 acres.

These cross sections suggest that the domes had reached isostatic stability at the time sulfur was formed, and if so, some of them have risen at a later date. At Hoskins Mound, the vertical difference in the position of the sulfur amounts to 1,000 feet; at New Gulf 2,000 feet; at Sulphur 1,000 feet; and at Lake Peigneur it is practically stable.

Lake Peigneur and Sulphur are definitely piercement type domes and small in size. The latter shows a sulfur zone extending completely across the top. This may have been the case at Lake Peigneur if isostatic adjustments had not taken place causing the dome to rise a second time so that the caprock was partially eroded.

If thickness of caprock depends on content of anhydrite in the salt and the rate of upward movement of the salt, then many of the domes that are completely covered by the caprock have had varied rates of upward movement. Smaller domes appear to have had a more rapid rise than larger ones, and the salt top is relatively flat. On larger domes, the surface of the salt is convex rather than flat.

There does not appear to be any particular relationship between thickness of the anhydrite-gypsum part of the cap and the sulfur zone above. The degree of alteration of the sulfate series is determined by stability of the salt dome, and stability also determines the thickness of the sulfur zone. The fact that many domes have more than one sulfur ore body at varying depths could mean that alteration took place at different times, or at the same time at different levels.

The sulfur zone lies in a matrix of limestone for the most part, but there is no genetic relationship between the sulfur zone and the barren limestone above and below. It is difficult to conclude that limestone is not a by-product of alteration of gypsum-anhydrite to limestone and sulfur, yet there is always barren limestone above the sulfur zone and rarely any below the sulfur zone. The uneven thickness of the barren limestone above the sulfur zone suggests faulting, unequal erosion of the limestone, or unequal reduction of the sulfate cap. If limestone and sulfur are formed at the same time, it appears that sulfur should be present on any part of the dome that has limestone, but on almost all domes this is not true. Limestone and sulfur, therefore, may not have been formed in the same environment, and perhaps not contemporaneously.

It appears that paragenesis of calcite and sulfur in salt domes is a direct result of the alteration of anhydrite, but not under the same conditions nor at the same time. Stratified limestone and sulfur suggest anaerobic bacterial reduction of sulfate to hydrogen sulfide. The presence of hydrocarbons may help to create the proper environment for bacterial action.

FUTURE RESOURCES

Frasch reserves of sulfur in the Gulf Coastal area of the United States and Mexico are kept confidential, but a general estimate is 170 million long tons.

Future resources of sulfur, in addition to the reserves of the present productive domes and those to be brought into production in 1968, must come from old domes that have undeveloped reserves, or from unprospected domes that occur offshore from the United States in the Gulf of Mexico. There is also the prospect of finding additional domes on the Isthmus of Tehuantepec, Veracruz, Mexico, that have not been proved or tested.

In respect to the prospects of finding additional productive domes offshore in the United States, it is a statistical fact that out of 37 domes in Texas and 41 in Louisiana onshore along the Gulf Coast, with the top of the salt encountered no deeper than approximately 3,000 feet below the surface, only 24 domes, or about 31 percent, have sulfur deposits, but one of these is noncommercial. There are 3 additional domes to be put into operation that have commercial deposits of sulfur, bringing the total to 27, or approximately 35 percent of the total number of domes. These domes are all located in the Gulf Coastal section of Texas and Louisiana.

If interior domes are included, it would add 14 in East Texas, 15 in North Louisiana, 34 in Mississippi, and 1 in Alabama, making a total of 64 additional domes that are nonproductive. This would make a total of 142 domes onshore, of which 26, or 18 percent, have commercial deposits. Most domes in Mississippi and East Texas have been tested, and probably those in North Louisiana.

There are 23 additional domes with possible caprock above 3,000 feet in Louisiana's offshore, and 5 domes in Texas. Freeport Sulphur Company drilled 19 wells on 4 additional salt domes in 1966, with negative results. How complete these domes have been tested is not known, but the offshore domes will probably not be tested as well as those onshore because of increased cost. If these 32 offshore domes are added to the

142 onshore, it makes a total of 174 domes, 26 of which have been found productive, or approximately 15 percent of the total. Thus, statistically, odds are against finding productive domes offshore beyond the present productive belt. Any prediction of future resources from Frasch sulfur production on salt domes must consider the results of past prospecting.

There is a geological factor which enters into the consideration of future prospecting. A glance at the map in Figure 1 shows that known sulfur-producing domes follow a narrow belt which roughly parallels the coastline. Does this mean that only those domes have sulfur deposits which have a common depth and environment for the generation of or formation of hydrogen sulfide gas or native sulfur? This could be a factor. Since structural contours on any horizon do not exactly follow the present shoreline of the Gulf Coast in Texas and Louisiana, and since this belt of sulfur domes does parallel the shoreline, it can mean that only certain conditions provide the environment necessary to the formation of sulfur. Due to isostatic adjustment, only certain ones are present at a given time and depth to provide this environment, and since the adjustment is greater in the immediate Mississippi River Delta, the belt follows shoreline rather than structural contours.

In general, the sulfur domes have limestone caprock around the shoulder of the dome, followed below by gypsum and below this impermeable anhydrite resting on rock salt. The sulfur body lies in the transition zone between limestone above and gypsum-anhydrite below. In a rich deposit, the sulfur body is entirely limestone and sulfur in equal amounts. Sulfur is crystalline, finely disseminated and even amorphous. Limestone is stratified in places, and in other places exhibits solution brecciation. It is only near the base of an ore body, or near the outer limits, that gypsum is present in any appreciable amount. Some deposits have a knife-edge contact between barren limestone at the top and the ore body of sulfur and limestone, and at the bottom, a clear contact either with barren, massive gypsum, or barren, granular anhydrite. Poorer deposits exhibit an ore deposit that is irregular in texture, being limestone and sulfur with some gypsum and anhydrite in irregular bodies with a lack of distinct division between the ore body and the surrounding caprock material. Thus, the sequence of limestone, to limestone and sulfur, to gypsum-anhydrite is repeated many times.

The carbonate content increases and the sulfate content decreases with an increase of sulfur. Formation of sulfur at normal temperatures and under conditions which confine the formation to land-locked basins, such as appears to be the case of the Upper Permian Ochoa Series of the Delaware Basin in West Texas, need not rule out similar conditions around salt domes or in other areas.

It appears that further drilling and testing of some old domes lying within the indicated productive belt might produce additional tonnages, if the ownership of the land and minerals is made available for prospecting.

Production of sulfur by the Frasch process cannot be done economically by individual leases or a small deposit, because a sufficiently thick deposit and one covering enough area to obtain adequate circulation of mine water is necessary to control the heating process, and therefore, the extraction of sulfur. In a sense, the Frasch process requires a fairly large deposit with mass production to make it economical.

It may be that another belt of sulfur producing domes will be found offshore, but if so, they will be found with fewer tests than those onshore. In the light of lack of sulfur on many of the interior domes and the few producing domes onshore in the present coastal belt, it seems difficult to imagine a repetition of another producing belt of domes offshore.

There does not appear to be any geological conditions existing in the Gulf Coastal area of the United States and Mexico that lends itself to the prospective formation of sulfur in commercial quantities, other than salt domes.

REFERENCES

Hawkins, M. E., and Jirik, C. J. (1966) Salt domes in Texas, Louisiana, Mississippi, Alabama and offshore tidelands, A survey: Bur. Mines Inf. Circ. 8313.

Kelly, P. K. (1925) The Sulphur salt dome, Louisiana: Bull. Amer. Assoc. Petr. Geol., vol. 9, pp. 479-496.

Marks, Archer (1936) Hoskins Mound, Brazoria County, Texas: Bull. Amer. Assoc. Petr. Geol., vol. 20, pp. 155-178.

O'Donnell, Lawrence (1935) Jefferson Island salt dome, Iberia Parish, Louisiana: Bull. Amer. Assoc. Petrol. Geol., vol. 19, pp. 1602-1644.

__________ (1959) Annual Report, Texas Gulf Sulphur Company.

THE LOUANN SALT OF THE GULF COASTAL REGION

George C. Hardin, Jr.
Kerr-McGee Corporation
Oklahoma City, Okla.

ABSTRACT

The more than 400 domes located around and under the Gulf of Mexico in which salt has been penetrated attest to the presence of one or more thick beds of salt underlying this large area. In 1945, Hazzard, Spooner, and Blanpied proposed the name Louann Salt for the salt bed penetrated by the Gulf No. 49 Werner well drilled at Smackover in southern Arkansas. The name Louann has since been extended to include all of the salt occurring in the U. S. Gulf Coast.

Salt has been mined from Avery Island, Louisiana, since 1862, and at present, the salt produced from the salt domes of the Gulf Coast account for about 44 percent of the salt production of the United States. In addition to rock salt, the salt domes of the Gulf Coast supply large quantities of sulfur and oil and gas.

The Louann Salt does not crop out, and our knowledge of Louann stratigraphic relationships is limited to what can be learned from drill holes and from a few rock salt mines. Consequently, the details of Louann stratigraphy remain obscure. The thickness, areal extent, geochemistry, and age of the salt, and whether it was deposited in one or several basins cannot be determined with certainty.

CAP ROCK GENESIS AND OCCURRENCE OF SULPHUR DEPOSITS

Ralph E. Taylor
Frontier Sulphur Company, Inc.
New York, N. Y.

ABSTRACT

The capping of distinctive mineral composition that covers upper contours of the rock salt of salt domes in the U. S. Gulf Coast, Mexico, Germany, and other regions of similar salt occurrence, is known as cap rock. At many of the shallower salt domes of the Gulf Coast, cap rock is a thick, indurated body of rock comprised of a series of layers of anhydrite occurring directly above the salt, overlain by secondary gypsum and calcite, with the calcite and gypsum in places being associated with significant amounts of native sulphur.

Detailed field and laboratory investigations have determined that the anhydrite zone of the cap rock is a residual accumulation of disseminated anhydrite grains leached from the salt, associated with minor amounts of other distinctive water-insoluble mineral grains and rock fragments that also occurred in the salt. Anhydrite has altered to gypsum where suitable depth and time factors have existed.

The native sulphur and calcite altered from the sulphates, anhydrite, and gypsum, in the presence of hydrocarbons. Earlier detailed petrographic analyses revealed that sulphur and calcite are direct replacements of the sulphates. More recent isotopic studies confirm this conclusion, both for cap rock and sedimentary sulphur deposits of Sicily. Work with isotopes also provided evidence that sulphate-reducing bacteria which utilized hydrocarbons were the alteration agent. Sulphur deposits in West Texas, northern Iraq, Poland, U. S. S. R., and elsewhere that are associated with sulphates and hydrocarbons also appear to be due to bacterial action.

Cap rock has been a principal source of native sulphur since the beginning of the century and mines in this country and Mexico provide two-thirds of production for the western world. Oil and gas have been produced in important amounts from cap rock in the Gulf Coast. Calcite and gypsum have been quarried at Pine Prairie and Winnfield salt domes in Louisiana and Gyp Hill and Hockley in Texas.

Many new data relating to cap rock have been developed through application of modern coring methods at more recently developed sulphur mines. Detailed studies of these data and publication of the results would add importantly to the knowledge of cap rock.

Deposits of native sulphur associated with the sulphates gypsum and anhydrite have been the principal world sources of elemental sulphur from the time that sulphur became a dominant industrial chemical raw material in the 19th century. First to be exploited were deposits of Sicily and southern Italy associated with sedimentary gypsum. At the close of the 19th century, Herman Frasch developed the hot water injection method of mining which brought the Gulf Coast deposits into production.

Gulf Coast salt domes (Fig. 1), their cap rock and sulphur were subjects of great interest to geologists, following discovery of salt at Avery Island, Louisiana, in 1862 and discovery of sulphur at Sulphur, Louisiana, in 1865. Various theories were advanced to account for these occurrences. They were first believed to be Cretaceous outliers, and then uplifts due to vulcanism, to gas, and to crystallization forces of salt from saline springs. Eventually comparison was made with comparable salt structures of the North German basin that were interpreted to be the result of intrusion of salt in plastic state from deep sedimentary beds (Fig. 2). German salt domes had been studied in detail and their origin well established through correlation of the shallow Zechstein rock salt and potash beds with the same beds in mines in the salt domes (Fig. 3). In this connection, it is of much interest that the German geologists G. Richter Bernburg and E. Hofrichter, who supervised extensive potash core drilling at Palangana salt dome, Duval County, Texas, were able to correlate rock salt and potash beds between core holes, establishing the stratigraphic sequence. Hofrichter (1968) recently reported on this work.

Cap rock, which has long been recognized at German salt domes, being referred to as "gyps hut" or gypsum cap, has been explained as an accumulation of residue leached from salt. Typically, gyps hut is of thin and irregular develop-

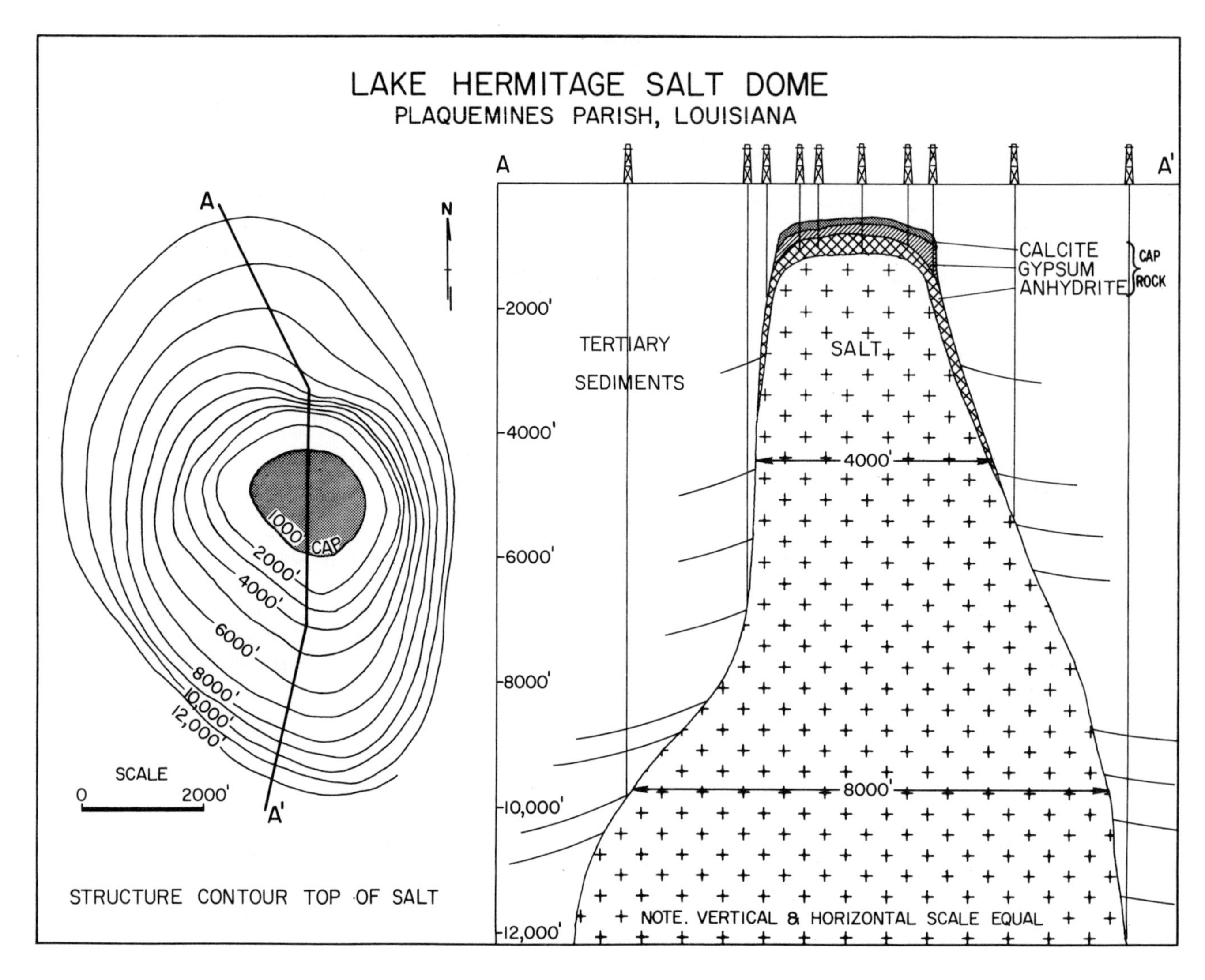

FIG. 1. Position of cap rock with respect to salt at Lake Hermitage salt dome, Louisiana.

ment. By contrast, typical cap rock at Gulf Coast salt domes is thick and massive, composed essentially of anhydrite, quite unlike gyps hut. As a result, Gulf Coast geologists originally did not think of Gulf Coast cap rock as a residual accumulation, but explained it as a block of sedimentary anhydrite carried upward as the salt was extruded. This idea has acceptance today among some geologists in spite of evidence from a number of field and petrographic studies. The remarkable thickness of some cap rock bodies, which require great thicknesses of salt to explain their formation as an accumulation of salt residue, as well as the occurrence of salt and related rocks in unconsolidated sediments of the Gulf Coast, are features that must be explained. For this reason a brief review of the geology of Gulf Coast salt domes is included at this point.

Cap rock at Lake Hermitage salt dome (Fig. 1) is typical of that at shallow salt domes. It rests on a virtually flat salt table and does not extend appreciably down the flanks of the salt plug. Fortunately, at several other localities cap rock is exposed at the surface and can be examined in detail. In quarries at Pine Prairie cap rock consists of massive, coarsely crystalline, fractured, secondary limestone, underlain by massive gypsum. Limestone and gypsum also have been quarried at both Winnfield, Louisiana, and Gyp Hill, Texas, salt domes. In addition, there are limited surface exposures of cap rock at Damon Mound and Hockley, Texas. At several localities in Louisiana and Texas, shallow cap rock is rimmed by reef limestone, as at Damon Mound and Nash, Texas, where coral of the *Heterostegina* zone has been cored. It is, of course,

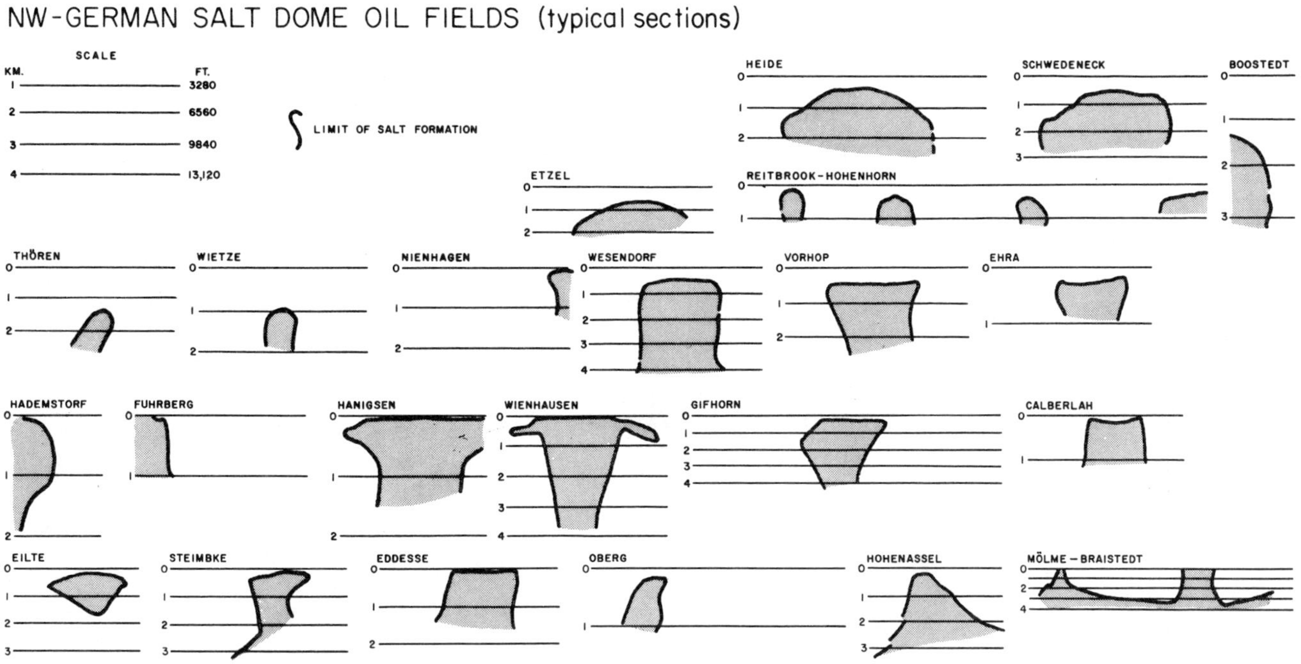

FIG. 2. Configuration of salt domes, North German salt basin.

evident that cap rock, which was exposed in the shallow Miocene Heterostegina seas, was formed at or near the sea floor well before that time. On the other hand, cap rock of appreciable thickness, 100 to 150 feet, has been reported at depths of 6,500 to 7,500 feet, overlying deep salt at domes in both Louisiana and Texas. Typically, however, thicker cap rocks from 600 or 700 to 1,500 feet in thickness normally occur at depths of a few hundred to 1,500 feet.

Geology and mineralogy of salt in salt domes is of course of primary importance in determining the genesis of cap rock. A sharp contact between cap rock and underlying salt occurs at both Lake Washington, Louisiana, and High Island, Texas. In both occurrences, high angle dark bands are quite distinctly apparent in the salt and are cut off abruptly at the sharp salt-cap rock contact. As will be shown in detail later, these dark bands are composed essentially of anhydrite grains, and are a distinctive feature of most Gulf Coast salt, as well as salt dome and bedded salt of some other regions. The cap rock at Lake Washington and High Island as revealed in cores is also banded parallel to the contact; this parallel banding resulted from development of secondary mineralization along partings or layers. In the Lake Washington dome in particular, an intricate pattern of minute faults occurs along the parallel bands.

At salt domes with thick cap rock, the top of the salt is generally a flat solution salt table. Where extensive drilling has penetrated this contact, as at the sulphur mines, a sharp contact is unusual, and in many cases a cavity containing brine with anhydrite sand is encountered.

Anhydrite bands in the salt were well illustrated in the now abandoned salt mine of Carey Salt Company at Winnfield. The concentration of the anhydrite grains in closely spaced parallel bands, the detrital appearance of the grains, and the elongation of the halite grains parallel to the anhydrite bands provide evidence of the intrusive origin of the salt domes by plastic flow of the salt from deeply buried salt beds. These anhydrite bands originated as "yargang" or year bands when the bedded salt was deposited.

Mineralogy of the salt, especially of the bands, is of particular interest in the genesis of cap rock. Extensive chemical and petrographic studies of the salt, both in Germany and in the Gulf Coast, provide a good understanding of the mineralogy. Petrographic thin sections of salt illustrate some fragmentation of anhydrite grains which probably occurred when the salt was intruded. Dolomite rhombs with nuclei are distinctive of Gulf Coast salt, although dolomite represents much less than one percent of the salt body.

The composition of sea water (Sverdrup, 1942) and the composition of whole samples of salt from a salt dome and a bedded deposit are very similar (table 1). In addition, the presence of significant amounts of calcium and sulphate ion provide an adequate source for anhydrite. Large volumes of almost pure anhydrite sand have been washed

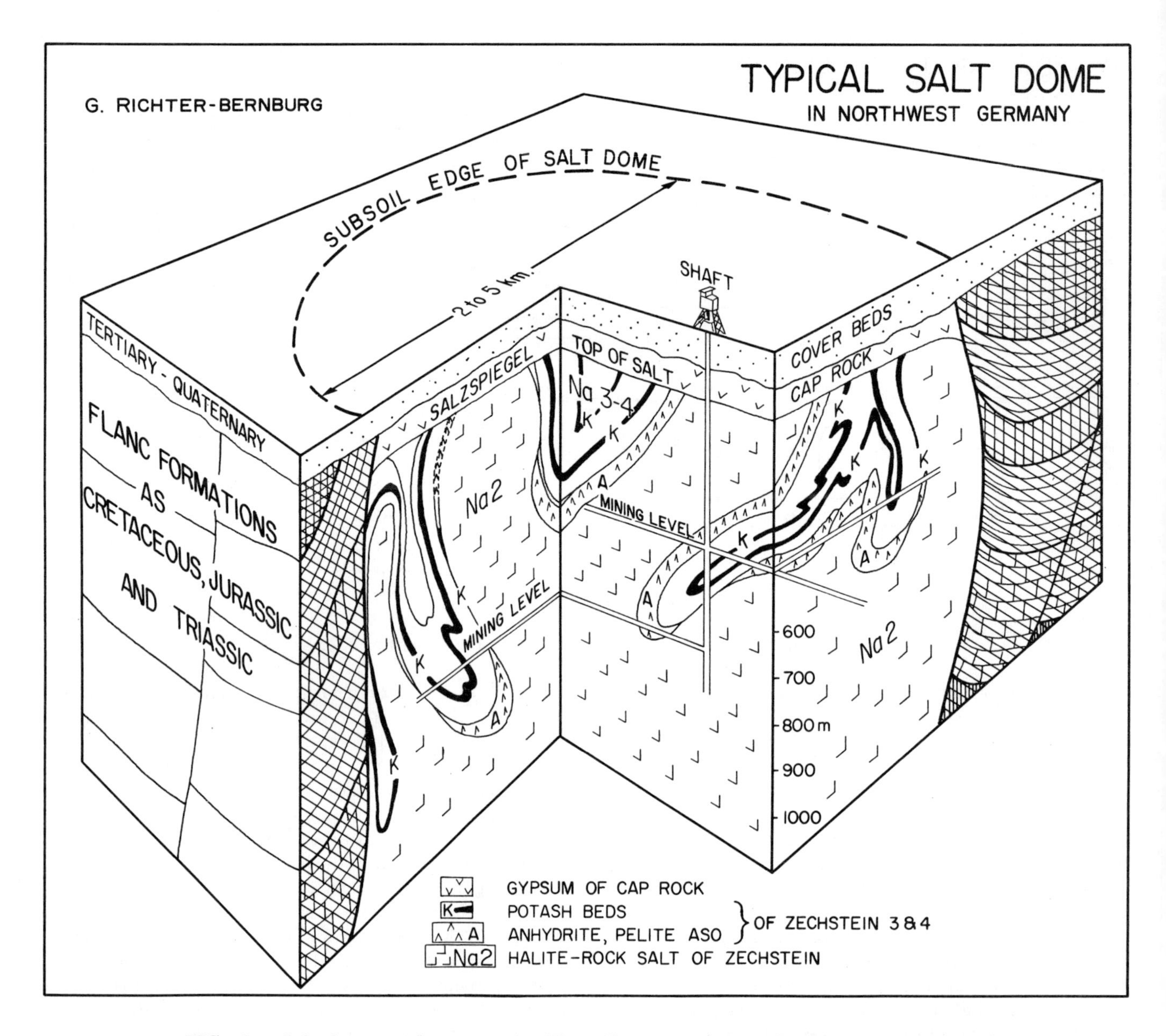

FIG. 3. Salt dome with cap rock, West Germany (after G. Richter Bernburg).

from brine wells of Dow Chemical Company at Stratton Ridge salt dome, Texas, and were piled in dumps of impressive size. Similar anhydrite sand dumps have accumulated at brine operations at Palangana, Pierce Junction, Bryan Mound and Sour Lake, Texas; Sulphur, West Hackberry and Choctaw, Louisiana; Petal, Mississippi, and McIntosh, Alabama, among others.

The amount of water-insoluble residue in salt is shown in Figure 4 for two salt domes, Bryan Mound and Stratton Ridge, as determined from cores of Dow Chemical Company brine wells. The amount of residue varies not only within a dome but from one dome to another, which suggests that the series of deeply buried salt beds--the "mother" salt--had a variable residue content and that different salt domes were supplied from different parts of the buried salt section. For example, in the Five Islands, all of which now have been opened for salt mining, certain of the mines are composed predominantly of hard, closely banded salt, whereas others are essentially a soft, friable salt with relatively sparse banding. Studies of these mines are being conducted by geologists at Louisiana State University (Kupfer, 1965) and eventually should shed light not only on the stratigraphy of the salt but on the thickness and, hopefully, the age. Spores (Jux, 1961), which are of rare occurrence and the only fossils so far recognized in salt, have not given definitive results but suggest Triassic or Jurassic age.

Table 1. Comparison of chemical analyses of sea water and brine from bedded salt and salt-dome salt.

	Salts in Sea Water (Sverdrup, after Dittmar and others)	Humble No. 1 Williams Clark Co., Alabama Depth 15,666-667 feet Bedded Salt	Clovelly Dome, Louisiana Salt
Cl	55.04	57.17	53.97
Br	0.19	---	---
SO_4	7.68	2.67	3.46
HCO_3	0.41	---	None
H_3BO_3	0.07	---	---
Mg	3.69	0.04	0.10
Ca	1.16	1.07	1.37
Sr	0.04	---	---
K	1.10	0.22	0.88
Na	30.61	37.09	35.02
Si	0.01	---	0.05
Fe	0.0001	---	0.24

Sampled intervals of salt from mines and cored wells have shown that water-insoluble residues are composed almost wholly of anhydrite grains, except those for short sections in wells at Choctaw, Bryan Mound, and Pierce Junction. Trace amounts of a number of minerals occur in certain of the localities, particularly Choctaw.

Distribution of the minerals identified in residues from 37 salt domes that were examined is illustrated in table 2. A limited mineral suite is present at most salt domes, typically including anhydrite, dolomite, quartz, calcite, and pyrite. Anhydrite grains, consisting of both crystal forms and cleavage fragments, are strikingly similar at all localities extending from McIntosh, Alabama, across Mississippi, Louisiana, Upper Gulf Coast of Texas, to Palangana in South Texas. A few salt domes are characterized by distinctive assemblages, such as Choctaw with rare and abundant borates: boracite, danburite, hilgardite, and parahilgardite. The only other known occurrence of these last two borates is at Heidelberg salt dome, Mississippi. Recently another new borate mineral, chambersite, was described from brine wells at Barbers Hill, Texas, and Venice, Louisiana. Further studies of water-insoluble residues from salt domes are desirable to aid in determining stratigraphy of the mother salt bed.

Large inclusions of rock occur in the salt or cap rock at several domes, including fragments of Lower Cretaceous limestone at McFaddin Beach, Texas, and Eocene shale at Bryan Mound and Pierce Junction, Texas, and Bayou Bleu, Louisiana. Large sedimentary sandstone inclusions at Avery Island and Jefferson Island, Louisiana, and quartzite at Belle Isle and Jefferson Island indicate plucked fragments of rocks contemporaneous with or older than the salt. Inclusions of anhydrite rock fragments are common.

An unusual series of anhydrite beds intercalated in the salt was encountered in 20 feet of core directly below cap rock in an oil test at an offshore locality (Block 30, West Delta, Louisiana). Imbrication and fragmentation of anhydrite bands in the core reflect great stresses encountered during salt intrusion. Petrographic thin sections of massive, indurated anhydrite cap rock just above the salt illustrate the detrital nature of anhydrite cap rock with its typical composition of anhydrite grains resembling those

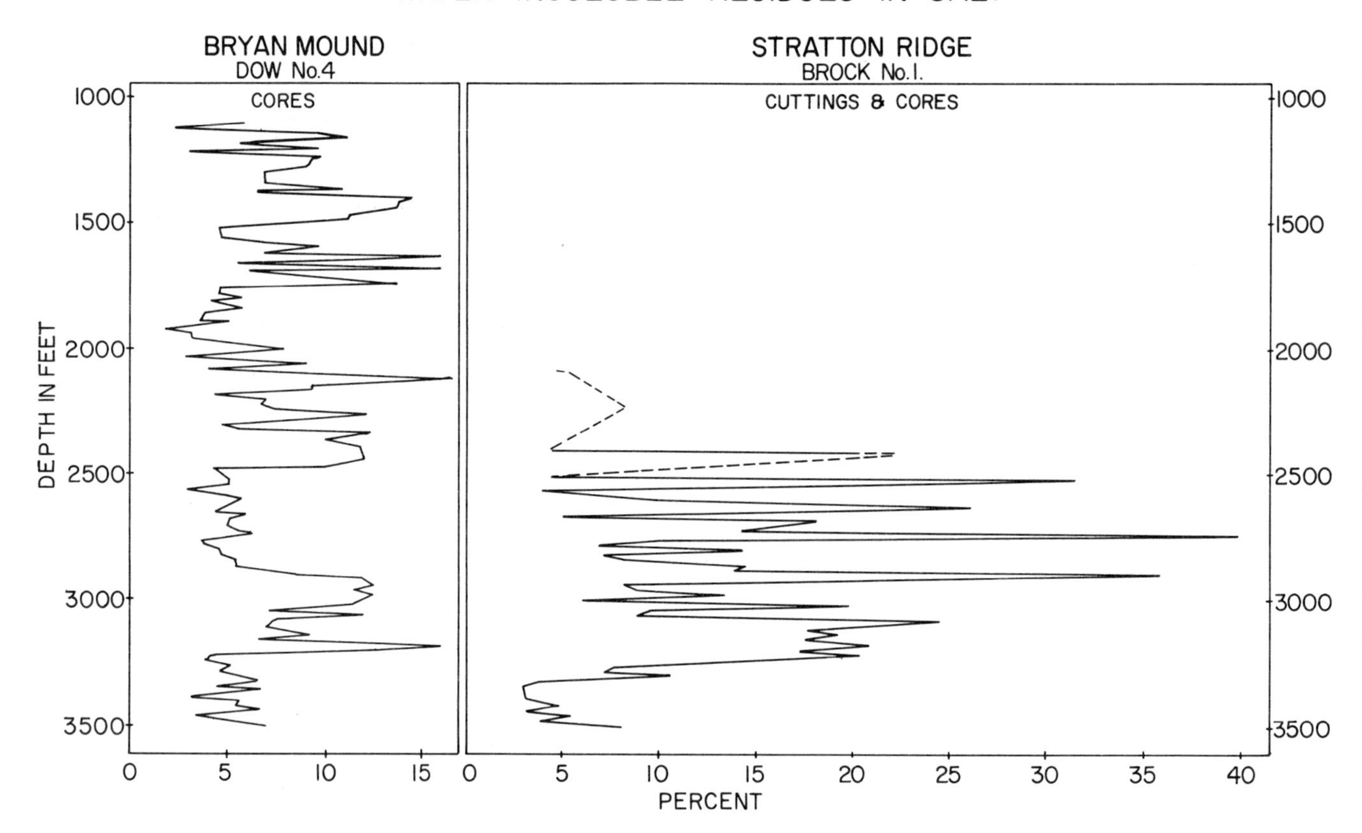

FIG. 4. Percentages of water-insoluble residues in salt from Bryan Mound and Stratton Ridge salt domes, Texas.

from salt residues.

In an earlier paper, the writer (Taylor, 1938) traced a distinctive suite of minerals from the salt residues into the cap rock. In particular, distinctive dolomite rhombs and quartz rosettes of the salt residues were traced into the anhydrite cap rock in detrital positions with anhydrite grains, upward within the cap rock into gypsum of the transition zone, and into the overlying calcite zone. Of particular significance also was replacement of the dolomite rhombs, with the retention of the distinctive nucleus, by both gypsum and calcite. On the basis of petrographic evidence, there seems to be no other tenable conclusions than that, like the gyps hut of the German salt domes, anhydrite cap rock of the Gulf Coast salt domes is a residual accumulation of the water-insoluble residue leached from the salt. The difference between the relatively thin gyps hut of German salt domes and the thick, massive anhydrite cap rock of Gulf Coast salt domes results from the difference in the inclusions in the salt of the two regions. Of interest in this connection is the discovery that in sulphur prospects drilled in the Texas Federal offshore lands of the outer continental shelf, only a thin capping of gypsum and shale overlies the shallow salt domes. Inclusions in the shallow salt are clay and fine-grained anhydrite. These outer shelf salt domes are separated from those near shore and on shore, which possess typical Gulf Coast anhydrite cap rock and salt, by the Heald Bank structural high, along which salt domes have not been found. The outer shallow salt domes undoubtedly drew their salt from a different sequence of salt beds, or possibly from a separate salt depositional basin.

Other minerals of significant occurrence in the cap rock have been derived by alteration of the anhydrite. Gypsum has been formed directly from anhydrite by hydration. Compilation of field data has determined that in Gulf Coast areas, such as Sulphur, Louisiana, where stable conditions have existed well into the Pleistocene, gypsum does not occur below a depth of 1,000 to 1,500 feet because of pressure-temperature relations that govern its stability. Anhydrite occurs below this depth, unless rapid burial or subsidence has occurred. As pointed out by Marcus I. Goldman, who conducted the original studies of the Sulphur, Louisiana, cap rock, a study of gypsum-anhydrite relations at salt domes across

Table 2. Chart showing distribution of minerals and other inclusions in water-insoluble residues from 37 Gulf Coast salt domes.

COMMON MINERALS (Approximate Order of Abundance)	SALT DOMES																			
	McIntosh, Ala.	Petal, Miss.	Midway, Miss.	Garden Is. Bay, La.	Lake Washington, La.	Venice, La.	Lake Hermitage, La.	Caminada Pass, La.	Clovelly, La.	Choctaw, La.	Darrow, La.	White Castle, La.	Bayou Bleu, La.	Weeks Island, La.	Avery Island, La.	Jefferson Island, La.	Anse La Butte, La.	Jennings, La.	Sikes, La.	Winnfield, La.
Anhydrite	X	X	X	X	X	X	X	X	X	X	X	X	X	X	X	X	X	X	X	X
Dolomite	X	X	X	X	X	X	X	X	X	X	X	X	X	X	X	X	X	X	X	X
Calcite	X	X	X		X		X	X	X	X	X	X	X	X	X	X	X		X	X
Quartz					X	X	X			X	X	X	X	X	X	X	X	X	X	X
Pyrite			X		X					X		X	X	X	X	X	X	X		X
Sulfur	X				X				X	X					X	X				
Limonite			X	X	X					X				X	X	X	X			X
Hauerite				X	X					X				X	X	X	X			
Manganosite?								X	X											
Hematite										X						X				X
Celestite					X									X						
Barite																X				X
Siderite																				
Other inclusions																				
Potash salts	X				X	T				X	T	T	X	X	X	X	T			T
Oil					X	X	X	X						X	X	X				
Gas				X	X	X	X	X			X	X		X	X	X				X
Quartz sand					X	X			X					X	X	X				
Anhydrite rock					X									X		X				X
Other rock	X									X						X				
Carbonaceous matter	X																			

COMMON MINERALS (Approximate Order of Abundance)	SALT DOMES																	BEDDED SALT		
	Arcadia, La.	Gibsland, La.	East Hackberry, La.	West Hackberry, La.	Sulphur, La.	Fannett, Tex.	Sour Lake, Tex.	Barbers Hill, Tex.	Hoskins Mound, Tex.	Stratton Ridge, Tex.	Bryan Mound, Tex.	Pierce Junction, Tex.	Blue Ridge, Tex.	Nash, Tex.	Hockley, Tex.	Grand Saline, Tex.	Palangana, Tex.	Humble-Williams #1, Ala.	Danciger-York #1, Ala.	Shell-Prince #1, Tex.
Anhydrite	X	X	X	X	X	X	X	X	X	X	X	X	X	X	X	X	X	X	X	X
Dolomite	X		X	X	X	X	X	X	X	X	X	X	X	X	X	X	X			
Calcite	X		X	X	X	X	X	X		X	X	X		X	X	X	X			
Quartz			X	X	X		X		X	X	X			X	X	X	X	X	X	
Pyrite			X	X			X		X						X		X	X		
Sulfur				X	X		X		X	X		X	X							
Limonite									X						X					
Hauerite				X						X								X	X	
Manganosite?				X																
Hematite										X					X		X	X	X	
Celestite					X		X		X					X						
Barite											X									
Siderite							X				X						X			
Other inclusions																				
Potash salts																		X		
Oil																				
Gas			X	X	X	X				X		X								
Quartz sand					X							X								
Anhydrite rock						X	X			X										
Other rock								X		X		X					X			X
Carbonaceous matter						X									X			X		X

the Gulf Coast can help provide accurate data on recent tectonics and sedimentation rates in the Gulf Coast.

Petrographic studies have revealed that sulphur has replaced anhydrite and seemingly to a lesser extent, gypsum; the corroded appearance of partly replaced anhydrite grains and the sulphur pseudomorphs after anhydrite as revealed in thin sections are particularly striking evidence of this relationship. Thin sections of the "disseminated" sulphur of the mines show direct replacement of the anhydrite by both sulphur and calcite. Important amounts of sulphur also occur as veins along partings and fractures, and as vug fillings.

Alteration of the anhydrite and gypsum to sulphur and calcite is a much less understood process than hydration to gypsum, although it has been accorded a great deal of attention at salt domes of the Gulf Coast and Mexico, as well as in Sicily and elsewhere (Feely and Kulp, 1957; Thode et al., 1954; Dessau et al., 1962).

Early workers, particularly in Sicily, considered all sulphur to be of volcanic origin. An inorganic origin also was suggested in which hydrocarbons reduced sulphates to hydrogen sulphide. Repeated laboratory experiments showed, however, that temperatures of 1000 degrees Fahrenheit, or higher, were required. In 1915, Walter Hunt published an extensive treatise on the origin of Sicilian sulphur deposits associated with gypsum and marl. He concluded that sulphate-rich brines were reduced to hydrogen sulphide by bacterial action in isolated lagoons at the time of deposition of the gypsum and marl. The sulphur either was deposited directly from hydrogen sulphide or in later stages after concentration as a calcium polysulphide. Recently published results of petrographic studies conducted on extensive suites of sulphur mine samples from Sicily, supplemented by spectrographic studies (Dessau et al., 1962), substantiate a bacterial origin. Results of similar studies in the south of the U.S.S.R. over a period of years by Murzaiev (1937) and Vinogradof (1956) seem to confirm the importance of bacterial action. Large deposits south of Mosul, Iraq, in the Miocene Lower Fars gypsum beds associated with bitumen, which were explored by the Russians, also are considered by them to be of bacterial origin.

The association of native sulphur and calcite with hydrocarbons, anhydrite, and gypsum certainly is a persistent one and characterizes all salt dome deposits. The possibility that bacteria have been the agent of alteration in the Gulf Coast sulphur deposits has been the subject of investigation over the past fifteen years. H. G. Thode and associates of McMaster University (Thode et al., 1954) conducted the first work along these lines as an outgrowth of attempts to use sulphur isotope ratios as means of determining ages of sulphide ores. Using isotope values of sulphur from meteorites as a standard, the ratio of heavy to light isotopes is consistently lower for Gulf Coast sulphur. Sulphur in anhydrite from salt and sea water also displays similar low ratios, but not as low as Gulf Coast sulphur. By contrast, volcanic sulphur has somewhat variable ratios, but ones consistently higher than for cap rock sulphur.

Experimental work by Thode and associates (1954) determined that known bacterial sulphur had distinctively lower isotope ratios, comparable to those of Gulf Coast sulphur. Further work indicated that bacteria have the faculty of fractionating heavy sulphur isotopes when they attack sulphates, reducing the amount of the heavy sulphur isotope in the resulting sulphur that they produce. From this it was concluded that Gulf Coast sulphur has been formed from anhydrite and gypsum--or sulphate brines--as a result of bacterial action, with probable utilization of hydrocarbons as a nutrient.

Isotope studies of Gulf Coast sulphur were subsequently conducted by Feely and Kulp (1957). These workers confirmed the findings of Thode and associates and also determined that the carbon isotope ratios for carbon in calcite from sulphur-bearing cap rock were very close in value to those for carbon in petroleum, and consistently lower than for carbon in chemically precipitated calcite. To check the sulphur isotope findings, Feely and Kulp (1957) cooperated in experiments at Rutgers University in which sulphur bacteria obtained from cultures secured from oil field waters were nurtured on petroleum derivatives. The more recent isotopic studies of Dessau et al. (1962) on Sicilian sulphur, previously mentioned, also indicate a bacterial origin.

In our present knowledge of the origin of sulphur in major sulphur deposits of the world, therefore, we must conclude that bacterial action provides the most acceptable explanation. The chemistry involved is fairly straightforward, calling for organisms to reduce the sulphate, producing hydrogen sulphide which then oxidizes to elemental sulphur, accompanied by the release of calcium which combines with carbon dioxide to form calcium carbonate. The environment in which the changes occur is readily explained for deposits where native sulphur is a contemporary sedimentary deposit, as believed, in Sicily and the south of the U. S. S. R. Modern localities are known where bacteria are reducing sulphate brines with release of hydrogen sulphide and deposition of sulphur and calcium carbonate.

In the cases of the Gulf Coast, West Texas, the Isthmus of Tehuantepec, Mexico, and Mosul, Iraq, where consolidated anhydrite and gypsum apparently are replaced by sulphur and calcium carbonate on a grand scale, the environment is not easy to reconstruct. Laboratory work has shown that hydrogen sulphide will replace anhydrite and gypsum with elemental sulphur, suggesting that bacteria have reduced the sulphate in brines to provide the hydrogen sulphide. It is not known, however, whether sufficient amounts of rich sulphate brines can be assumed to be circulating in the consolidated formations to account for large sulphur deposits. Analyses of subsurface brines and sulphur formation from sulphur mines by Professor Lehman,* University of California, revealed living sulphate-reducing bacteria; however, the subsurface samples in all probability were contaminated by surface waters which are rich in sulphate-reducing bacteria. Brines from deep oil wells were reported to contain sulphate-reducing bacteria by Zobell (1951); here again, the question of contamination arises to cast doubt on this finding.

There is good evidence that large Gulf Coast sulphur deposits were formed at very shallow depths, quite possibly at or near the bottom of the sea or connecting shallow bays or lagoons. The evidence includes the common occurrence of shales interbedded with sulphur formation around the periphery of the cap rock, the existence of fringing coral reefs as previously mentioned, and the eroded upper surface of the cap rock at a number of the shallow Texas salt domes. At the time that Dessau (Dessau et al., 1962) was conducting studies of Sicilian sulphur, lengthy discussions were held with him by the writer, attempting to interpret conditions for the Gulf Coast sulphur deposits similar to those indicated for Sicilian sulphur deposits. Professor Dessau plans to study core samples from Grand Isle Block 18, offshore Louisiana, where good cores were obtained of the shales interbedded with sulphur-bearing rocks.

Improved methods of coring in recent years

*Sample collection observed by writer at Lake Washington mine of Freeport Sulphur Company.

have provided much better samples and hence more information about cap rock has been obtained, particularly at the more recently developed sulphur mines. Studies of these newer mines, including detailed petrographic and spectrographic analyses of cores, as well as consideration of the stratigraphic and structural relationships of both the interbedded and adjacent sediments, can not only improve our understanding of the genesis of the sulphur deposits but also of the environment in which the entire cap rock was formed, and of the history of the uplift of the salt domes.

The economic importance of salt-dome cap rock as a source of sulphur is, of course, of greatest significance. In addition to potential discoveries in the offshore area of Louisiana and Texas, there are similar types of saline structures that are currently under exploration in the Isthmus of Tehuantepec, Mexico. Recent developments in Poland, Iraq, and the U. S. S. R. lead to speculation that there must be other areas of evaporites with possibilities for discovery of large Frasch-type deposits. Figure 5 shows distribu-

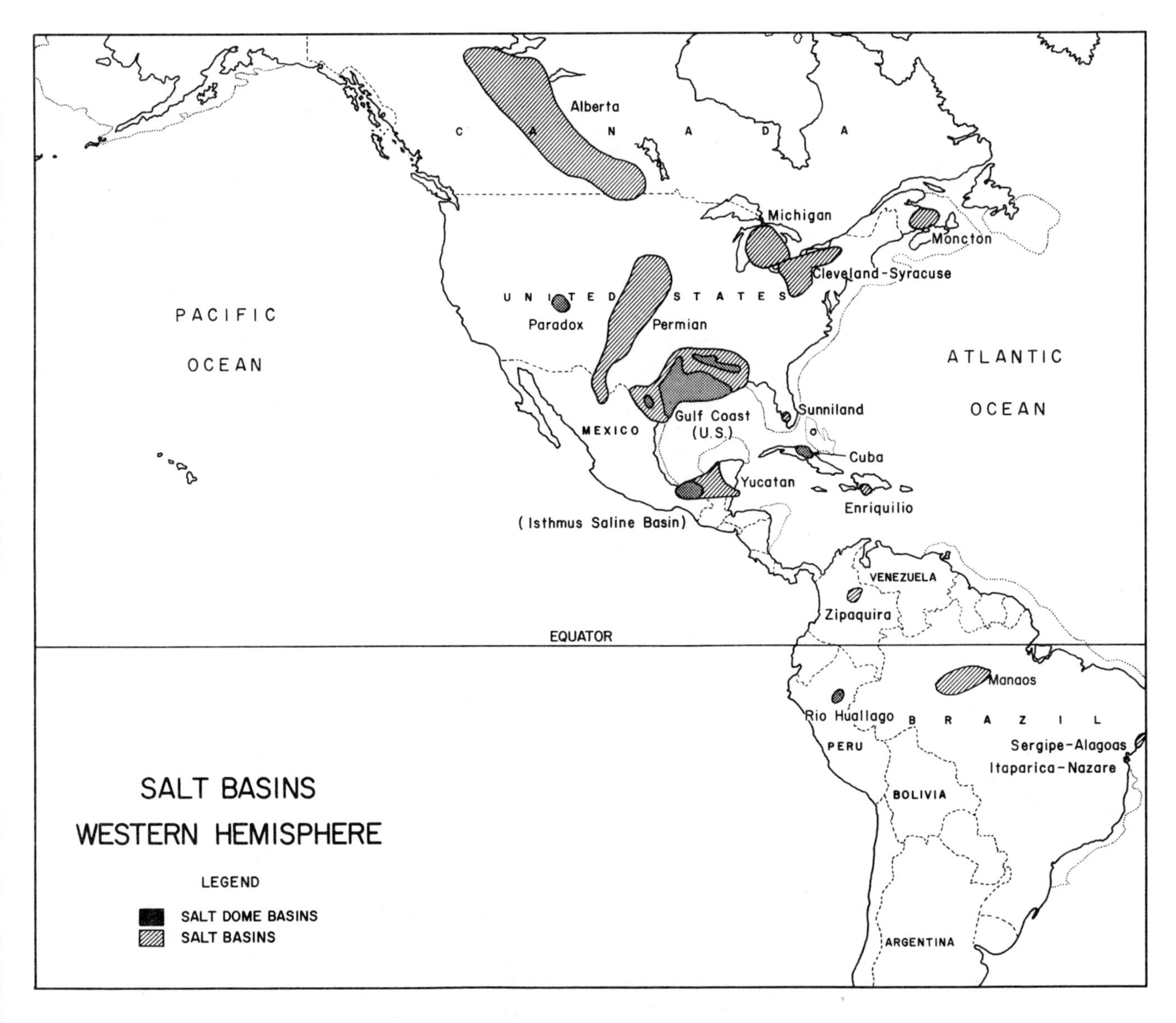

FIG. 5. Salt basins and salt-dome areas of Western Hemisphere (from Dow Chemical Company).

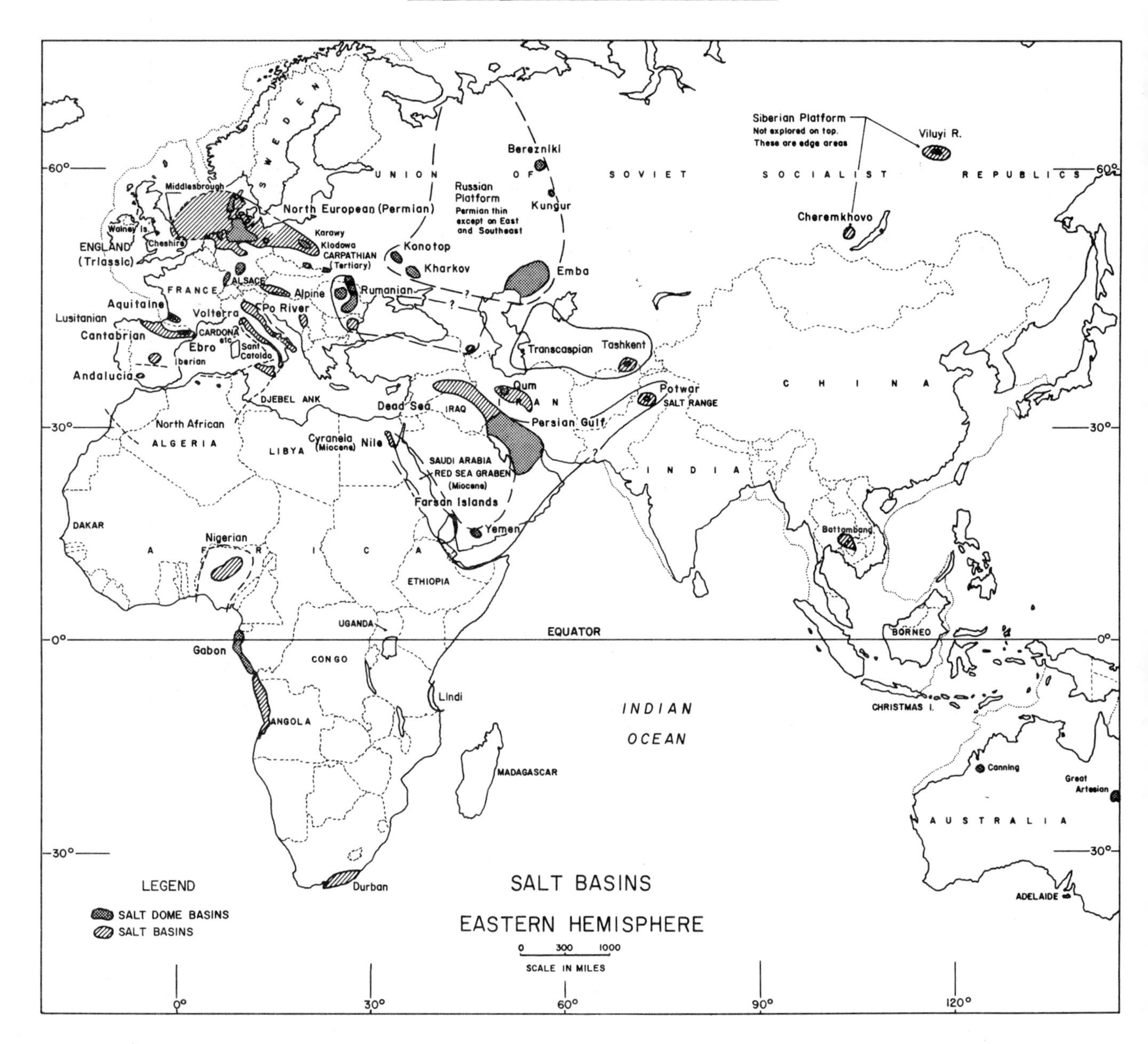

FIG. 6. Salt basins and salt-dome areas of Eastern Hemisphere (from Dow Chemical Company).

tion of saline deposits in the western hemisphere, and Figure 6 illustrates distribution for the eastern hemisphere.

Cap rock has been an important petroleum reservoir, providing the production at Spindletop, the first salt-dome oil field discovered on the Gulf Coast. Although there was large production from cap rock at a few other salt domes, production from this source at the present time is quite minor. As previously mentioned, both limestone and gypsum have been quarried from cap rock at outcrops in Louisiana and Texas. Currently, some gypsum and anhydrite are being quarried from cap rock at Winnfield, and recent scouting reports state that Freeport Sulphur Company has acquired the gypsum deposit at Gyp Hill--ostensibly for recovery of the sulphur from gypsum.

REFERENCES

Ault, W. U., and Kulp, J. L. (1960) Sulfur isotopes and ore deposits: Econ. Geology, vol. 55, pp. 73-100.

Dessau, Gabor (1962) Indagini sull'origine dei giacimenti solfiferi delta: Sicilia Consiglio Nazionale Delle Recherche, an. 2, ser. 2, pt. 1, vol. 2, pp. 13-22.

Dessau, G., Jensen, M. L., and Nakai, N. (1962) Geology and isotopic studies of Sicilian sulfur deposits: Econ. Geology, vol. 57, pp. 410-438.

Feely, H. W., and Kulp, J. L. (1957) Origin of Gulf Coast salt dome sulfur deposits: Bull. Amer. Assoc. Petrol. Geol., vol. 41, pp. 1802-1853.

Hofrichter, Erich (1968) Stratigraphy and structure of the Palangana salt dome, Duval County, Texas: Geol. Soc. America Spec. Paper 88, pp. 365-380.

Hunt, W. F. (1915) The origin of the sulphur deposits of Sicily: Econ. Geology, Vol. X, pp. 543-579.

Jensen, M. L. (1968) Isotopic geology and the origin of Gulf Coast sulfur deposits: Geol. Soc. America Spec. Paper 88, pp. 529-536.

Jux, U. (1961) The palynological age of diapiric and bedded salt in the Gulf Coast provinces: Louisiana Geol. Survey Bull. 38, 46 pp.

Kupfer, D. H. (1965) The relationship between internal and external structure in Gulf Coast salt domes: Bull. Amer. Assoc. Petrol. Geol., vol. 49, p. 349.

Murzaiev, P. M. (1937) Genesis of some sulphur deposits of the U. S. S. R.: Econ. Geology, vol. 32, pp. 69-103.

Sverdrup, H. U., Johnson, M. W., and Fleming, R. H. (1942) The oceans: Prentice Hall, New York, N. Y.

Taylor, R. E. (1938) Origin of the cap rock of Louisiana salt domes: Louisiana Geol. Survey Bull. 11, 191 pp.

Thode, H. G., Wanless, R. K., and Wallouch, R. (1954) The origin of native sulphur deposits from isotope fractionation studies: Geochim. et Cosmochim. Acta, vol. 5, pp. 286-298.

Vinogradof, A. P. (1956) Isotopnyi sostav servy v svyzi s voprosom o vozraste piritov osadochnogo geneziu: Geochim. No. 1, pp. 96-105, English summary.

Zobell, C. E. (1951) Contributions of bacteria to the origin of oil: World Petr. Cong., 3d, The Hague, Proc., sec. 1, pp. 414-420.

INTERNAL STRUCTURE AND PETROFABRICS OF GULF COAST SALT DOMES*

William R. Muehlberger and Patricia S. Clabaugh
Department of Geology
The University of Texas at Austin
Austin, Texas

ABSTRACT

Internal structures and preferred orientation of halite crystals in Grand Saline salt dome, Texas, and Winnfield salt dome, Louisiana, indicate emplacement of the stocks by successive differential movement of spines and lobes. Differences in intensity of shear folding and varied degrees of preferred orientation in these and other Gulf Coast salt domes suggest that mines within the domes are at different structural levels and that movement of the salt has, in some cases, been very complex. Study of the preferred orientation of bedded salt from Oklahoma showed that it increases with depth.

INTRODUCTION

Of the more than 300 salt domes in the Gulf Coast region of the United States, only eight have mines in which the internal structure of the salt can be studied. Two of these are in Texas; the rest are in Louisiana (Kupfer, 1963). Inasmuch as the mines themselves occupy only small parts of the domes, mapping within them affords only a limited sample of the structure, and the writers' extrapolations and conclusions admittedly are based on fragmentary data. Nevertheless, it appears that individual Gulf Coast salt domes have been intruded as pencils, or spines, and individual lobes. Furthermore, differential uplift, that is, uplift of one part of the dome at one time and of another part at another time, also has been demonstrated by study of the sedimentary section on and surrounding the domes (Atwater and Forman, 1959). The writers' purpose is to discuss some of the data accumulated from the interior of the domes and the conclusions derived therefrom. Much of this is published information; therefore, this paper will review the earlier work and present some newer data believed to support the case for differential uplift.

---

*Reprinted with permission from "Diapirism and Diapirs," pp. 90-98, Memoir No. 8, American Association of Petroleum Geologists.

The writers are indebted to D. E. Feray, who took the mine photographs; T. E. Longgood, Jr., who assembled the photomosaic (Fig. 3); and L. W. Bridges, W. T. Haenggi, T. E. Longgood, Jr., B. E. St. John, and E. L. Trice, Jr., who assisted in the mapping at Grand Saline; Haenggi also assisted in the work at Winnfield dome. Louise Jordan furnished the cores of bedded salt from northwestern Oklahoma. The writers also are grateful to the mine officials who permitted access to the mines and helped in many ways, and to the Houston Geological Society and the Texas Bureau of Economic Geology for permission to republish certain illustrations.

PREVIOUS STUDIES

Robert Balk published the first major study of internal structures of Gulf Coast salt domes. His comprehensive paper on Grand Saline (1949) was followed by a companion paper on Jefferson Island (1953). These papers formed a solid foundation of basic data and cautious interpretation for all later studies.

The doubling in size of the mine in the Grand Saline salt dome furnished an opportunity to expand on Balk's studies there. Muehlberger and students produced a map, a short text, and photographs of features exposed in the mine (Muehlberger, 1959). Some of these photographs are included in this paper to illustrate the main structural features observed. Maps of salt structure in Winnfield dome (Hoy et al., 1962) and Weeks Island (Kupfer, 1962) have been published, and structure maps now are available for all salt mines (Kupfer, 1963).

Petrofabric study of halite and anhydrite by Clabaugh was begun in Grand Saline dome (Clabaugh, 1962) and later expanded to Winnfield and Hockley, unoriented cores from Tatum dome, Mississippi, and bedded salt from western Oklahoma. Data and conclusions from these studies, as well as more recent unpublished work by the writers, are presented in this paper.

GRAND SALINE SALT DOME

The most extensive work by the writers was done in Grand Saline salt dome in northeast Texas. At the mine level, about 800 ft below ground level, the dome is about 8,000 ft across. Gravity measurements suggest that the dome is nearly cylindrical, although there is probably an overhang along the eastern side (Peters and Dugan, 1945). The mine itself is about 1,800 by 3,000 ft (inset map on Fig. 4 shows location of mine in dome at mine level). It consists of rooms 60-80 ft wide alternating with unmined pillars of salt of about the same size. Heights of different rooms range from 25 ft in some of the newest workings to 100 ft in the oldest.

FIG. 1. View of folds exposed in ceiling near northeast edge of Morton Salt Co. mine, Grand Saline salt dome, Van Zandt County, Texas. Ceiling more than 100 ft above camera. Short straight lines are 6-8-ft drill shot holes. From Muehlberger (1959, Pl. V).

DESCRIPTION OF STRUCTURAL FEATURES

The walls of the mine typically show alternate dark and light layers of uniform thickness standing vertically. The ceilings show the same bands in a series of complex folds. Figure 1 is a photograph of a single fold exposed on a 100-ft-high ceiling. The short, straight, dark lines are shot holes, each of which is from 6 to 8 ft long. It should be noted that the shape of the fold changes abruptly. The layers become parallel as the fold dies out just out of view. The folds themselves plunge nearly vertically, and the effect is similar to that seen by looking upward at the folds of curtains or drapes.

The tops and bottoms of individual beds can be recognized in Figure 1, even though these beds are now nearly vertical. The evaporation of sea water produces a chemically graded bed that is rich in anhydrite near the bottom and grades upward into nearly pure salt. The anhydrite-rich layers are seen as dark bands and the nearly pure salt layers are light. Thus each bed should grade upward from dark to light, with a sharp return to dark in the next overlying bed. This implies a deposition in a closed basin where recurrent floods replenished the water supply. Obviously this idealized situation will be realized rarely. Reversals of grading have been observed in the mines (as well as in bedded salt at Hutchinson, Kansas), and this criterion for determining "tops," especially in such highly disturbed bedding, must be used with caution (Kupfer, 1962, p. 1466).

Shear folding is the most common structural feature in the mine. Individual layers can be traced for various distances--some as far as 1,000 ft--before attenuation, dust cover, or intervening unmined pillars of salt prevent detailed correlation.

In a few places the layers close on themselves. The closure illustrated in Figure 2 can be interpreted as being either the top of an anticline (or the bottom of a syncline) or the pinched-off end of an isoclinal fold that reopens out of the field of the illustration. Large pillars nearby conceal the critical data needed to resolve this problem. The fold shown in Figure 3 has been refolded, as indicated by the distortion of the axial-plane cleavages visible on each limb. Most of the closures recognized in Grand Saline are in or near structural discontinuities.

GIANT SALT CRYSTALS

One of the most interesting features in Grand Saline salt dome is the giant salt crystals. Crystal faces 1-3 ft across are common in the giant crystal beds. The largest single crystal (or crystal group with almost identical crystallographic orientation) that was identified is about 4 ft wide, 20 ft high, and 100 ft long, as traced along the ceiling. These giant-crystal layers, which are beds of nearly pure salt, are bounded by anhydrite-rich layers. The correlation between purity and grain size is excellent throughout the mine; the coarsest salt is that which has the least included anhydrite. (These beds are indicated by a special symbol on the map, and one group of giant crystals is shown on Pl. III C and Pl. IV A of Muehlberger, 1959.)

INTERPRETATION OF STRUCTURE

Figure 4 is an interpretive map of the bedding and structure mapped in Grand Saline salt dome. No single bed can be traced throughout the mine. The lines drawn on the map are parallel with the observed bedding. Figure 4 illustrates the sweep of the folds and is more indicative of the entire structure than the detailed map (Muehlberger, 1959), which shows myriads of individual beds and folds. Two prominent lines of discontinuity, or at least abrupt changes in fold-trend orientation, are found along nearly east-west lines in the south-central and northwestern parts of the mine. These lines are nearly parallel with the regional fault pattern and suggest that faults in the sedimentary overburden may be related to uplift of this part of the dome. Movement on a particular fault may have produced a local stress minimum which then localized the salt uplift to that zone. These changes in trend also can be seen in the central map of Figure 5, which shows the traces of the axial planes at the mine level.

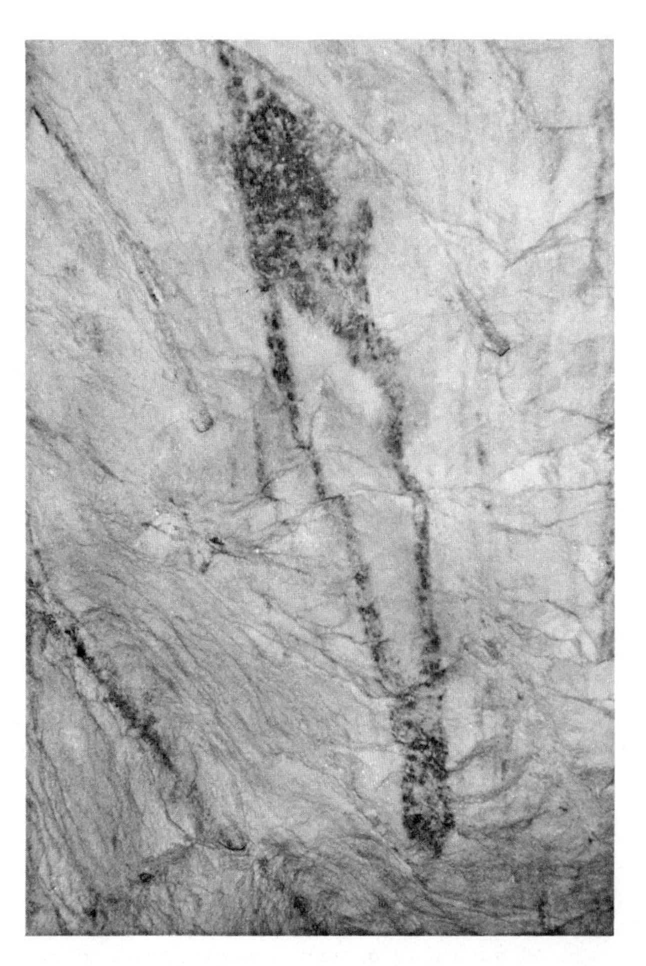

FIG. 2. Small closed fold exposed in ceiling in south-central part of Morton Salt Co. mine, Grand Saline salt dome. Length of closure shown, about 10 ft. From Muehlberger (1959, Pl. II D).

PETROFABRIC STUDY

The petrofabric diagrams in Figure 5 show that a significant number of the salt crystals are oriented similarly, indicating that the salt movement occurred along crystal glide planes. Because the salt moved in a direction parallel with the axial planes of the folds, the traces of these planes are shown. In contrast to normal petrofabric plotting, all three axes of each salt crystal must be plotted because isometric minerals have no single axis that uniquely defines the orientation of a crystal. The contouring is based on the percentage of total points, not crystals. Thus a 5-per cent point maximum really represents a concentration of axes from 15 per cent of the crystals. The area of the largest maximum in diagram D (Fig. 5) actually includes axes from about 25 per cent of the crystals measured in that sample.

Study of the crystal glide planes of salt shows that where one of the three maxima is in the axial plane of a fold, gliding took place parallel with dodecahedral planes. Where two maxima are within the axial plane, gliding was parallel with cubic planes. Possible glide directions can be located from these orientation diagrams with reference to the axial planes of the folds; these directions are not parallel in different parts of the dome. The writers believe that this further confirms the idea of differential uplift of the dome.

Some of the diagrams show better preferred orientations than others. The poorer orientations could be explained as resulting from simultaneous movement along both cubic and dodecahedral glide planes in different crystals within the mass of the

FIG. 3. Refolded closed fold exposed in ceiling in south-central part of Morton Salt Co. mine, Grand Saline salt dome. Axial-plane cleavage shown by shear folds has been folded into an arc parallel with elongation of fold. Length of fold shown is about 35 ft. From Muehlberger (1959, Pl. VI).

sample measured. However, it seems more likely that mixed patterns represent successive movements of the salt in different directions. A diagram of the mass of salt measured to produce diagram A, Figure 5, for example, probably would be more like diagram C if the mass of salt were moved in a different direction for a long enough time. The diagrams which show the best preferred orientations (A and D) are from the northern part of the mine. In diagrams of the southernmost part, closer to the margin of the dome, the preferred orientations are relatively poorly defined. The salt in the southern area is also finer grained and visibly more strained than the coarse-grained, better-oriented salt farther from the dome margin, which probably indicates a higher strain-rate differential nearer the margin.

WINNFIELD SALT DOME

The writers have done similar work in Winnfield salt dome in northwestern Louisiana. The interpretation of the gross structural features, derived from the map by Hoy et al. (1962), is shown in Figure 6. Some of the samples from this mine show a well-developed preferred orientation of the salt.

Diagram D, Figure 6, is the only example of highly preferred orientation, although the concentrations in the other diagrams are too great to be considered statistically random. The lack of geometric arrangement of these maxima seems to suggest movement in several different directions by the masses of salt which were sampled, inasmuch as uniform directional stress over a long period of time should produce a general alignment of the salt crystals affected. In fact, movement probably has occurred relatively recently. This is suggested by a survey by Muehlberger of a water-etch line that was created accidentally 27 years ago when the mine was flooded for several months. This line is no longer flat, although the maximum change has been only a few inches.

WEEKS ISLAND SALT DOME

Figure 7 is the writers' interpretation of the map published by Kupfer (1962) of the Weeks Island salt mine. The figure shows characteristic folds with smooth, rounded shapes which give the impression that the entire mass has flowed. This pattern is very different from that of the shear folding typical of the structures in Grand Saline and Winnfield domes. The abrupt changes in strike resulting from the refolding of older folds during differential uplift of the salt are clearly visible (and also are shown in slightly different style by Kupfer). The topographic prominence over the dome, in conjunction with the lack of

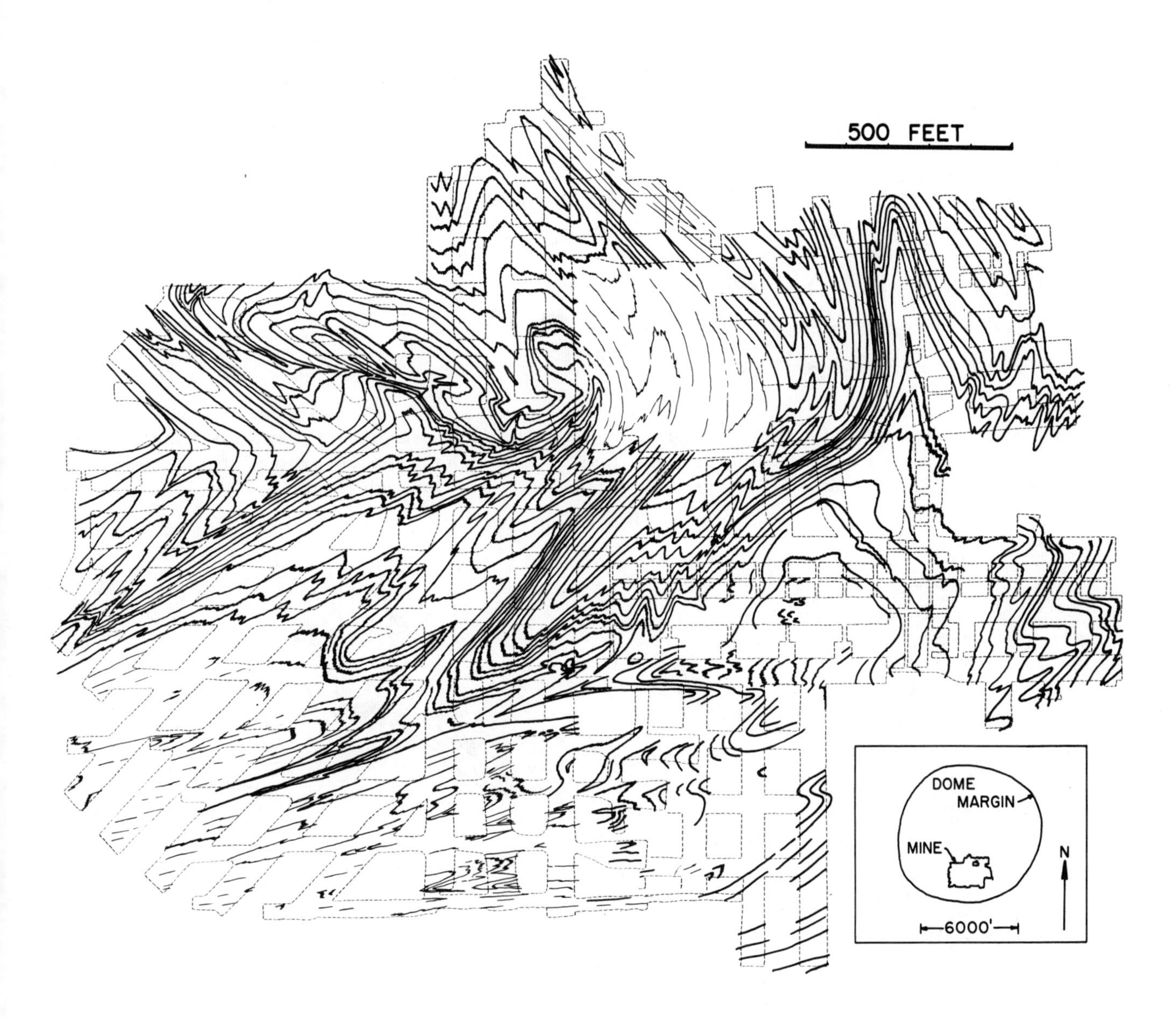

FIG. 4. Interpretation of salt structure, Morton Salt Co. mine, Grand Saline salt dome. Lines drawn parallel with observed bedding in mine. Mine workings shown by dashed lines. Location of mine within dome at the mine level is shown in inset map, lower right. From Muehlberger et al. (1962, p. 274).

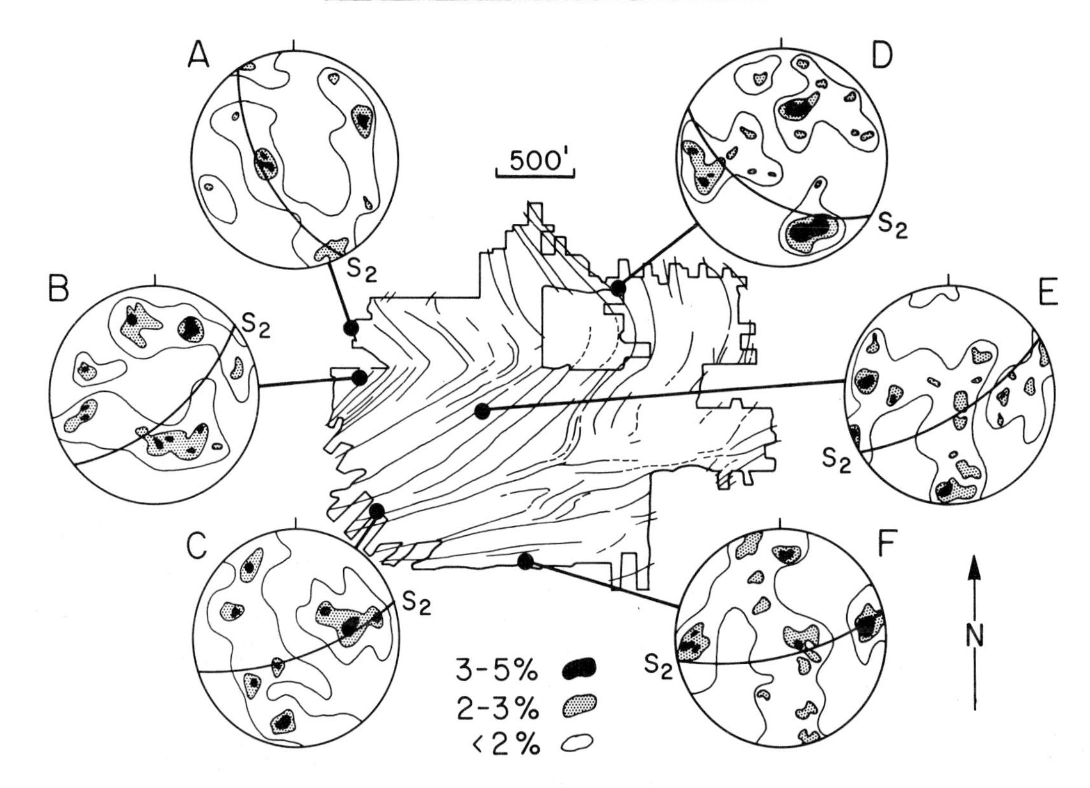

FIG. 5. Outline map of mine in Grand Saline salt dome and petrofabric diagrams of halite showing trace of axial planes (S_2) of folds at mine level. Lower hemisphere projection. A, 600 points (200 crystals); B-F, 300 points (100 crystals). Modified from Clabaugh (1962).

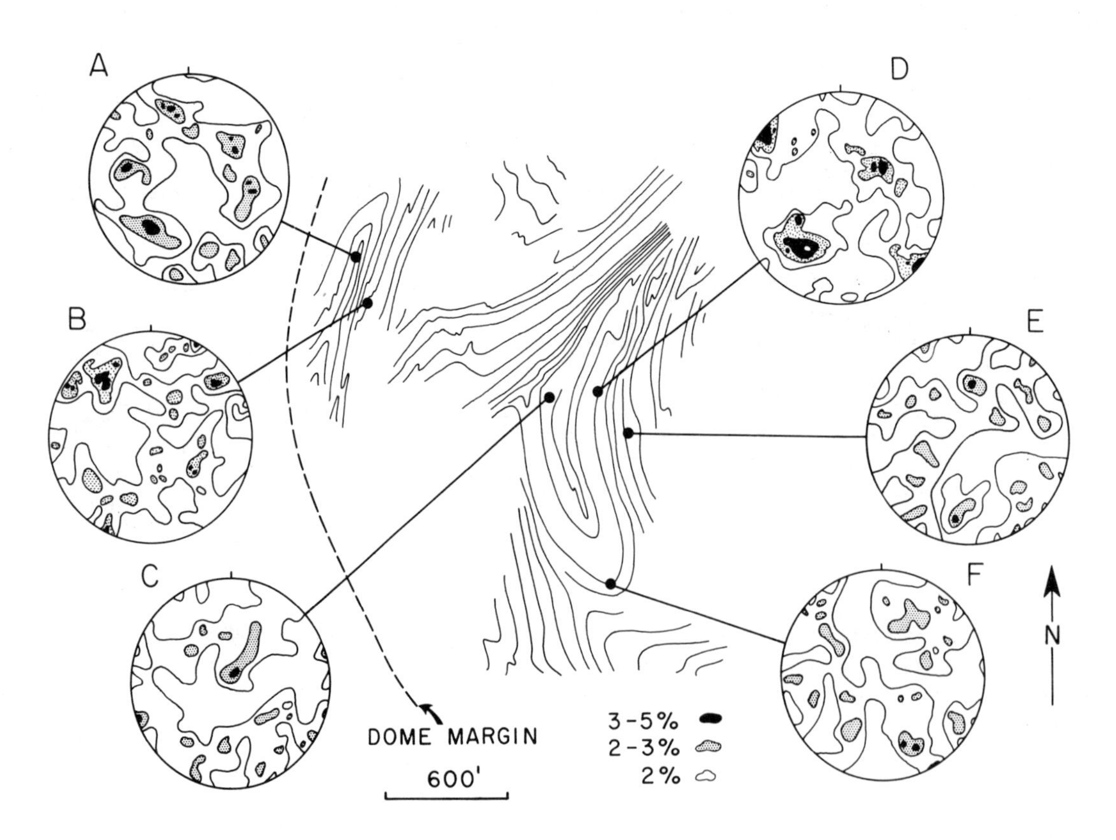

FIG. 6. Petrofabric diagrams of halite and map showing interpretation of salt structure (derived from Hoy et al., 1962, p. 1450-1451) in mine in Winnfield salt dome, Winn Parish, Louisiana. A-F are 300-point diagrams (100 crystals).

FIG. 7. Interpretation of salt structure in mine in Weeks Island salt dome, Louisiana. Derived from Kupfer (1962, p. 1462-1463).

cap-rock development, indicates that uplift has occurred recently. The appearance of these internal and external features of the dome also suggests recent movement, and the writers predict that future salt-fabric studies will confirm it (by finding mixed cubic-dodecahedral patterns or a lack of clearly defined maxima).

The complex fold structures of the salt in the Weeks Island mine seem to indicate that the mine is in a structurally high horizontal position in the dome where interference between adjacent parts of the moving salt mass has produced innumerable distorted and sheared-off folds. The horizontal positions of the mines in Winnfield and Grand Saline salt domes apparently are structurally lower or deeper, as shown by broader scale, more consistent fold patterns, as well as thick cap rock. Topographic depressions over Grand Saline and Winnfield domes indicate that uplift, if it still is going on, is proceeding more slowly than solution. Unpublished maps made by the writers of the few structures visible in the mine at Hockley dome suggest that it, too, is a structurally low cross section.

OTHER EXAMPLES

Additional petrofabric diagrams have been made of samples from Tatum dome in Lamar County, Mississippi, and from Hockley dome northwest of Houston. Well-oriented fabric patterns were found, but the writers have little additional structural data to which to relate them. The simplicity of the internal structures and the excellent fabric orientation of the one petrofabric sample from Hockley dome seem to indicate stability and lack of recent differential movement.

Petrofabric studies also were made of a core of undeformed bedded salt from Beaver County, Oklahoma (Fig. 8; see also Jordan, 1962). This salt is coarsely crystalline and apparently was recrystallized during compaction. The fabric patterns indicate a consistent tendency of the bedded salt crystals to be oriented with a cube corner up (octahedral plane parallel with bedding). A more systematic orientation is evident in the diagrams made from samples taken at progressively greater depths.

CONCLUSIONS

Structural data from the interiors of salt domes are of two kinds. First, mapping of the layering and the structural interpretations derived from this mapping show refolded folds and discontinuities, indicating that the domes have histories of uplift far more complex than those suggested by Nettleton's fluid model (1934, 1943), in which a low-density oil flows upward in a teardrop shape through a more dense oil. Second, petrofabric studies have demonstrated empirically that the salt moves on crystal glide planes, and that there are two glide symmetries. Furthermore, certain areas within the domes have much more poorly developed orientation patterns than others, probably as a result of more recent differential movements of the salt. These recent movements may have destroyed older preferred orientations caused by steady movements of the salt that produced well-oriented fabrics such as those found in the deeper or more central parts of the domes. Petrofabric analysis also demonstrates that bedded salt may show remarkably good preferred-orientation patterns, which are increasingly better with depth.

REFERENCES

Atwater, G. I., and Forman, M. J. (1959) Nature of growth of southern Louisiana salt domes and its effect on petroleum accumulation: Bull. Amer. Assoc. Petrol. Geol., vol. 43, pp. 2592-2622.

Balk, Robert (1949) Structure of Grand Saline salt dome, Van Zandt County, Texas: Bull. Amer. Assoc. Petrol. Geol., vol. 33, pp. 1791-1829.

__________ (1953) Salt structure of Jefferson Island salt dome, Iberia and Vermilion Parishes, Louisiana: Bull. Amer. Assoc. Petrol. Geol., vol. 37, pp. 2455-2474.

Clabaugh, P. S. (1962) Petrofabric study of de-

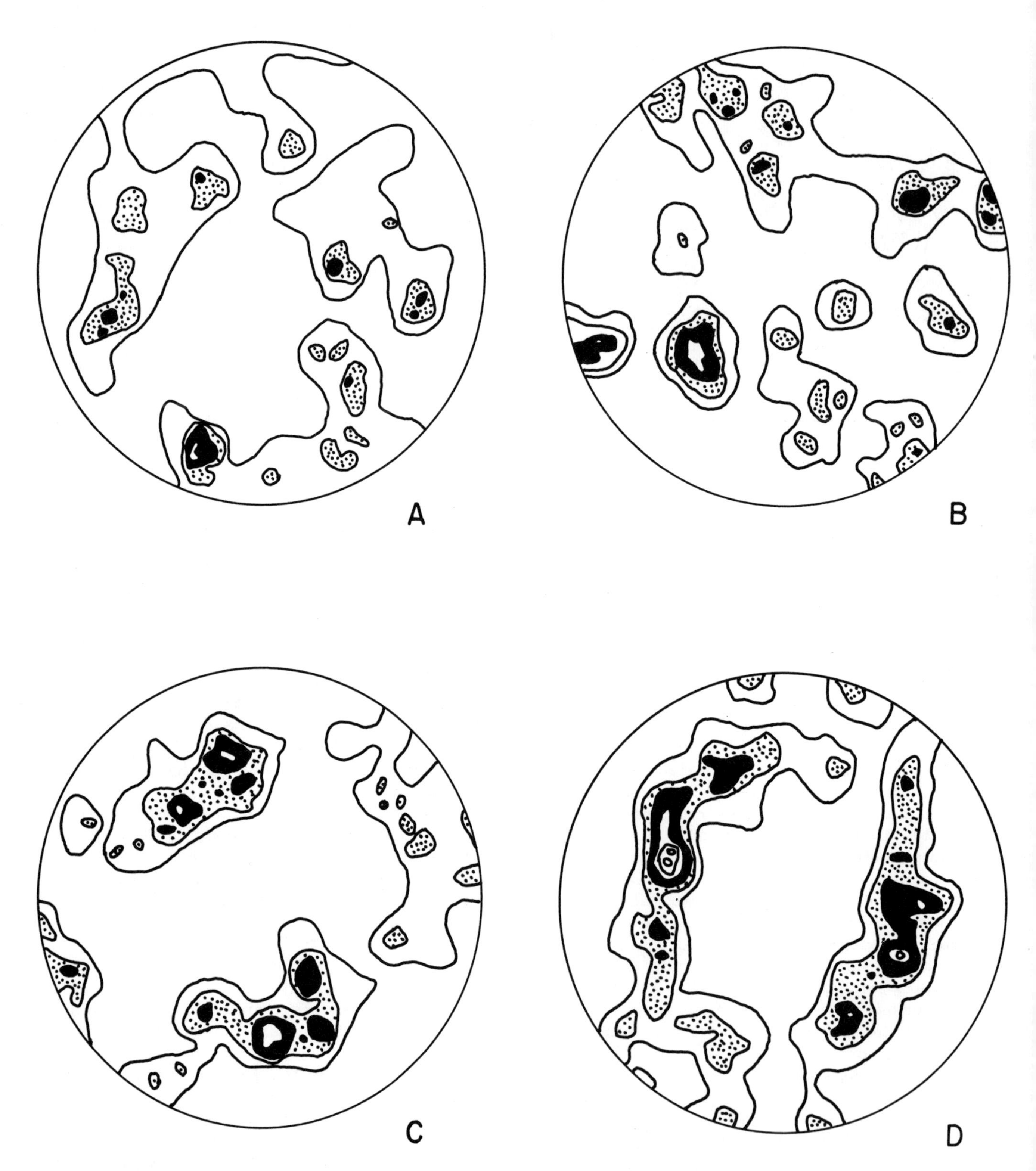

FIG. 8. Petrofabric diagrams of halite from four sections of a vertical core in Cimarron Anhydrite, western Oklahoma. Depths to samples: A, 1,490 ft; B, 1,530 ft; C, 1,630 ft; D, 1,715 ft. Direction of north in core is unknown. Contours in per cent: 1, 2, 3, etc., of total points measured. Solid areas are 3-4 per cent; areas within solid zones are higher--in diagram D, amounting to 6 per cent of points measured (18 per cent of crystals). Diagrams A, B, and C contoured on 150 points (50 crystals); diagram D, 300 points (100 crystals).

formed salt: Science, vol. 136, no. 3514, pp. 389-391.

Hoy, R. B., Foose, R. M., and O'Neill, B. J., Jr. (1962) Structure of Winnfield salt dome, Winn Parish, Louisiana: Bull. Amer. Assoc. Petrol. Geol., vol. 46, pp. 1444-1459.

Jordan, Louise (1962) LPG storage in Cimarron Salt, Beaver County, Oklahoma: Oklahoma Geol. Notes, vol. 22, pp. 23-26.

Kupfer, D. H. (1962) Structure of Morton Salt Company mine, Weeks Island salt dome, Louisiana: Bull. Amer. Assoc. Petrol. Geol., vol. 46, pp. 1460-1467.

__________ (1963) Structure of salt in Gulf Coast domes, in Symposium on salt: geology, mining, evaporated salt, solution mining, underground storage, A. C. Bersticker, ed.: Northern Ohio Geol. Soc., Cleveland, Ohio, 1962, pp. 104-123.

Muehlberger, W. R. (1959) Internal structure of the Grand Saline salt dome, Van Zandt County, Texas: Univ. Texas, Bur. Econ. Geology Rept. Inv. 38, 22 pp.

__________ (in press) Internal structures and mode of uplift of Texas and Louisiana salt domes: Geol. Soc. America, Memoir on Proceedings of International Conference on Saline Deposits.

__________, Clabaugh, P. S., and Hightower, M. L. (1962) Palestine and Grand Saline salt domes, eastern Texas, Field excursion no. 6, in Geology of the Gulf Coast and central Texas and guidebook of excursions, Geol. Soc. America, 1962 Ann. Mtg.: Houston Geol. Soc., pp. 266-277.

Nettleton, L. L. (1934) Fluid mechanics of salt domes: Bull. Amer. Assoc. Petrol. Geol., vol. 18, pp. 1175-1204.

__________ (1943) Recent experimental and geophysical evidence of mechanics of salt-dome formation: Bull. Amer. Assoc. Petrol. Geol., vol. 27, pp. 51-63.

Peters, J. W., and Dugan, A. F. (1945) Gravity and magnetic investigations at the Grand Saline salt dome, Van Zandt County, Texas: Geophysics, vol. 10, pp. 376-393.

GEOLOGICAL EXPLORATION WITH A MAJOR CHEMICAL COMPANY

W. N. McAnulty, Sr.
Department of Geology
The University of Texas at El Paso
El Paso, Texas

ABSTRACT

Almost every major chemical company employs geologists and conducts some kind of geological problem; however, the manner in which geologists are used and the geological programs conducted vary greatly among the companies. A company originally established to manufacture chemicals is not as likely to use and rely on geologists to the extent that oil and mining companies do. Policies and philosophies developed to fit the original enterprise often strongly influence approaches to problems and methods of operation in other fields. It appears that most large corporations whose principal products are based on initial- and intermediate-stage processing of geological raw materials are integrating backward and forward and that the time is near when so-called Oil, Mining, and Chemical companies could better be classified as Mineral Resource companies.

A great variety of minerals are used as raw materials by the chemical industry in enormous amounts. Several companies use their geologists only to keep up with the general supply situation. Most chemical companies undertake mining or geological exploration and acquisition of reserves only if (1) a company-developed process or surplus products produced by the company can be profitably applied to mining and/or processing of a given mineral deposit, or (2) dependable sources of supply, at predictable costs, are needed to back up production or research.

The major chemical companies are well equipped to conduct geological exploration, and several of them have conducted many successful projects. There are advantages and disadvantages for exploration geologists working for a major chemical company. Among the principal advantages are: (1) chemical companies are strongly research-minded; (2) they have excellent analytical facilities; and (3) their various research departments provide expertise in basic and applied sciences and technology. Disadvantages include: (1) narrow scope of geological interests--insofar as exploration is concerned; (2) general lack of knowledge and understanding, on the part of management, of geology and the geological approach; and (3) their economic valuation practices are generally not applicable to geological projects.

While employed as an economic geologist with a major chemical company (1953-64) the writer was involved in investigations of minerals ranging from alunite to zircon. Successful exploration projects were conducted for fluorspar, limestone, beryllium, thorium and rare earths, and tellurium. Reserves of several minerals were acquired, and a major fluorspar mining and milling operation in Mexico was started. The major chemical companies will have increasing need for well-trained, energetic, and imaginative geologists.

Almost every major chemical company employs geologists. Hundreds of geologists are employed in the so-called chemical industry. However, the manner in which geologists are used and the geological programs vary greatly among the companies, principally because no two companies are alike. How does a major chemical company differ today from a major oil company, or a major mining company, etc.? Once it was easy to classify a company on the basis of its activities: oil companies produced, refined, and marketed oil and gas; chemical companies manufactured and sold certain chemicals and chemical products; mining companies mined, beneficiated, and sold certain minerals or mineral products. Today, oil companies not only produce and refine crude oil, but many are engaged in various mining and chemical activities; some mining companies are producers of chemical products and oil and gas, as well as minerals and mineral products, and many major chemical companies own and operate oil and gas and mining divisions. It appears that all large corporations whose products are based on initial- and intermediate-stage processing of mineral raw materials are integrating backward and forward, and that the time is near when oil, mining, and chemical companies should be classified as mineral resource companies.

Where strictly geological products are concerned, a company originally established to manufacture chemicals, a chemical company, is not as likely to use and rely on geologists to the extent that oil and mining companies do. Policies and philosophies developed to fit the original enterprise often strongly influence methods of operation and approaches to problems in other fields.

Most chemical companies are not interested in mining per se. They generally employ geologists and become involved in geological activities for one or more of the following reasons:

1. To provide planners, production managers, and purchasing departments up-to-date information on reserves, availability, location, and costs of certain mineral raw materials; for example, the principal assignment of several geologists employed by one major company is to keep up with sources of supply, production costs, etc., on fluorspar, titanium, and a few other commodities.

2. To gather information on reserves, availability, location, costs, etc., of certain minerals as back-up support (insured supply) for chemical process research, planned or in progress; for example, one major chemical company conducted an extensive survey of fluorspar occurrences for this reason.

3. To locate and possibly acquire reserves of minerals or ores, the processing of which would consume excess or waste products generated in existing or planned manufacturing processes; for example, to find ores or mineral deposits, the beneficiation of which might use surplus HCl.

4. To gather information on reserves, or to find, explore, and recommend acquisition of reserves, of certain mineral assemblages amenable to beneficiation or refinement by company-developed, patented processes.

5. To find, explore, and assist in development of deposits of certain mineral raw materials basic to major product lines of the company; for example, deposits of rock salt, brines, and calcium and magnesium carbonates, or hydrocarbons (for use in the manufacture of organic chemicals) for a company like the Dow Chemical Company.

6. To find, explore, and assist in development of deposits of certain minerals consumed in appreciable quantities, the price of which commonly fluctuates greatly because of cartel manipulations, or for other reasons; for example, mercury, large quantities of which are used in mercury cells in the manufacture of caustic.

The major chemical companies are well equipped to conduct geological exploration--better in many ways than some large, conventional ("old line") mining companies. There are advantages and disadvantages for exploration geologists working for a chemical company. Among the advantages are:

1. Chemical companies are strongly research-minded; research is vital to their existence.

2. They maintain excellent analytical laboratories and constantly work to improve analytical techniques.

3. The various research departments of chemical companies provide expertise in basic and applied science and technology.

4. Chemical companies are, in general, progressive organizations.

Disadvantages include:

1. General lack of knowledge and understanding, on the part of management, of geology and the geological "point of view" or the uncertainties of exploration.

2. Narrow scope of real geological interest.

3. Expertise is scattered among too many departments; difficult to mobilize.

4. Economic valuation practices commonly used by chemical companies are usually not applicable to geological projects--short-term process obsolescence vs. long-term exploitation of large mineral reserves.

I was employed by a major chemical company as an economic geologist for nearly 12 years. My assignments included planning and supervising several geological projects ranging from literature surveys through exploration and development. Our group was not concerned with oil and gas because the company owned and operated a separate oil and gas division. The company with which I was associated manufactures a large number of organic and inorganic chemicals and chemical products, ranging from acetylene to xanthates, with sales in excess of one billion dollars per year. Enormous amounts of a great variety of minerals and mineral products are consumed by a major chemical company. Purchasing agents have a heavy responsibility to keep production and research departments supplied with mineral raw materials, and geologists play an important role in keeping buyers informed about these materials, especially those in short supply.

Although the company with which I was employed started by "mining" brines from subsurface formations, and still "mines" brines, as well as rock salt, oil and gas, sea water, and fluorspar, it has little interest in mining as a means of making money--only as a means of supply. Geological exploration and acquisition of reserves and mining were not automatically ruled out, but such projects were undertaken only if:

1. A company-developed process or products produced by the company could be applied profitably to mining and/or processing of a

particular mineral deposit, and

2. Sure sources of supply of certain minerals, at predictable costs, were needed to back up production or research planned or in progress.

Serving the needs of purchasing, research, and production departments, the geological group which I headed studied, investigated, and reported on many different commodities, including: alunite, beryllium, cesium, lithium, iodine, uranium, thorium, rare earths, rutile, ilmenite, mercury, molybdenum, manganese, bismuth, tungsten, nickel, tellurium, clays, rock salt, brine, oyster shell, limestone, dolomite, magnesite, lignite, fluorspar, and ground water. Needless to say, we were spread pretty thin much of the time, but it was excellent experience. Many of the investigations never went beyond literature surveys; some proceeded through field reconnaissance surveys; and a few advanced through various stages of exploration and development. Exploration work was done on fluorspar, lithium, beryllium, mercury, limestone, dolomite, oyster shell, uranium, thorium, and tellurium. Reserves of several minerals were acquired, and one full-scale mining and milling operation was started.

The normal geological program of the company did not include exploration, and, in a sense, exploration was an "extra-curricula" assignment for the geological group. Each exploration project had to be "sold" to top-level management by the geologists, with support from interested research or production departments. Many factors were involved, and interest in exploration waxed and waned. In general, there was lack of understanding on the part of management of the time required to plan, organize, and conduct an exploration project. In spite of the many inherent difficulties, however, exploration projects were approved from time to time, and our group never lacked for interesting and challenging assignments.

One project in particular stands out as a classic example of the right way to approach exploration. Interest in fluorspar developed when the company undertook an intensive program of research in fluorine chemistry. As this research proceeded successfully, it was decided that at least 500,000 tons of fluorspar reserves should be acquired, and the problem was turned over to the geologists. As there was no immediate need for fluorspar, we were permitted to plan and carry out the project in a systematic manner. After checking known occurrences in the United States and Canada, Mexico was chosen as the most promising area. The first stage included a reconnaissance survey of all known fluorspar districts in Mexico. The second stage focused on three of the seemingly most favorable areas seen during Stage I. The third stage consisted of detailed mapping and drilling of properties selected in northern Coahuila. The project was successful beyond expectations, and the company had more fluorspar than it needed. Consequently, it was decided to enter the fluorspar market, and a mining and milling operation was started. The geologists stayed with the project through the development stage.

Geological exploration is dependent to a high degree on complete, reliable analytical data, and the excellent analytical laboratories maintained by major chemical companies contribute greatly to geological exploration. For example, careful analytical studies of fluorspar from one of the ore bodies discovered in Mexico revealed the presence of potentially commercial amounts of beryllium in the fluorspar (in bertrandrite). Beryllium mineralization had not been suspected and was not recognized in the field. This discovery of beryllium prompted additional exploratory drilling which resulted in proving an important commercial occurrence of non-pegmatite beryllium ore. This led to checking other fluorite occurrences in the United States and Mexico for beryllium and finally to another exploration project for beryllium in Utah.

Geologists working for a chemical company are given interesting and sometimes very difficult assignments. For example, our group was asked to find commercial deposits of tellurium. Nearly all tellurium produced comes from electrolytic refining of copper; there are no tellurium mines. Of course, there are many tellurium minerals, particularly of copper, lead, gold, and silver. Some gold and silver telluride deposits are important sources of gold and silver, but such deposits are normally small and not important sources of significant amounts of tellurium. Our assignment was to find tellurium ore--not ores of lead, copper, gold, or silver. At first this seemed impossible, but after we began to get a grasp of the problem it became apparent that there was a great deal more tellurium than was suspected.

The company's interest in tellurium waned and this project was gradually curtailed and finally terminated, but the project was a geological success. Sizeable deposits rich in native tellurium, along with a host of tellurites, tellurates, and tellurides, were found in Sonora, Mexico. More than 15 new tellurium minerals were discovered in these deposits. Again, much credit for the success of this project must go to company analysts who provided us with simple field tests for tellurium and analyzed hundreds of field samples.

Being strongly research-minded, as are most chemical companies, the company urged the geologists to use any and all modern tools and techniques which might prove helpful in locating and valuating mineral deposits. In fact, far too much emphasis was placed on the use of tools and techniques, and not enough on basic geology.

We were encouraged to develop new tools and techniques, and research groups concerned with instrumentation and analytical methods were often called on for assistance. From time to time use was made of gravity, magnetic, seismic, and electrical geophysical methods; berylometers, geiger counters, scintillators, mercury detectors, ultraviolet lamps, and well-logging instruments were used. Several geochemical methods were tried. Detailed mineralogical studies were made in the laboratory. Small aircraft were used often for aerial reconnaissance and for geophysical or radiometric surveying. Aerial photographs, both black and white and colored, were used. Detailed mapping and sampling was done where appropriate. Mobile laboratories were used on certain projects. We owned and operated our own core drilling equipment, although some drilling was contracted.

Working as a geologist for a major chemical company is not without problems. Communication is a major problem. The view commonly held by many chemical engineers (and too many other engineers, for that matter) is that anyone with a book and a map can be a geologist, and their reluctance to recognize geology as a profession, and the geologist as a professional, gets to be annoying.

Lack of technical knowledge and of mining philosophy on the part of economic evaluators makes selling of mining projects to chemical company management very difficult. It is not easy to find a mine or prospect which has the potential of becoming a mine; the odds against it are very high. It is practically impossible to find a mine or prospect which still looks good after having been subjected to the valuation procedures employed by chemical company economists. Geological considerations often carry too little weight in the chemical company valuation methods. Geological extrapolations have little value.

In my opinion, the greatest handicap to successful geological exploration by a chemical company is the general lack of support from top-level management--at the Board level. Without top-level support the "follow through" required to succeed may not be approved. A successful minerals exploration program must be a continuing program--it cannot be turned on or off on the basis of current issues of E. & M. J., Mining Engineering, Mining World, or the Wall Street Journal.

With mining firms merging and diversifying into a variety of ventures, and petroleum, chemical, and other industries becoming actively engaged in mining, the mining industry, and especially minerals exploration, is undergoing revolutionary changes. Entry of petroleum and chemical companies into the field of mineral exploration and exploitation is forcing the conventional mining companies to adopt more progressive methods and to make better use of geologists. Each new industry coming into mining has a contribution to make, and, when the transition to true mineral resource companies has been accomplished for all, geological exploration will operate much more efficiently.

Chemical companies have made significant contributions to geological exploration, and, with more experience and slight changes in managerial attitudes, they promise to become much more competitive in this field.

THE STRUCTURE OF THE GULF COAST CHEMICAL INDUSTRY

Stanley A. Arbingast
Department of Marketing Administration and Bureau of Business Research
The University of Texas at Austin
Austin, Texas

ABSTRACT

Since 1940 one of the largest concentrations of the chemical industry in the world has developed in the coastal areas of Louisiana and Texas. In these two States, chemicals rank first in value added by manufacture among all manufacturing industries. Major attractive factors for this development have been the low assembly costs for raw materials, the availability of relatively cheap power, and location on the warm Gulf, midway between the Atlantic and Pacific coasts with easy access to the nation's inland waterways, particularly the Mississippi-Ohio-Missouri system. Natural gas, the major source of power for the chemical industry, is found in all the coastal counties and in deposits located offshore. In addition to natural gas, area raw materials in demand by the chemical industry include petroleum, sulfur, salt, lime (oyster shell), and water.

Integrated oil companies such as Humble began early the manufacture of chemicals, particularly petrochemicals. Nevertheless, it has been chemical companies such as DuPont, Union Carbide, and Monsanto that have been responsible for most of the capital investment in the area. Further, there is a strong tendency for chemical companies and oil companies to merge or to enter into joint ventures in petrochemical production.

Many chemical manufacturers find major markets for their end products within a few miles of their plants, for the end product of one chemical plant is often the raw material for another. For this reason the chemical plants are commonly connected with one another or with refineries by pipelines. A comparatively recent development is the establishment of industrial districts, such as Humble's Bayport project, designed for plants devoted to manufacture of chemicals and for plants specializing in manufacturing end products needed by the chemical industry.

The outlook for the industry in the area is for continued expansion. Suitable plant sites are numerous. Supplies of raw materials and power should be more than ample for industry needs for the next 25 years.

Since 1940 one of the largest concentrations of the chemical industry in the world has developed in Louisiana and Texas, especially in the coastal area which extends from Baton Rouge and New Orleans on the east to Brownsville on the west. In these two states, which might well be described as the heartland of the United States chemical industry, chemicals rank first among all manufacturing industries in value added by manufacture. Value added by chemical manufactures in Texas amounted to $4.15 billion in 1965, almost half of the state's total value added by manufacture. This is even more impressive when it is considered that as late as 1939 the Census of Manufactures did not report organic chemical production anywhere in Texas and the word petrochemical was not even in the dictionary.

Major attractive factors for this development have been the low assembly cost for raw materials, the availability of relatively cheap power, and tidewater location on the warm Gulf of Mexico, midway between the Atlantic and Pacific coasts with easy access to the nation's inland waterways, particularly the Mississippi-Ohio-Missouri system. The chemical industry of the Southwest is based primarily on raw material availability. Natural gas, the major source of power for the industry and one of its major raw materials, occurs in substantial quantity in all the coastal counties and in offshore deposits. Of course it is also available by pipeline from inland fields. In addition to natural gas and its liquid derivatives, area raw materials in demand by the industry include petroleum, sulfur, salt, lime (oyster shell), and water. This fortunate combination of raw materials for chemical manufacture does not occur elsewhere in the United States.

Capital investment for new chemical manufacturing plant and equipment in Texas for 1965, the most recent year for which there is reliable information, was approximately $457 million, a little over 40 percent of that year's total new manufacturing investment in the state. In Louisi-

CHEMICAL PLANTS, WESTERN GULF AREA

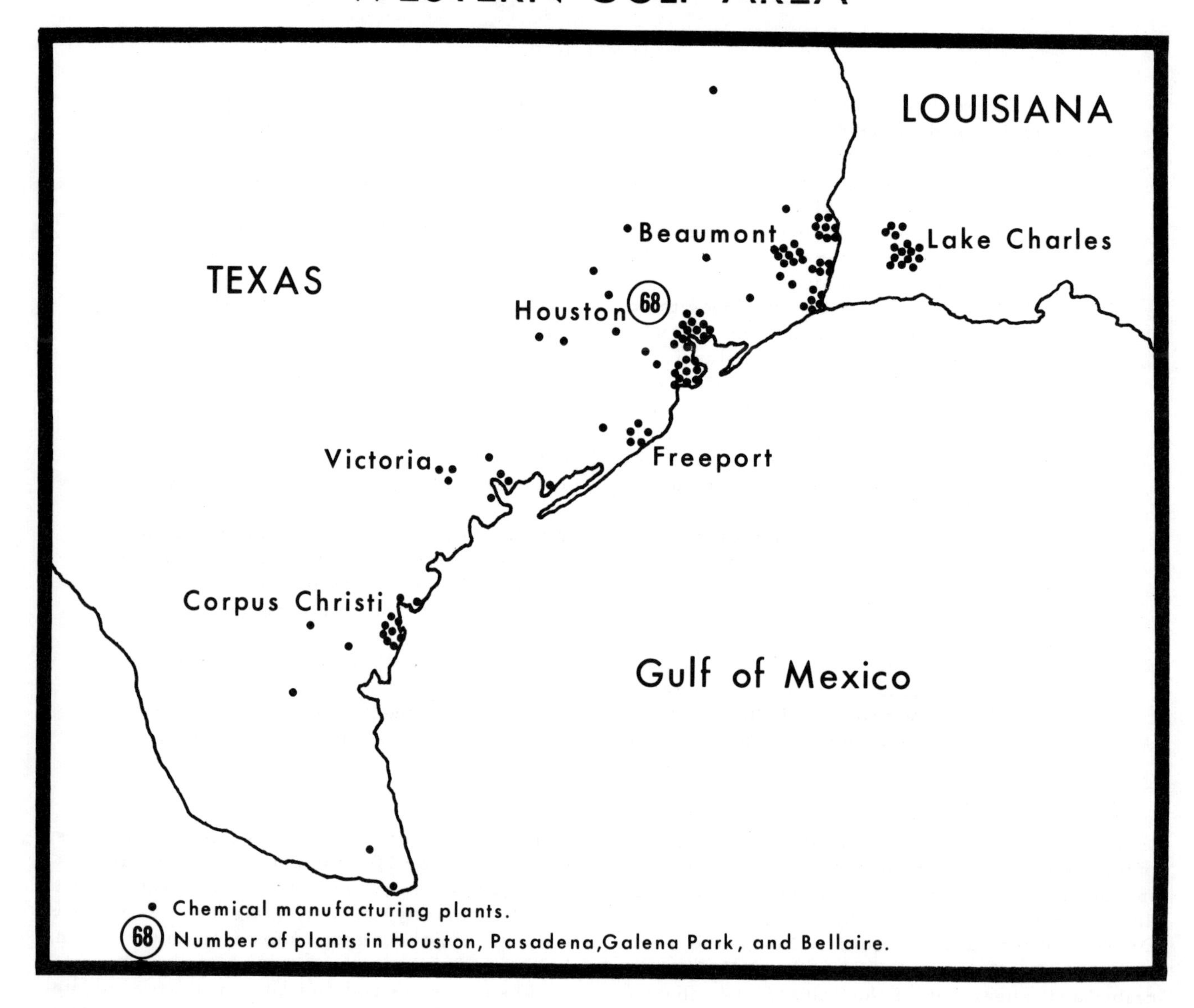

ana chemical industry investment during the same year totaled $133 million, or about 35 percent of total new manufacturing plant investment. The Texas Employment Commission estimates that in January, 1968, the Texas chemical industry employed 60,200 workers, about nine percent of the state's manufacturing labor force. Transportation equipment and food processing were the only other categories of manufacturing which employed more workers. During the same month in Louisiana, the chemical industry was estimated to be employing 21,300 workers.

Major chemical centers in the region, from west to east, are Brownsville, Corpus Christi, Victoria, Seadrift, Bay City, Freeport, Texas City, Houston, Beaumont-Port Arthur-Orange, Lake Charles, Baton Rouge, and New Orleans. The greatest concentration is found within 150 miles of Houston; investment is estimated conservatively to total about $7 billion in the area. Inland Texas locations which have sizable chemical manufacturing operations are Big Spring, Odessa, Borger, Denver City, Pampa, and Longview. Except around Baton Rouge, there are few inland chemical manufacturing locations in Louisiana.

Twenty-eight chemical firms are included in the listing of the nation's 100 largest industrial corporations. All of the twenty-eight firms have plants in the area: Mobil, Texaco, Gulf, Standard Oil of New Jersey (Humble and Enjay), Standard Oil of Indiana (American Oil), Goodyear, Union Carbide, Firestone, Continental Oil, Eastman, Phillips, Monsanto, Borden, Sinclair, Dow, Cities Service, W. R. Grace and Company, Allied Chemical, Tenneco, Olin Mathieson, Atlantic Richfield, B. F. Goodrich, Celanese, Food Machinery Corporation, General Tire and Rubber, American Cyanamid, Pittsburgh Plate Glass Industries, and Signal Oil and Gas.

This roster of twenty-eight firms reads like a Who's Who of the Chemical Industry. But, impressive as it is, it does not reveal that many of these firms have two or three or as many as five plants in Texas and Louisiana. For example, Dow has plants in each of the two states and is beginning to construct another Texas facility at Oyster Creek, near its Freeport installation. There are probably 100 other companies with plants in the region. Further, one integrated chemical complex owned by a single company may produce fifty or more products.

The first real push for establishment of the chemical industry in the Southwest on a large scale came with the beginning of World War II. Prior to that time inorganic products had been made at several locations, particularly at Houston and Corpus Christi. As the war began, demand sharply expanded for such hydrocarbon-based commodities as synthetic rubber. By-product gases from local refineries and the natural gas resources of the coastal area provided the requisite raw materials for two of the major synthetic rubber ingredients, butadiene and styrene. As a result, almost two dozen wartime chemical plants were built in Texas and Louisiana, many of them by the federal government. When the war was over private investors acquired the government-built plants, expanded them and constructed others.

According to the Census of Transportation for 1963, Texas and Louisiana were their own best customers for chemicals. Almost 40 percent of industrial chemicals and slightly over 50 percent of the miscellaneous chemical products produced in the two states were sold in the Southwest. Other major markets were the Middle Atlantic, the East North Central, and the East South Central states. When the routes that Louisiana- and Texas-produced chemicals must travel to reach the market are analyzed, the importance of access by manufacturers to the Gulf and to the inland waterways can readily be seen; chemicals and chemical raw materials are typically heavy and bulky.

Largest volume petrochemical produced in the area is ethylene, which in turn is a major raw material for the production of dozens of other major chemical products, including acetaldehyde, organometallics, ethylbenzene, polyethylene, diethyl ketone, and vinyl acetate. Ethylene and many other chemical products are shuttled from plant to plant by an elaborate system of pipelines described as "The Spaghetti Bowl." Plants as much as 100 miles apart supply each other with raw materials; some plants are even connected with those of competitors. End products and raw materials are often piped to nearby salt domes where they are stored until needed. As the industry expands and becomes more complex, the network of pipelines becomes ever more intricate. Cost of the lines is usually borne by the supplier, but occasionally the buyer shares the cost. It has been said that no other area in the world is laced with such a concentration of piping.

It is this interchange of raw materials that has given rise to a comparatively new development, which, for lack of a better name, might be called a "chemical industrial park." Most elaborate of these is Bayport, a tract of about 9,000 acres, owned by Friendswood Developing Company, a subsidiary of Humble Oil. (Humble operates a large refinery and the Enjay Chemical plant just across Galveston Bay.) Not only do the tenants in Bayport benefit from ease of access to feedstocks from Humble and Enjay nearby, but at least fifty and perhaps as many as seventy crude oil, gas, and product lines cross the area. These pipelines are not just the sources of raw materials; they are also sources of energy materials and provide avenues to market. In addition the port has been designed to provide the most up-to-date accommodations for long ocean vessels and for barges. Highways, rail facilities, storage terminals, and utilities have also been provided.

Also, Bayport is a fine example of the "new town" so much discussed today. It is a community that provides large tracts for residential and commercial use as well as for factories. Though not as well known as Columbia in Maryland or Reston in Virginia, Bayport is nearer completion and probably deserves more notice than it has received among economists, sociologists, and psychologists. Already it has 5,000 or more residents, and 40 or 50 new homes are being built each month. It is anticipated that Bayport will have about 150,000 residents by 1980. The Bayport complex is probably the forerunner of several others like it along the coast.

Integrated oil companies such as Humble began early the manufacture of chemicals, particularly petrochemicals. Nevertheless, it has been the chemical companies, such as DuPont, Union Carbide, Monsanto, Dow, Firestone, Goodrich, Goodyear, and Celanese, that have been responsible for most of the capital investment in the area. Furthermore, there is a strong trend currently for chemical companies and oil companies to merge or to enter into joint ventures in chemical production. Examples of this tendency are the recent purchase of Champlin Refining by Celanese, the merger of Beaunit into El Paso Natural Gas Products, the acquisition of Texas Butadiene and Chemical by Sinclair, and the purchase of Cosden Petroleum Corporation by American Petrofina. Another pronounced trend is for refineries to move into chemical production, particularly the making of fertilizers, on a much larger scale than in the past.

Chemical plants must typically be large so as to effect economies in production costs. Constant expansion and rebuilding programs are also typical, because the pace of technology in the industry causes plant and equipment to become obsolescent rapidly. Smaller and older plants find it difficult to compete in this highly competitive industry.

Because the industry is relatively new in Texas and Louisiana, plants in this area are highly automated. Many persons, unfamiliar with the industry, are surprised to learn that the maintenance force in a $200 million plant may be larger than the operating force. Investment per worker is the highest for any major industry in the area.

The outlook is for continued expansion. Suitable sites for plants and industrial districts are still numerous. The marketing geography is favorable. Supplies of raw materials and energy should be more than ample for at least the next 25 years. It is certain that Texas and Louisiana can readily produce the necessary building blocks for chemicals that are yet to be produced commercially. Southwestern markets for fabricated plastics and other end products should expand, but the most important contribution of the chemical industry in the near future will likely be that of supplier to manufacturers in the Southeast and in the American Manufacturing Belt.

COORDINATE EVALUATION OF EVAPORITE DEPOSITS BY DYNAMIC MODELINGS AND SIMULATION

Louis I. Briggs and Darinka Zigic-Toshich
Department of Geology and Mineralogy
The University of Michigan
Ann Arbor, Michigan

ABSTRACT

Quality, form, distribution, and relation of stratigraphic components are identified by systematic analysis of distributive patterns of deposit types and their associative elements.

Integration of desired stratigraphic variables by dynamic modeling and simulation serves as an effective tool, and gives a better understanding of distribution of evaporite deposits.

Evaporite deposits of commercial value in the Michigan Basin include Silurian (Salina) salt, Devonian (Detroit River) salt, gypsum and dolomite, and Mississippian (Michigan) gypsum. There has been significant exploitation of the Salina salt and the Michigan gypsum deposits.

The Salina Group evaporites were deposited within a basin rimmed by Niagaran carbonate reefs, across which ocean brine flowed to connect the open ocean with the central, moderately deep water basin. In the central part of the basin, rock salt comprises more than 75 percent of the total evaporite section of salt, anhydrite, and carbonate rocks. The central basin model best fits the depositional conditions, which have been simulated using a digital computer.

The Detroit River Group evaporites are characterized by two distinct phases of sedimentation, a lower carbonate-sulfate phase which covers the entire Michigan Basin, and an upper complex carbonate-sulfate-chloride phase largely restricted to the central part of the basin. A combined central basin-sebhka model best fits the texture and distribution patterns of the evaporites.

The Michigan Formation is primarily a shallow marine shale-dolomite-anhydrite sequence characterized by strong stratigraphic facies zonation. The facies indicate deposition in a marginal shallow marine basin, having pseudo-marine and brackish water zonation adjacent to the distal margins of the basin subject to periodic influx of terrestrial silicate silt and mud.

Exploitation of evaporite deposits can be aided by dynamic model simulation, defined by appropriate structuring of the stratigraphic system analysis network.

INTRODUCTION

Evaporite deposits in the Michigan Basin of commercial value include Silurian Salina salt, Devonian Detroit River salt, gypsum and dolomite, and Mississippian Michigan gypsum deposits. Specific characteristics and geologic settings of the three groups represent three distinctively different types of evaporite depositional environments. The Salina Group was deposited as a carbonate-evaporite sequence within a relatively deep reef-rimmed basin. This sequence represents the central basin type evaporite environment. The Detroit River Group is also a carbonate-evaporite sequence which combines the basin environment with the shallow marginal sebhka evaporite environment. Geometric configuration of the basin was controlled largely by regional structural features and differential tectonic adjustment; boundary reefs were of relatively little importance in controlling the evaporite distribution and sequence. In contrast to these, the Michigan Formation represents a shallow marine marginal basin characterized by alternate deposition of evaporite sulfates and carbonates. This deposition was interrupted by occasional influx of terrestrial muds, silts, and sands into the basin. Strong lateral lithologic facies characterize the Michigan Formation distribution and sequence.

Stratigraphic system analysis based on association of stratigraphic components uniquely defines the three models of evaporite deposition and the criteria for simulation of the evaporite models with the digital computer. Experimental variation of stratigraphic components that control parameters of the simulation models imply the optimal criteria that define the geologic settings for each evaporite sequence. Dimensions of the optimal stratigraphic components, their spatiotemporal variation, and their association in terms of defined paleogeographic and economic concepts serve as the basis for geologic and economic decisions in exploitation of mineral deposits.

STRATIGRAPHY

Michigan Formation

The Michigan Formation is a Lower-Middle Mississippian stratigraphic unit which occurs only in the Lower Peninsula of the State of Michigan. It overlies the Coldwater Shale and underlies the Bayport Limestone over most of this area. The formation is a complex sequence of sandstone, siltstone, shale, dolomite, limestone, and gypsum having strong lateral and vertical facies changes that have been clarified by outcrop and subsurface studies of Moser (1963). Stratigraphic relationships are shown on the correlation chart (Fig. 1), and subsurface members are defined for the purpose of detailed stratigraphic analysis in Figure 2.

Of eleven subsurface units defined by Moser from electric log responses and well cuttings and core studies, nine contain beds of anhydrite and gypsum. Two of these, the National City Gyp (Q-R) and the Triple Gyp (M-N), are currently exploited for gypsum, the first near Saginaw Bay, and the latter in the vicinity of Grand Rapids on the western side of the state. With current economic practice, these are the only areas where abundant gypsum of high purity occurs at shallow depth near inexpensive lake transportation. Elsewhere, problems of transportation, thick glacial drift overburden, bed thickness or anhydritic composition make commercial exploitation unfeasible.

Upper "Marshall" sediments are exclusively sandstone and were deposited in a regressing sea (Newcomb, 1933; Lane and Seaman, 1909). Positive tectonic elements enclosed the basin on all sides except northeast, where the "Marshall" seas joined the open ocean.

Description.--Return of the sea marked the beginning of deposition of the Michigan Formation. Conditions of sedimentation which favored deposition of evaporites developed contemporaneously. Circulation of ocean brine between the basin and open sea was probably restricted by a sub-aqueous sill which had a northeast trend, characteristic of structural elements in the Michigan Basin. There were probably changes

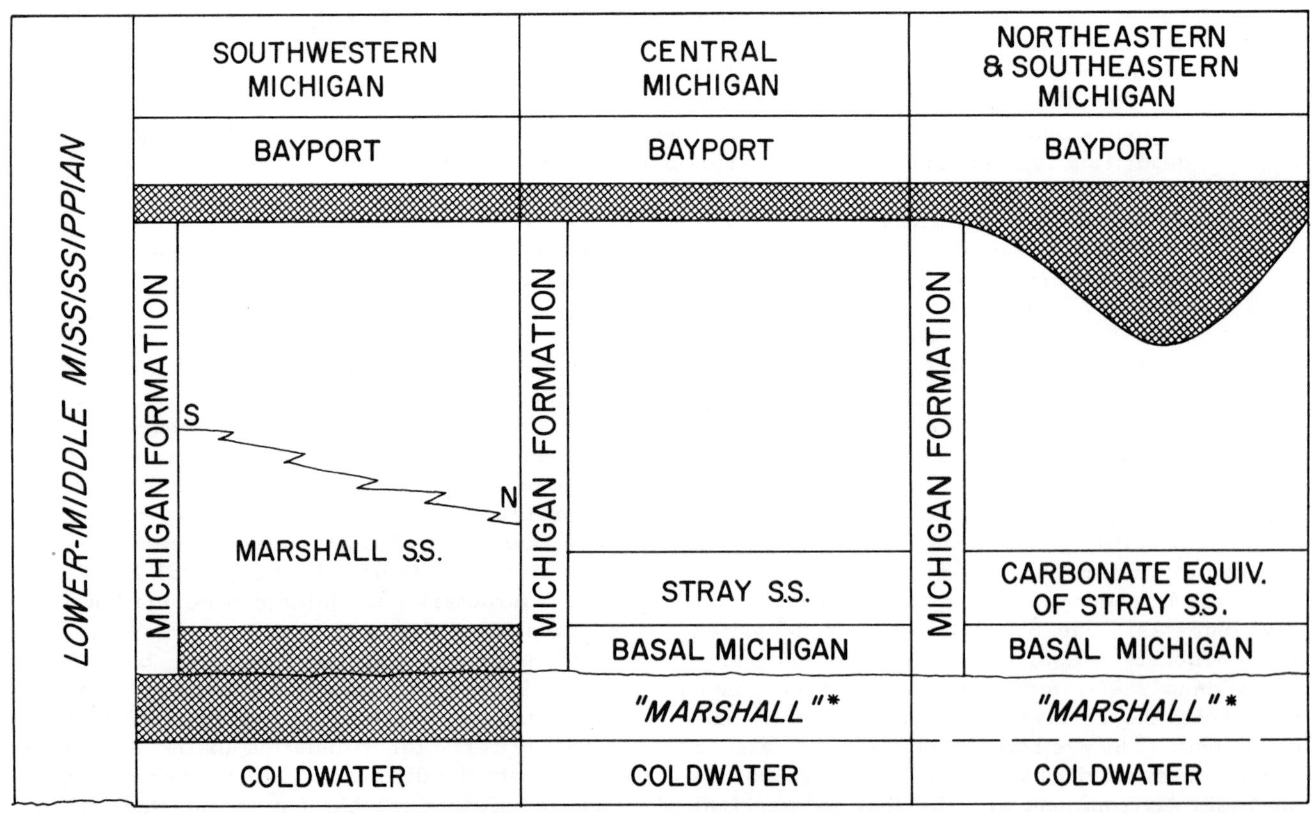

FIG. 1. Lower-Middle Mississippian correlation chart, Michigan.

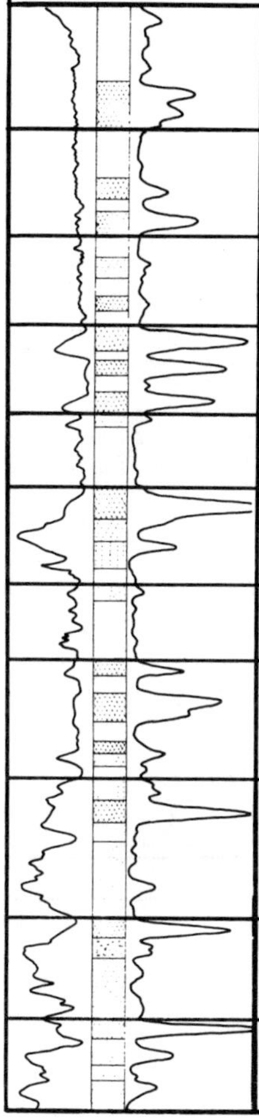

COMPOSITE ELECTRIC LOG	ELECTRIC LOG MARKER AND UNIT NO.	UNIT NAME	MAXIMUM THICKNESS (FT.)	DESCRIPTION
	UNIT 9	TOP OF MICHIGAN-K	120	ANHYDRITE, DOLOMITE AND SHALE
	K			
	UNIT 8	K-L	60	ANHYDRITE, DOLOMITE AND SHALE
	L			
	UNIT 7	PENCIL GYP L-M	50	RED AND WHITE ANHYDRITE WITH SHALE AND DOLOMITE
	M			
	UNIT 6	TRIPLE GYP M-N	70	ANHYDRITE AND SHALE WITH SOME ANHYDRITIC DOLOMITE
	N			
		N-O	40	MOSTLY SHALE WITH SOME DOLOMITE
	O			
	UNIT 5	BROWN LIME O-P	50	TAN SUCROSIC DOLOMITE, ANHYDRITE, LIMESTONE AND SHALE
	P			
		STRAY-STRAY P-Q	60	MOSTLY SANDSTONE AND SHALE
	Q			
	UNIT 4	NATIONAL CITY GYP Q-R	65	ANHYDRITE AND SHALE WITH MINOR DOLOMITE
	R			
	UNIT 3	STRAY R-S	200	SANDSTONE, DOLOMITE, LIMESTONE & ANHYDRITE
	S			
	UNIT 2	SUGAR DOLOMITE S-T	60	COARSE SUCROSIC DOLOMITE, CRINOID LIMESTONE & ANHYDRITE SOME SANDSTONE AND SHALE
	T			
	UNIT 1	T-MARSHALL	50	DOLOMITE, ANHYDRITE AND SANDSTONE

FIG. 2. Division of Michigan Formation in subsurface by electric log markers.

in the climatic regime as well, accounting in part for the sequence of evaporites and terrestrially derived clastics.

A series of maps showing relative percentages of main lithologic types for the interval from the K marker (Fig. 1) to the bottom of the Michigan Formation is shown in Figure 3. The strong northwest trend of all percentage lines is a reflection of the dominant orientation of strand lines that persisted during most of Michigan Formation deposition. The source of clastics was mainly from the southwest while the direction to the open sea was northeast. The zone of highest anhydrite deposition was the approximate center of the subsiding basin.

The lowermost unit in the Michigan Formation was deposited during the initial advance of the Michigan sea over an eroded "Marshall" landscape. The unit is limited to a small area in the north-central part of the state.

Clastics of Unit 1 are fine-grained, shaly sandstone and gray shale. Overlying the clastic sediments is a bed of dark brown, coarse-grained dolomite. There is, however, an anhydrite bed on top of the dolomite, and the juxtaposition of dolomite and anhydrite indicates that at least part of the dolomite is a primary precipitate.

The areal extent of the Sugar Dolomite is much greater than that of the previous unit because of continuing transgression of the Michigan sea. The lower part of the unit is composed of light and dark brown sucrosic dolomite and black, fissile, dolomitic shale. The upper section consists of a thin, continuous bed of fossiliferous limestone overlain by a bed of anhydrite (Fig. 4). Principal fossil components are thin-shelled brachiopods and crinoids but only the crinoids can be distinguished in cuttings. The two fossils are nearly mutually exclusive, and brachiopods are concentrated toward the bottom of the bed. There is a thin transition zone between the limestone and anhydrite, consisting of

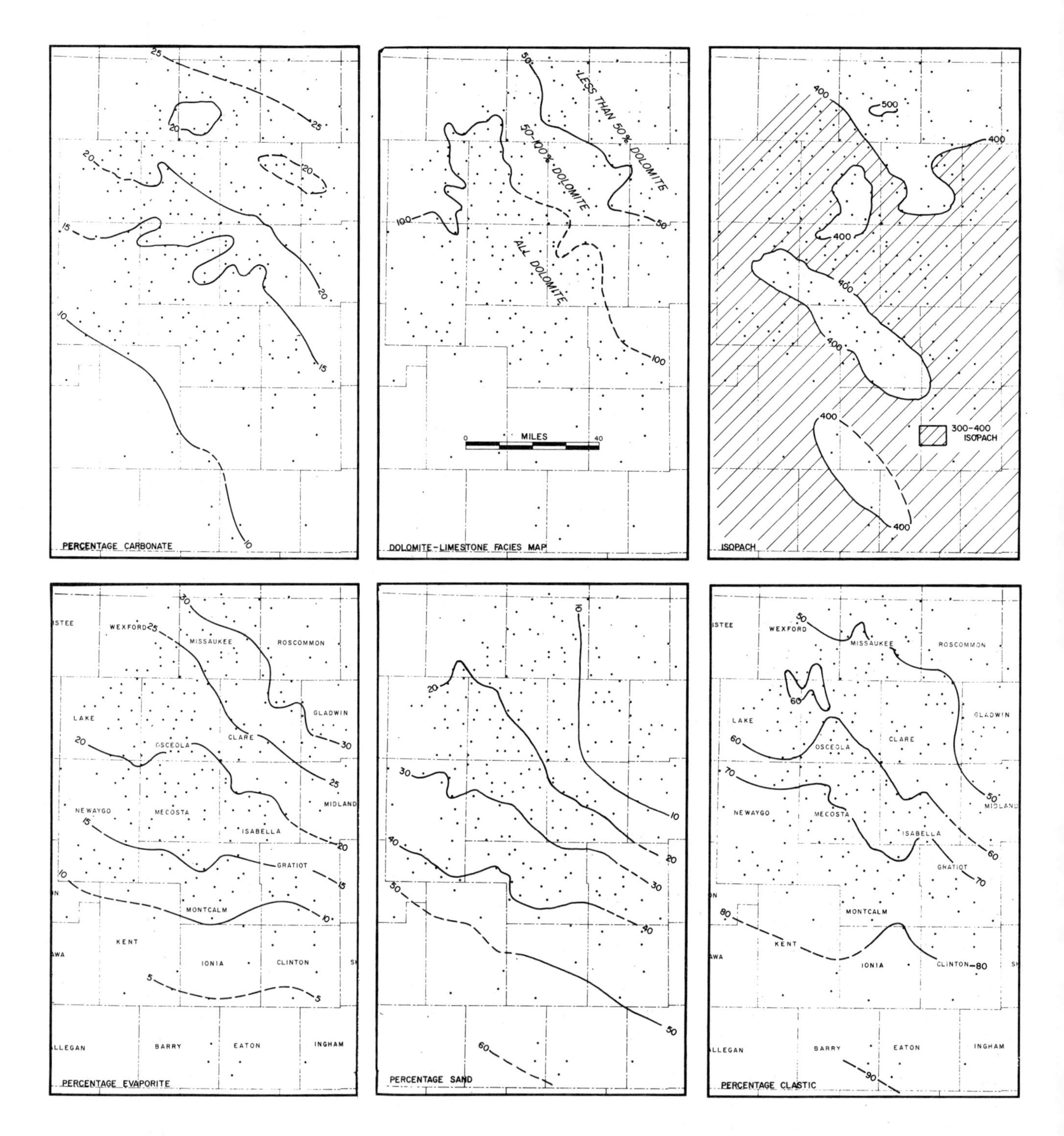

FIG. 3. Isopach and lithofacies map of total Michigan Formation.

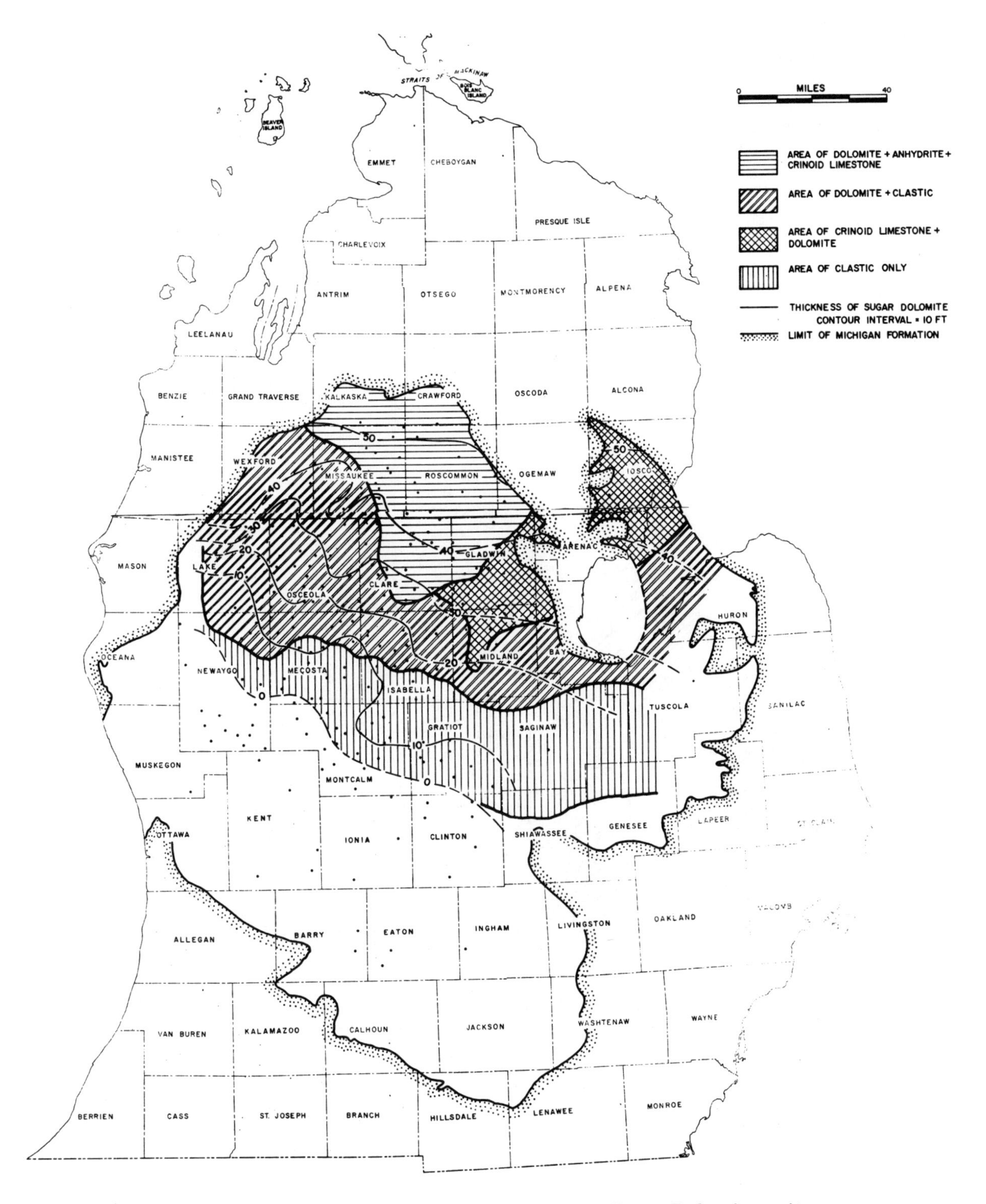

FIG. 4. Isopach and lithofacies map of Sugar Dolomite unit.

a black, calcareous shale containing pockets of crinoids.

The upper boundary of the Sugar Dolomite is placed at the end of another cycle of evaporite deposition and the beginning of a new regressive phase in which the Stray Sandstone was deposited.

Clastic sediments predominate in the Stray. They are everywhere more than half the section, and over a large area they are the only component. By far the largest clastic component is sand. The non-clastic equivalents of the Stray in the northeast consist of dark, shaly, fine-grained limestone, white massive anhydrite, and light brown, somewhat sandy dolomite. There is a complex relationship between these three lithologic types (Fig. 5).

The Brown Lime unit is an excellent example of vertical and horizontal development of an evaporite facies. The vertical sequence is shale to carbonate to sulfate. Each of these beds can be traced horizontally over a large area, indicating very stable conditions of sedimentation. In the nearshore environment they all grade gradually into a sandstone. The top of the unit is a bed of

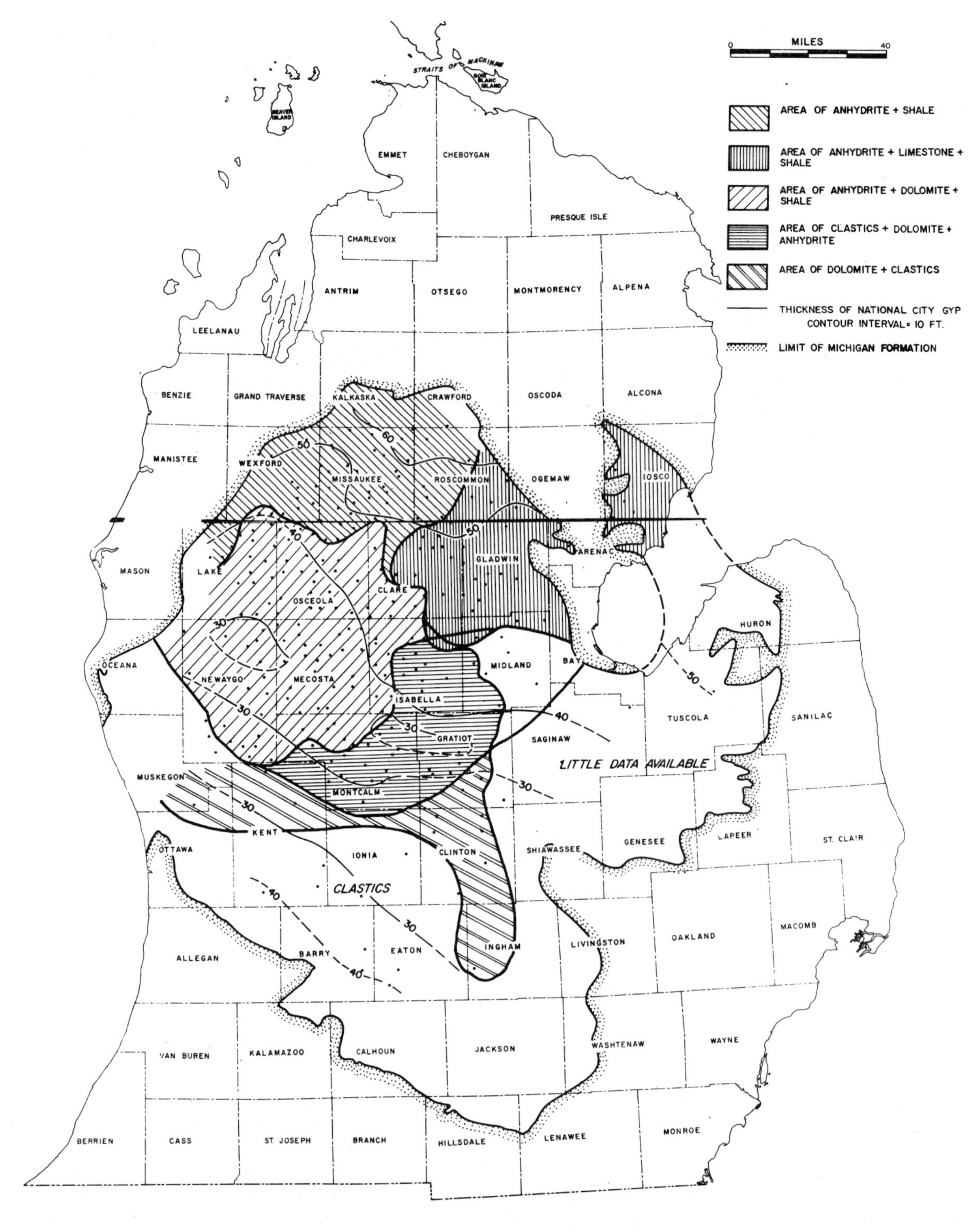

FIG. 5. Isopach and lithofacies map of National City Gyp unit.

massive anhydrite which is thickest in the north where it reaches a maximum of 15 feet.

The ingression of the Michigan sea onto the land culminated with the deposition of the Triple Gyp consisting of three massive anhydrite beds interbedded with two shale members. The anhydrite beds in the Triple Gyp have the greatest areal extent of any precipitate member in the Michigan Formation. In the direction of the paleoshoreline, some of the anhydrite grades into a brown, anhydritic dolomite (Fig. 6).

The fortuitous find of a fossil shark spine in

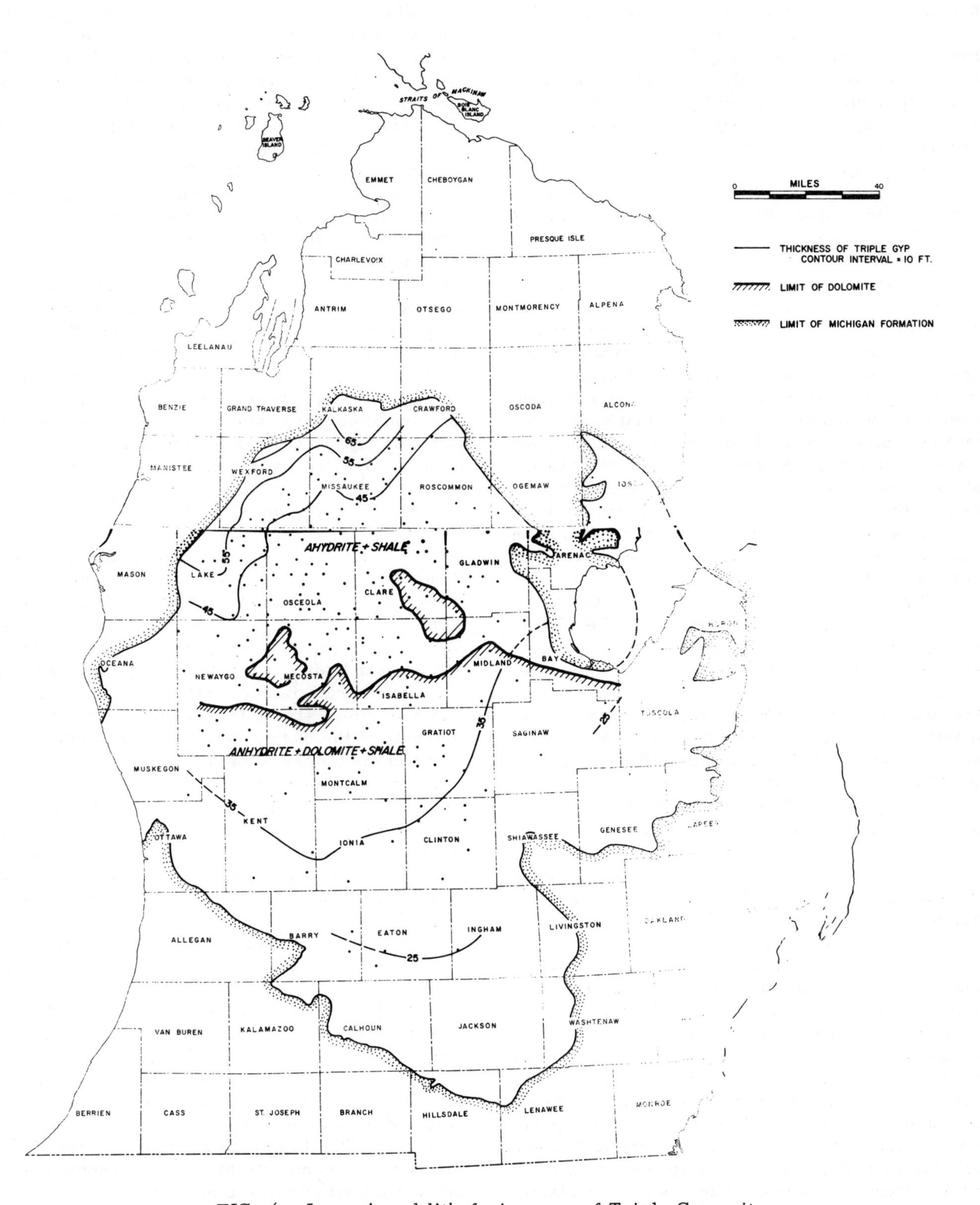

FIG. 6. Isopach and lithofacies map of Triple Gyp unit.

the Triple Gyp provided additional evidence for the hypothesis that climate was the main cause for interbedded shale and anhydrite (Dorr and Moser, 1963). The spine was found in an 8-inch sandy shale layer within the main gypsum bed at the mine of the Bestwall Gypsum Company in Grand Rapids. The locality of the mine was probably not far from the distal edge of the sea during deposition of the Triple Gyp. During precipitation of the anhydrite the salt-saturated waters would have been intolerable for sharks. However, during the time represented by the shale partings, when climatic changes brought an increased influx of fresh water, animals could enter the basin and live along the margins.

The Pencil Gyp unit consists mainly of gray shales intercalated with red anhydrite and brown dolomite. To the northeast, dolomite grades into white anhydrite.

The basic shale-anhydrite-dolomite alternations, typical of the upper part of the Michigan Formation, continued during deposition of Unit 8 (K-L). There are, however, important signs both in lithology and in isopach patterns of the impending termination of those environmental conditions which gave rise to the clastic-precipitate cyclicity. Intervals of evaporite deposition were becoming much shorter. The percentage of shale in this unit increases rapidly from 40 percent in the north to over 70 percent in southern areas. The northern edge was not far from the present day outcrop area and the center had shifted southward.

The top unit of the Michigan Formation can be subdivided into two members. The lower precipitate member is composed of red and white anhydrite which grades into a light brown dolomite. Because this unit is below an unconformity, the anhydrite has been hydrated to gypsum in many wells. The upper member is predominantly gray and green shale with localized development of anhydrite and dolomite.

Conclusions.--The period of deposition which preceded the Michigan Formation had its climax in a widespread regression of seas from the basin which produced a profound unconformity between the Michigan and the Coldwater everywhere in the basin and a complete hiatus between the two units in southwest Michigan.

During deposition of the Michigan Formation the basin was enclosed by land on all sides except to the northeast. In this direction lay the connection with the open ocean. A tectonic sill in the vicinity of the connection was the cause of restricted water exchange between basin and ocean. The sill, combined with climatic changes, resulted in clastic-precipitate cycles of deposition. During the arid phase, evaporites were precipitated while during humid phases sands and shales were deposited. The source of clastic sediments lay mainly in the land margins of the basin.

Evaporite deposits of the Michigan Formation consist predominantly of dolomite-anhydrite sequences with some limestone-anhydrite and dolomite-limestone cycles. No salt was found and the available evidence indicates that waters did not reach the salt precipitating stage.

The Michigan sea transgressed gradually into the basin from the northeast with only minor regressive movements during deposition of the Stray and Stray-Stray Sandstone units. The transgression reached its apex during deposition of the Triple Gyp unit. This slow shelfward expansion caused the sand-evaporite margin to lie progressively southward. Center of the depositional basin during Michigan times was in the area of the present northern edge of the formation and only the southern half of the deposited sediments remain.

Deposition of the Michigan Formation terminated as a result of epeirogenic activity which was strongest in the eastern parts of the state. This eventually resulted in a complete withdrawal of the seas from the Michigan Basin for a short period. Along a series of en échelon folds in eastern Michigan much of the Michigan Formation was removed prior to return of the sea and deposition of the overlying Bayport Formation.

Detroit River Group

The Middle Devonian Detroit River Group, described by Ehman (1964), is represented in the subsurface at the center of the Michigan Basin by a thick sequence of evaporite and carbonate rocks. This sequence grades laterally into a thick carbonate and sandstone unit at the marginal outcrops.

Lithostratigraphic analysis is difficult because of these strong facies changes toward the margins of the basin. Stratigraphic subdivisions made in the center of the basin where evaporites have accumulated and lithostratigraphic units are more distinct were defined and correlated by means of mechanical logs, with sample cuttings for control. In this way the stratigraphy in the middle of the basin was extrapolated through lithofacies changes to the margins of the basin (Fig. 7).

Four divisions of the section, in ascending order, in the Central Basin area have been delineated.

(1) Base of Group to Marker A, a fossiliferous carbonate unit, the Amherstburg Formation

(2) Pre-Salt Evaporite, from Marker A to Marker B, an essentially non-fossiliferous carbonate and evaporite unit

(3) The Lower Salts, from Marker B to Marker G, composed chiefly of carbonates with significant anhydrite and halites

(4) The Upper Salts, from Marker G to

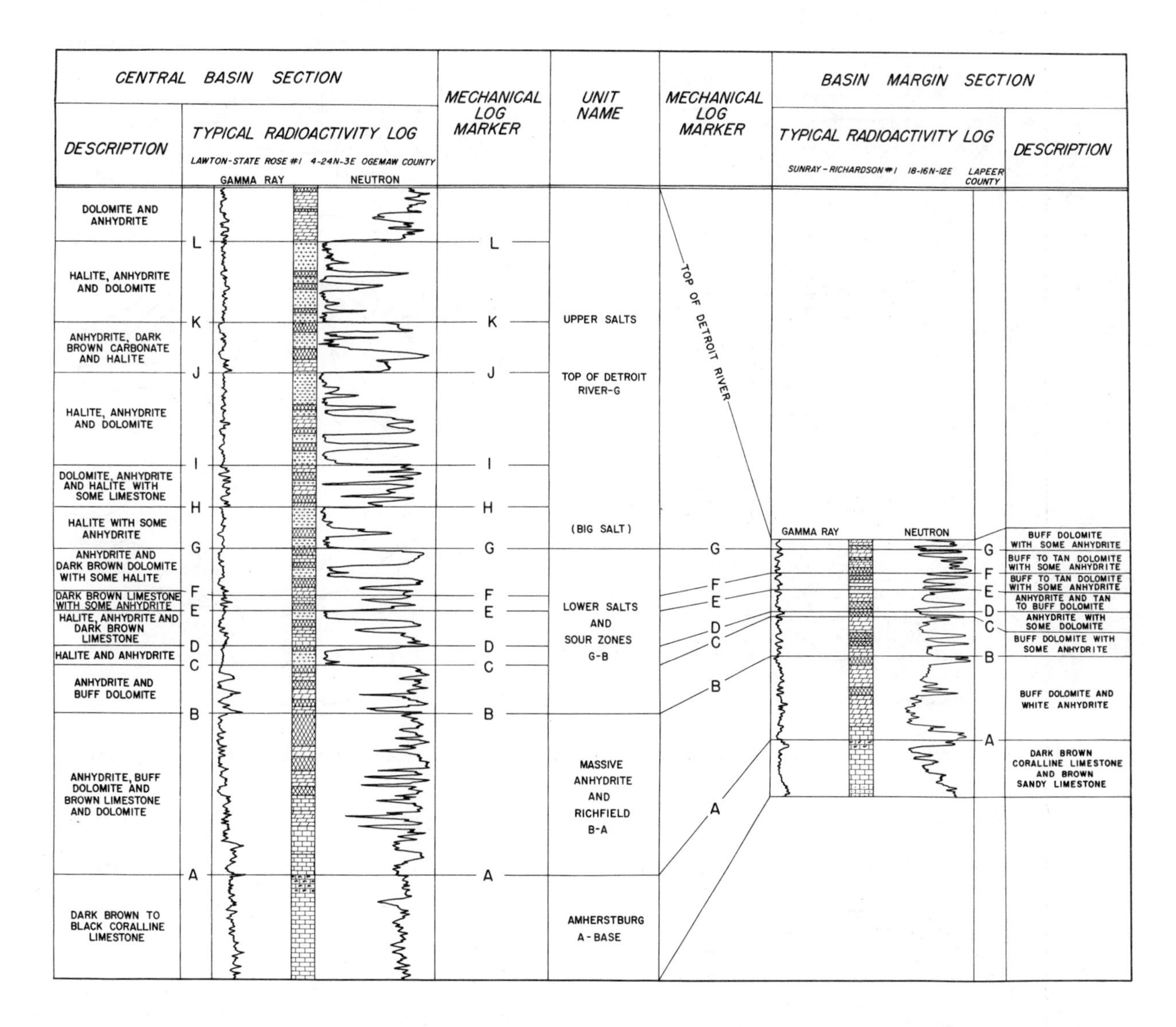

FIG. 7. Division of Detroit River Group by mechanical log markers.

top of Detroit River marker, composed chiefly of halite and anhydrite with some carbonate stringers.

Base of Group to Marker A: Amherstburg Formation.--This lowest unit of the sequence is a fossiliferous non-evaporitic carbonate rock containing appreciable amounts of silt and sand. The rocks grade from a thick sequence of fossiliferous, crystalline, dark limestone in the north and east, to dark crystalline dolomite in the extreme western part of the state. In the southern part of the area, the rock is a light gray to brown, silty, sandy dolomite.

Marker A to Marker B: Pre-Salt Evaporites.--The Pre-Salt Evaporite is comprised of two subsurface units long used by Michigan petroleum geologists. The Richfield Member is predominantly a carbonate rock unit, but contains numerous thin anhydrite stringers. The Massive Anhydrite, as the name denotes, is composed chiefly of anhydrite, but also contains interbedded carbonate rock stringers in some parts of the basin (Fig. 8). The rocks are thickest in the northwestern quarter of the basin, but thin rather uniformly toward the margins.

A color change similar to that of the underlying Amherstburg Formation occurs in the carbonate rocks of the Pre-Salt Evaporite. Basinal carbonate rocks of this unit are commonly brown to dark brown. The rocks become lighter in color as the unit is traced marginward; the thinner the unit becomes, the lighter its color.

Marker B to Marker G: Lower Salts.--Basal salts and carbonate Sour Zones of the Lucas Formation are included in the sequence defined between the B and G markers (Fig. 7). The area of

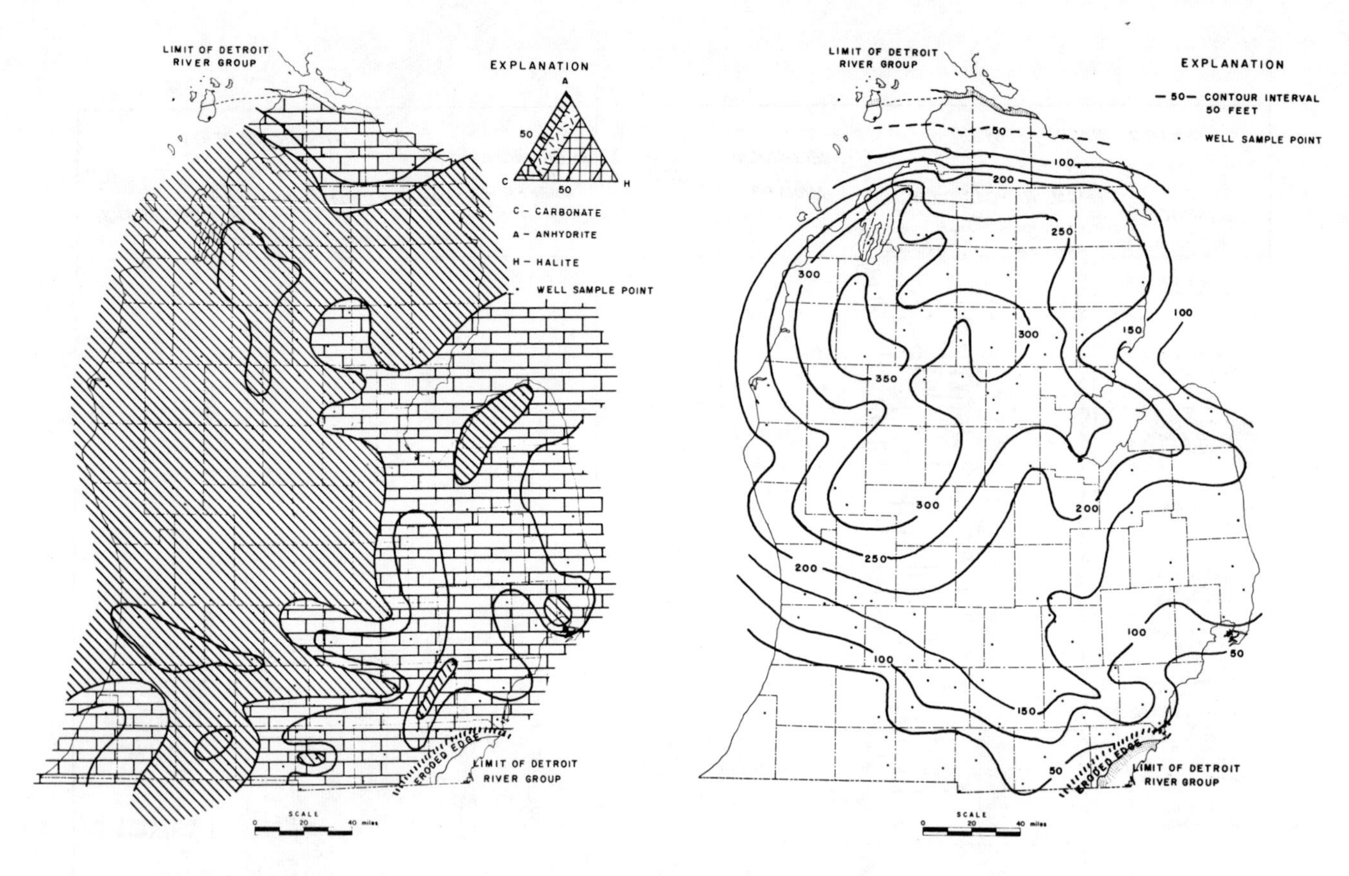

FIG. 8. Evaporite facies and isopach maps, Marker A to Marker B, Pre-Salt Evaporite.

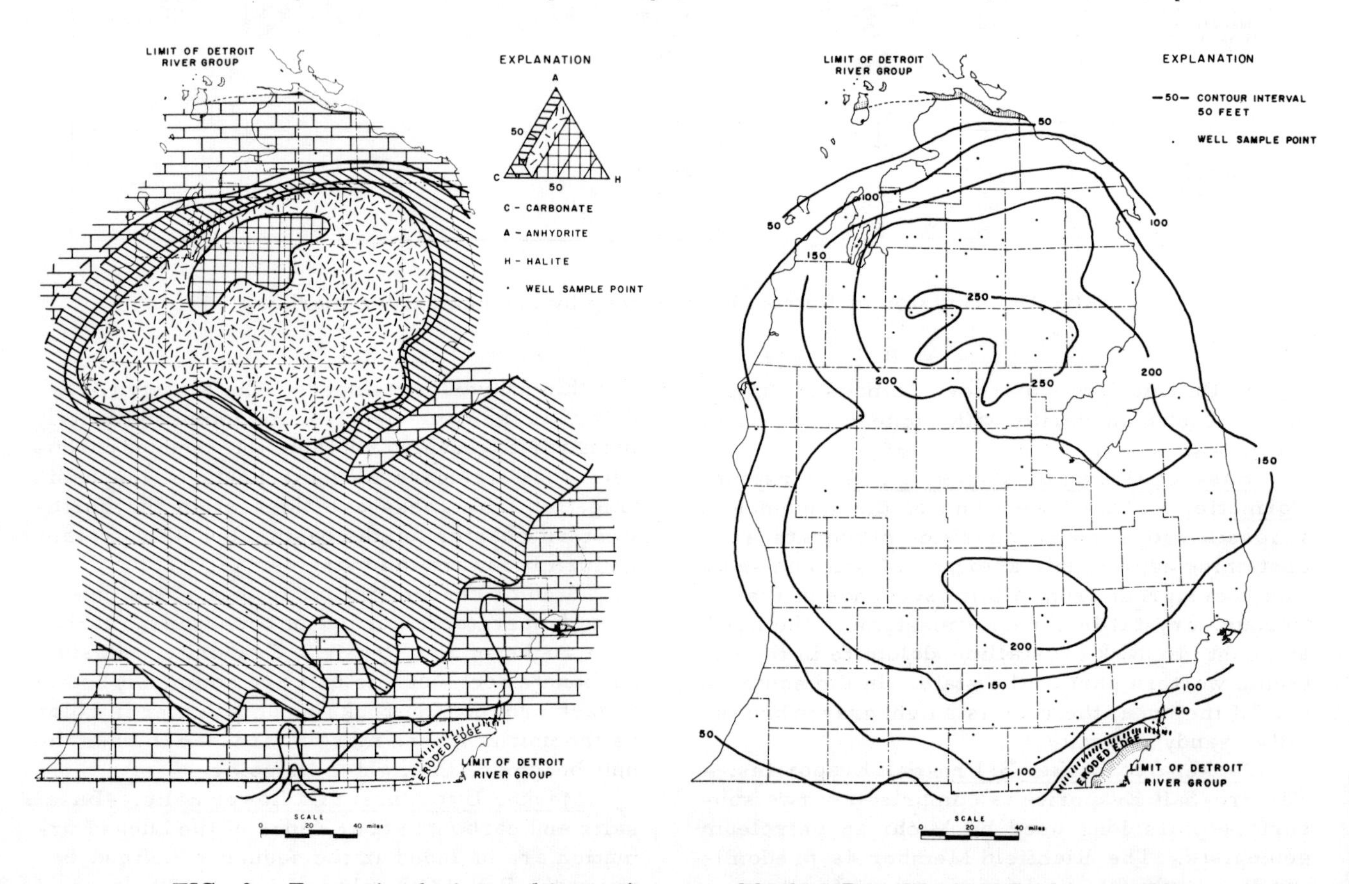

FIG. 9. Evaporite facies and isopach maps, Marker B to Marker G, Lower Salts.

salt accumulation was restricted to the northern half of the Southern Peninsula of Michigan. Maximum thickness of the entire Detroit River Group, as well as the Lower Salts, is found in this area of salt accumulation (Fig. 9).

Subdivisions of Lower Salts were traced as far as possible from the Central Basin area. The member between the B and C markers is defined as the lowest Sour Zone. The B-C member is typically a buff, commonly sucrosic carbonate rock with thickly interbedded anhydrite layers. The carbonate rock is dolomite except in the area south of Saginaw Bay, where it is extremely calcareous.

The C-D member is the basal Detroit River salt. In the center of the salt basin, the rocks attain a thickness of 90 feet and are more than 80 percent salt in Antrim County (Fig. 10). The

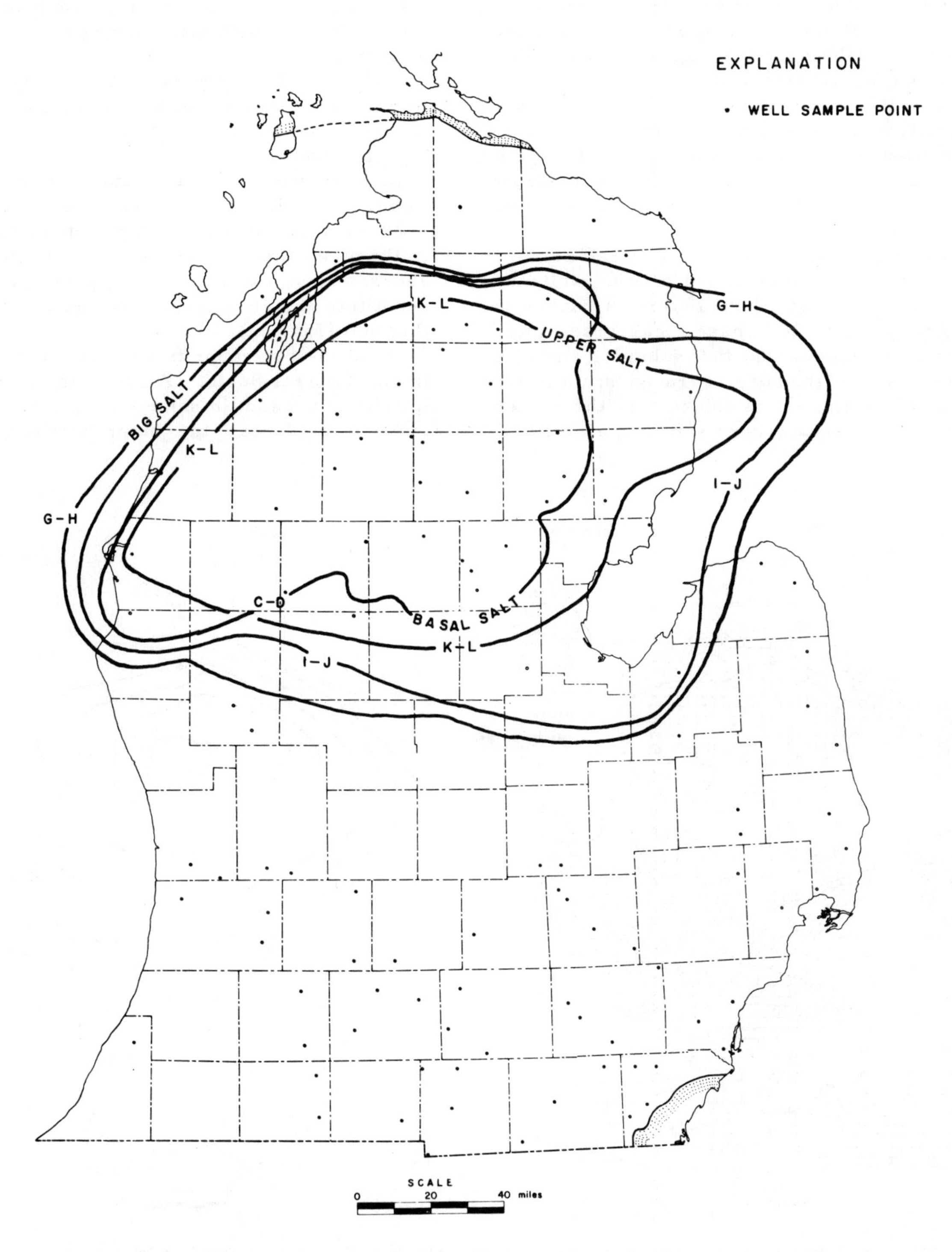

FIG. 10. Map showing the distribution of rock salt in the Detroit River Group.

sequence grades from halite and gray anhydrite in the salt basin to gray and white anhydrite and buff dolomite toward the southern basin margin.

The D-E member is composed of an upper salt and a lower, dark brown limestone layer in the Central Basin. This sequence grades into a unit composed of an upper, white anhydrite and a lower, buff, sucrosic dolomite toward the southern margin of the basin. It has a uniform thickness of 50 to 70 feet in the Central Basin, and decreases gradually in thickness toward the margin to the south and southeast. Salt increases in thickness toward the northwest, reaching a maximum thickness of 40 feet in Antrim County.

The E-F member is composed of rocks grading from a dark brown limestone in the Central Basin area to a buff-colored dolomite on the southern margin of the basin. It has a uniform thickness of 20 to 40 feet over all the area except the extreme basinal margins.

The F-G member in the Central Basin is characterized by interbedded halite, white and gray anhydrite, and buff to brown carbonate, principally dolomite. These rocks grade into a sequence of interbedded buff dolomite and white anhydrite as the member is traced south toward the basin margin. It is thickest in the eastern Central Basin area, where salt is present.

Regarding the Lower Salts as a whole, the following conclusions can be made: (1) Area of maximum thickness is in the Central Basin. (2) Salts of this unit increase in thickness to the northwest, then abruptly decrease in thickness. (3) The dark brown color of carbonate rocks in the Central Basin area changes to a light brown or buff marginward. (4) Limestone-dolomite patterns closely follow that of underlying Detroit River units. Limestone is concentrated in the Central Basin and in an area south of Saginaw Bay, whereas dolomite dominates the basin margin.

Marker G to top of Detroit River: Upper Salts.--The Upper Salts of the Detroit River Group is a sequence of interbedded halite, anhydrite, and carbonate rocks. Some limestone is present in the salt basin and in northern Huron County, but dolomite is the typical carbonate rock of this sequence. Gypsum is present in wells of northern counties, and of Allegan and Jackson counties. Evaporites (principally halite) constitute the greatest percentage of the rock in this unit (Fig. 11).

The G-H member is 80 to 100 percent halite in the Central Basin. A gray, argillaceous anhydrite is present in some wells, particularly in those located near the outer boundary of salt

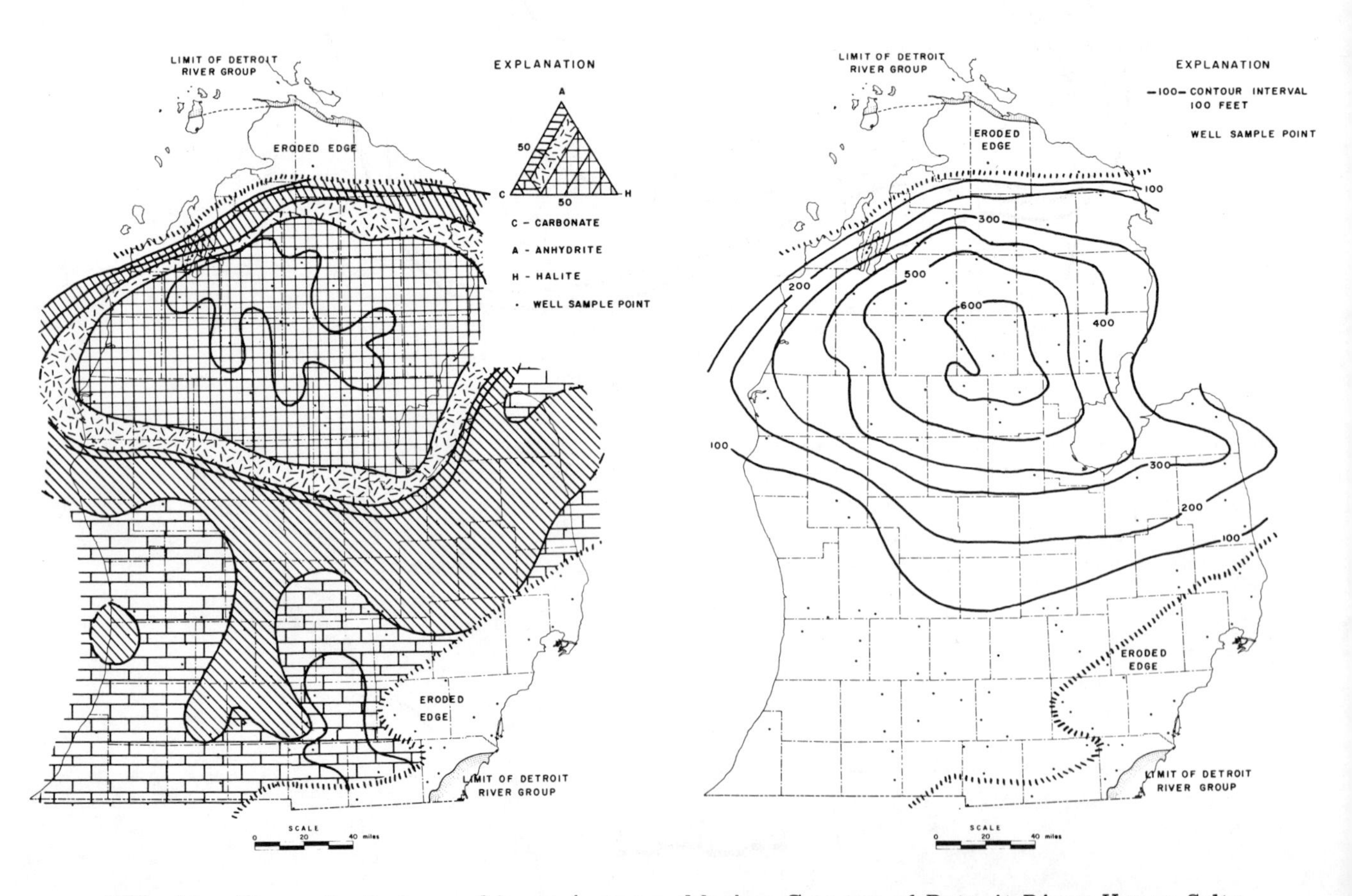

FIG. 11. Evaporite facies and isopach maps, Marker G to top of Detroit River Upper Salts.

accumulation (Fig. 10). Salt grades to an anhydrite facies outside the Central Basin area. The anhydrite facies in turn becomes more dolomitic still farther from the center of the basin.

The sequence of rocks above the H marker, the H-I member, represents two complete evaporite cycles. The vertical upward sequence of rocks in this member is anhydrite, carbonate, anhydrite, halite, anhydrite, and carbonate.

The I-J member is a thick salt sequence with interbedded gray and white anhydrite and light brown dolomite beds. This sequence thins gradually from the basin center toward the margins where it is characterized by anhydrite and dolomite.

One of the best marker beds of the Upper Salts is a dark brown carbonate bed in the J-K member. This rock, highly dolomitic except in the northern "thumb" area, underlies an anhydrite-salt and anhydrite sequence. The J-K member has a fairly consistent composition of 40 percent anhydrite, 30 percent salt, 30 percent dolomite, and a thickness of 60 to 90 feet throughout the Central Basin area. Eastward, south of Saginaw Bay, limestone is the predominant lithology.

The uppermost salt layer in the Detroit River Group is the K-L member of the Upper Salts. This sequence is confined to the Central Basin, and is composed predomimantly of halite with minor amounts of brown to buff dolomite and gray argillaceous anhydrite.

This section extending from the L marker to the top of Detroit River marker, the uppermost member of the Upper Salts, is also confined to the Central Basin area. It is composed of white and gray anhydrite and brown dolomite. In general, upper and lower anhydrite beds are separated by a dolomite layer.

Patterns emerging from the stratigraphic descriptions of the three lower Detroit River units appear in general to hold true for the Upper Salts as well. Color, composition, and textural changes of the carbonate rocks of this sequence, when traced from the Central Basin toward basin margins, are the same as those of underlying units.

Salina Group

The following description closely follows Alling and Briggs (1961, p. 519). In the Michigan Basin the system of stratigraphic classification of the Salina Group devised by Landes (1945) and modified by Evans (1950) is widely used by subsurface stratigraphers in Michigan and Ontario. The terminology marks the first comprehensive application of basinal evaporitic facies to the regional stratigraphy of a depositional basin of the Cayugan. The following stratigraphic units were defined from top to bottom:

Unit H: Bass Islands Dolomite, 175-570 feet thick, largely buff dolomite, some gray dolomite near base, anhydrite and salt beds near center of basin.

Unit G: Uppermost Salina Group, 4-100 feet thick, characteristically gray shaly dolomite with green and red shales near Mackinac Straits (Pte. aux Chenes Formation).

Unit F: Uppermost salt in Salina section, 0-1,230 feet thick. Thick beds of salt separated by shale, shaly dolomite, dolomite, and anhydrite.

Unit E: 30-122 feet of gray or red shale with some dolomite, shaly dolomite, and anhydrite.

Unit D: 25-65 feet of nearly pure salt, with thin partings of buff dolomite.

Unit C: 60-160 feet of largely shale or shaly dolomite with anhydrite and buff dolomite in places.

Unit B: 240-400 feet thick, almost pure salt with minor dolomite.

Unit A: 30-1,105 feet thick, limestone, dolomite, salt, and anhydrite, which was subdivided into:

A_2--Buff to brown dolomite or limestone
Dark gray dolomite
Salt or anhydrite

A_1--Fine brown to grayish brown dolomite or limestone
Fine gray dolomite or limestone
Salt or anhydrite

Around the rim of the Michigan Basin, the basal salt or anhydrite rests on and grades into light brown crystalline dolomite of the Guelph or upper Lockport whereas in the center of the basin, basal evaporites rest directly on reddish argillaceous bioclastic limestone. Evaporite facies and paleogeographic concepts are illustrated in Figure 12.

Dellwig (1955) described the mineralogy, textures, and structures of the Salina Salt in the Michigan Basin, comparing the basin facies with marginal facies. Stratigraphic components differ in characteristic properties and their dimensions. Dellwig attributed this to deposition of the salt sequence of the basin facies in relatively deep water, and in the marginal facies within the wave-affected zone.

CONSTRUCTION OF THEORETICAL STRATIGRAPHIC MODEL

Construction of a theoretical model necessitates defining each stratigraphic unit with its characteristic components appropriately defined and dimensioned. Each stratigraphic unit in the stratigraphic system analysis network consists of numerous defined characteristics, which are interconnected to the highest degree of complexity by their functional relationships. In order that characteristics be defined, they must be modified in terms of their characteristic properties, and scaled in terms of their characteristic dimensions. Both property and dimensioning functions are prescribed in terms of the stratigraphic concept to which the characteristic component has information potential value.

Each defined stratigraphic component consists of essential characteristics and accidental characteristics (general and specific respectively). Essential characteristics are represented as elements of a bounded set, and are the common ones. Accidental characteristics are contained in the bounded sets. If the two types of characteristics overlap, the intersection represents the subset of identified stratigraphic components (c, Fig. 12).

For example, grain size, crystallinity, and granularity are undefined terms. They become defined by the introduction of measurement scales of size, crystal morphology, and granularity. The association of dimensioned components produces a derived measured concept of the texture defined in terms of the characteristic properties of each rock component. Thus, the texture concept is developed from a set of characteristic components with defined associative properties.

The procedure can be summarized by the following example:

grain size: undefined term - unassociated to the stratigraphic system.

medium grain size: defined term - unassociated to the stratigraphic system, but associated to the number order system.

medium euhedral subangular grain: defined term - associated to the stratigraphic system by means of the scales of grain size, degree of perfection of crystal form, and grain sphericity.

Thus, definition of the stratigraphic concept (in the example, texture) is formulated by a process of observation, description, and scaling. Observation identifies the stratigraphic component, description defines its characteristic property, and scaling determines its characteristic dimension. In this manner the intensity value of properties of the rock components is determined. To have stratigraphic significance, the associative properties related to the stratigraphic concept must have stratigraphic significance in terms of their extensity value. The development of the intensity and extensity of all stratigraphic components in a similar manner prescribes the stratigraphic analysis network. Only at this stage of analytical development can the stratigraphic model be formulated. The analytical methodology of classification and pattern development by this process is illustrated in Figure 12. The characteristic component groups with their related

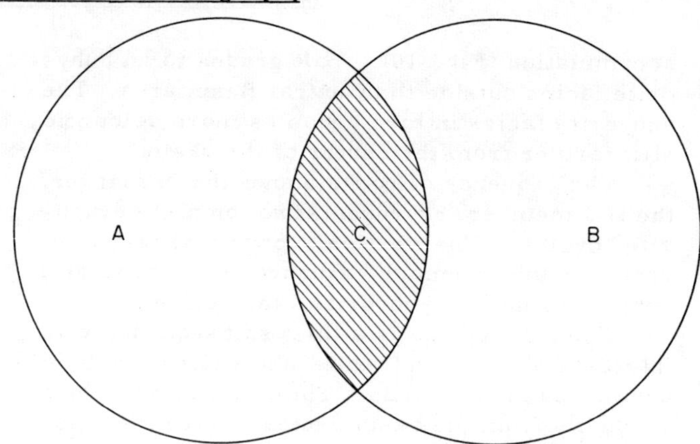

FIG. 12. Association of essential and accidental characteristics of stratigraphic components.

characteristic properties are composed of sets of observation components and properties, one set for each subgroup of components and properties. Examples are given below for the evaporite concept model of the Silurian Salina deposition.

Fossil Grains - Matrix Components	Characteristic Properties shape	size
Detrital		
Skeletal		
Pellets		
Algal		
Growth in situ		
Algal stromatolite		
Dead talus of algae		
Lump & coated grains		
Reef material		
Sphaerulites		
Lithographic		
Gravels		
Sands		
Clays		
Silts		
Composition		
Biological		
Biochemical		
Physical		
Physicochemical		
Chemical		

In a similar manner, texture, structure, and environmental component sets are defined relative to the specific stratigraphic concept model. If the stratigraphic concept model were not known or specified, component sets would include all possible stratigraphically significant terms, and extension of the observations would define the concept model from the stratigraphic system analysis.

The characteristic dimensioning properties (Fig. 13) are the functional dimensioning qualities that classify the defined associated terms and develop the associative patterns. Diagenesis modifies the classification and the associative components by means of a set of stratigraphic analytic functions (Fig. 14).

The stratigraphic sequence, such as described previously for the Michigan Formation, the Detroit River Group, and the Salina Group, is constructed from analysis of each stratigraphic unit so as to produce a sequence of changing paleogeographic settings. Integration of the set

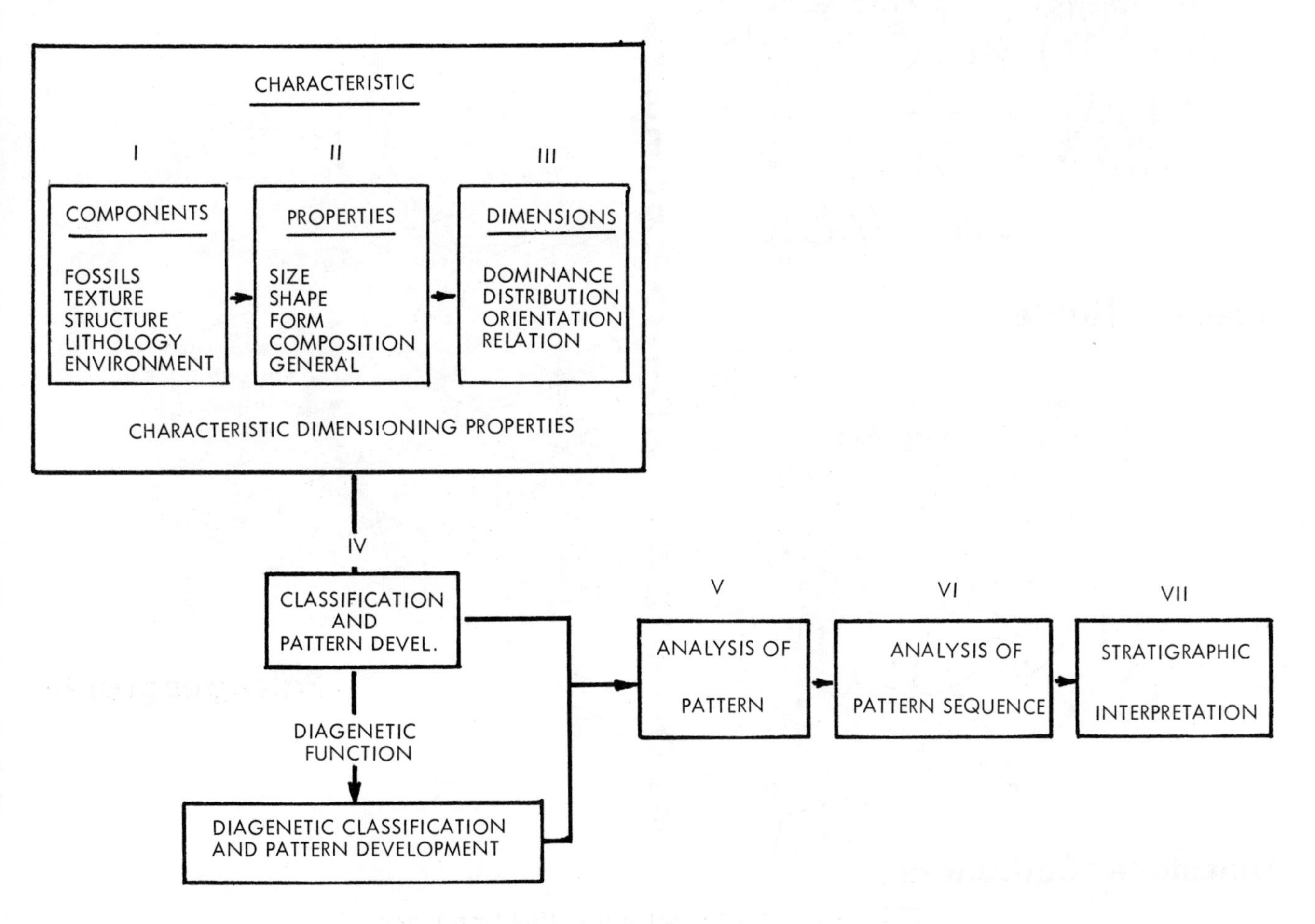

FIG. 13. The stratigraphic information system.

patterns. The analytic methodology prescribes feedback and recycling until all stratigraphic terms are classified and their associative patterns defined.

The analytic process of classification and pattern development produces sets of identified stratigraphic components distributed by location coordinates. These are stratigraphic facies which constitute the basis for stratigraphic analysis and model interpretation. The evaporite depositional models are designed to produce facies by means of classification of the identified stratigraphic components and their distributive patterns. Integration of the facies provides the elements of defined environmental components of the paleogeography. Thus, association of the facies components is transformed into paleogeographical of paleogeographic sequences constitutes the stratigraphy of the evaporite formation or group of formations.

In summary, construction of the theoretical stratigraphic model (Fig. 13) includes the following steps for each stratigraphic unit:

(1) Identification of characteristic components of fossils, texture, lithology, and environment.

(2) Definition of components by means of their characteristic properties (size, shape, form, composition, general).

(3) Dimensioning of defined components in terms of dominance, distribution, orientation, and relationship.

(4) Classification of defined associated components by their characteristic dimensioning

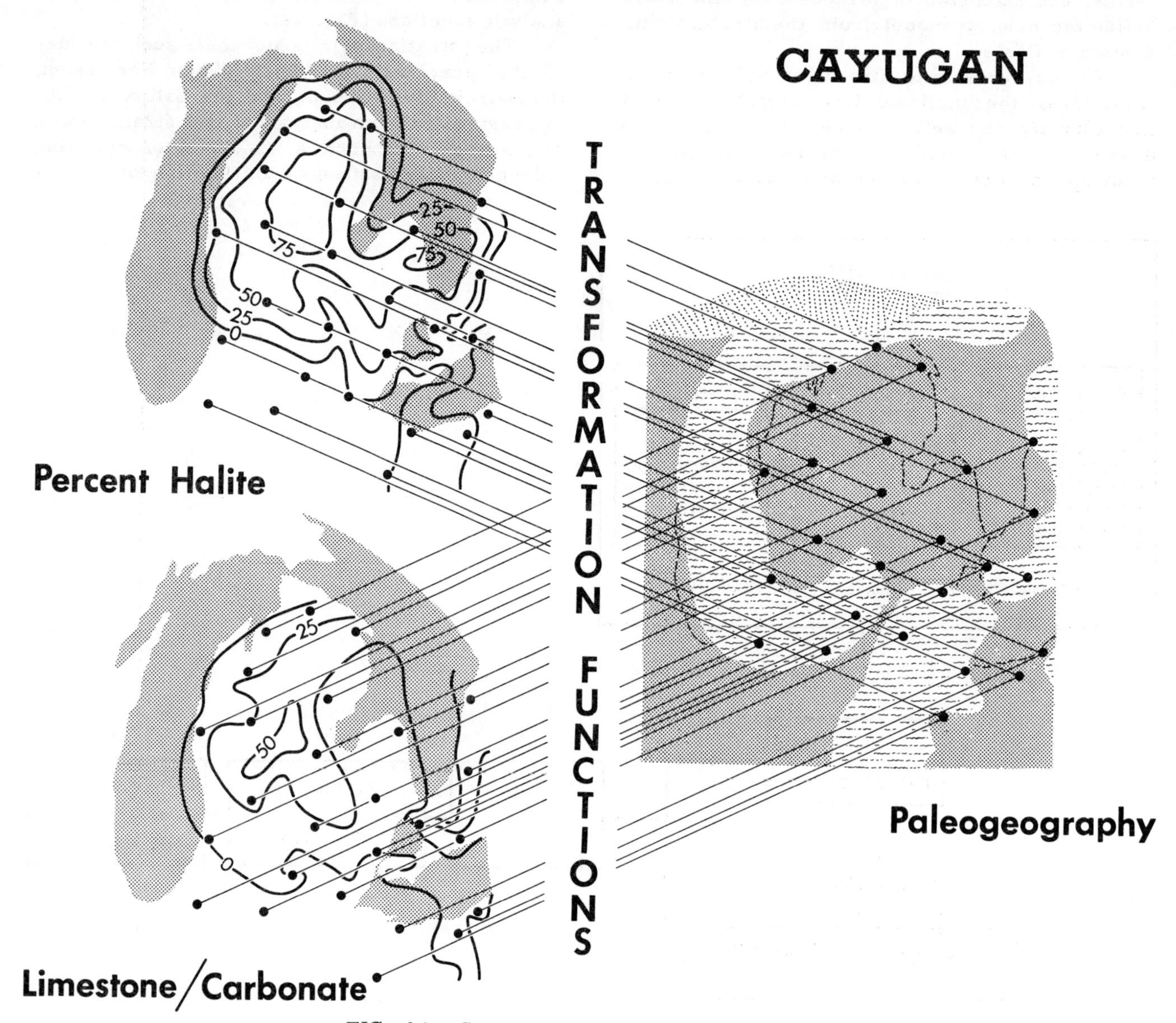

FIG. 14. Cayugan transformation functions.

property functions and development of their facies patterns.

(5) Transformation of classification and facies patterns by diagenetic processes.

(6) Association and dimensioning of facies components to produce paleogeographic settings at each stratigraphic level.

(7) Definition of stratigraphy from the sequence of paleogeographic settings corresponding to the stratigraphic succession.

DEVELOPMENT OF EVAPORITE SIMULATION MODELS

The theoretical stratigraphic model provides the framework for construction of the evaporite simulation model in that identified stratigraphic components are logically related through the stratigraphic system network, and the components on each level are appropriately defined and dimensioned.

In the simulation model relative values of the defined characteristic components of the environmental types are placed into a matrix form, and analyzed through their distribution sequence. The model is a dynamic steady state of the relative distribution of environmental patterns. The dynamic system is characterized by a finite number of parameters, a finite group of parametric values, and finite stratigraphic categories determined by the conceptual formulation.

For example, the following are environmental types and their associated lithologic components in Michigan evaporites.

Normal marine: fossiliferous limestone, shale
Reef zone: dolomitized biostromal limestone
Penesaline: dolomite, anhydrite
Saline: anhydrite, halite
Pseudomarine: limestone, dolomite, shale, sandstone

The succession of types includes various combinations of the general sequence from normal marine to evaporite to normal marine. The Michigan Formation sequence includes, in the evaporite phase, penesaline, and penesaline and pseudomarine. The Detroit River Group sequence includes the reef zone, penesaline, and saline types. The Salina Group sequence has a major development of the reef zone and the saline types, with the penesaline and pseudomarine of lesser importance.

Evaporite simulation models can be classified into deep central basin type, marginal marine type (sometimes called "megasebhka"), and sebhka type. Each has particular facies distributive patterns and component associations. Salina evaporites belong to the central basin type, Michigan evaporites to the marginal basin type, and Detroit River evaporites to combined central basin and sebhka types.

REFERENCES

Alling, H. L., and Briggs, L. I. (1961) Stratigraphy of Upper Silurian Cayugan evaporites: Bull. Amer. Assoc. Petrol. Geol., vol. 45, pp. 515-547.

Dellwig, L. F. (1955) Origin of the Salina Salt of Michigan: Jour. Sed. Petrology, vol. 25, pp. 83-110.

Dorr, J. A., and Moser, Frank (1963) Ctenacanth sharks from the mid-Mississippian of Michigan: Michigan Acad. Sci., Arts and Letters, Papers.

Ehman, D. A. (1964) Stratigraphic analysis of the Detroit River Group in the Michigan Basin: Univ. Michigan, MS thesis, p. 63.

Evans, C. S. (1950) Underground hunting in the Silurian of southwestern Ontario: Geol. Assoc. Canada Proc., vol. 3, pp. 55-85.

Landes, K. K. (1945) The Salina and Bass Island rocks in the Michigan Basin: U. S. Geol. Survey, Prelim. Map 40, Oil and Gas Inv. Ser.

Lane, A. C., and Seaman, A. E. (1909) Notes on the geological section of Michigan for geologists, teachers, and drillers: Tenth Ann. Rept. of State Geologists Bd. of Geol. Survey for 1908, pp. 23-120.

Moser, Frank (1963) The Michigan Formation; a study in the use of the computer oriented system in stratigraphic analysis: Univ. Michigan, Ph. D. dissertation, p. 96.

ENVIRONMENTS OF PHOSPHORITE DEPOSITION IN THE CENTRAL FLORIDA PHOSPHATE DISTRICT

D. H. Freas and S. R. Riggs
International Minerals & Chemical Corporation, and
Department of Geology, East Carolina College
Skokie, Illinois, and Greenville, North Carolina

ABSTRACT

Detailed mapping in the Noralyn and Achan mines of International Minerals & Chemical Corporation has provided new information on the stratigraphy, petrography, and environment of phosphate deposition. A revised stratigraphic section has been proposed in which the Hawthorn Formation has been raised to group status so as to include all phosphatic sediments of the district. The Bone Valley Formation is the principal phosphorite unit of the Hawthorn Group; it is composed of a lower unit, the Noralyn Member, which is primarily clayey and/or gravelly phosphorite, quartz sand, and is of marine origin. The upper unit, the Homeland Member, is mostly phosphorite gravel containing a heterogeneous mixture of phosphorite grains, quartz pebbles, and a prolific articulated land fauna with mixed and worn marine vertebrates and invertebrates; it is principally a fluvial or estuarine deposit. Underlying the Bone Valley Formation is the clayey, sandy phosphatic carbonate of the Arcadia Formation, previously called the Hawthorn Formation.

The upper surface of the Arcadia Formation in the Noralyn mine consists of a series of low ridges thought to be depositional carbonate mounds and banks and to which many sediment characteristics of the overlying Bone Valley Formation can be related. These banks and mounds formed shoal environments during the main period of phosphorite deposition.

As climatic and chemical conditions changed, carbonate deposition of the Arcadia Formation gave way to phosphorite deposition of the Noralyn Member. Phosphorite was deposited in this shallow-water, marine environment, which was possibly partially intertidal in the eastern part.

The marine phosphorite of the Noralyn Member was reworked and redeposited in fluvial and estuarine environments following regression. The reworked deposits, the Homeland Member, are locally thick and normally occur at the top of, and cutting into, the marine phosphatic section; in some places, the member rests directly on the Arcadia Formation.

Primary deposition of phosphorite and phosphate replacement of the carbonate substrate occurred in the Noralyn Member over the banks. This primary phosphorite, designated <u>microsphorite</u> because it is an <u>in situ</u> micro-crystalline phosphorite bed, was partially indurated, disrupted, and fragmented. In this manner, abundant clastic particles of phosphorite were produced which are the most common form of phosphorite in the sediments.

The Noralyn Member of the Achan area contains little or no microsphorite. Here phosphorite clasts, formed in areas to the north and east, were transported into an environment where intermittent carbonate and fine clastic sedimentation were the norm. The resulting sediments are, therefore, strongly bimodally sorted since they contain abundant granule and pebble size phosphorite clasts in a matrix of clay. Carbonate becomes increasingly abundant toward the base of the Noralyn Member and occurs in mound-shaped lenses.

INTRODUCTION

Phosphate deposits of the Central Florida Phosphate District (Fig. 1) are among the world's largest and have produced more than 70% of the U. S. requirements of phosphate rock. These deposits are, however, geologically unlike many of the other large phosphate deposits of the world. They do not have the typical black shale-chert lithologic association and appear to be very shallow-water deposits. The phosphorite has a characteristic coarse-grained component in the abundant gravel-sized phosphorite grains (pebble) mixed with quartz and phosphorite sand and clay. The stratigraphy of the deposits is complicated by rapid facies changes and post-depositional

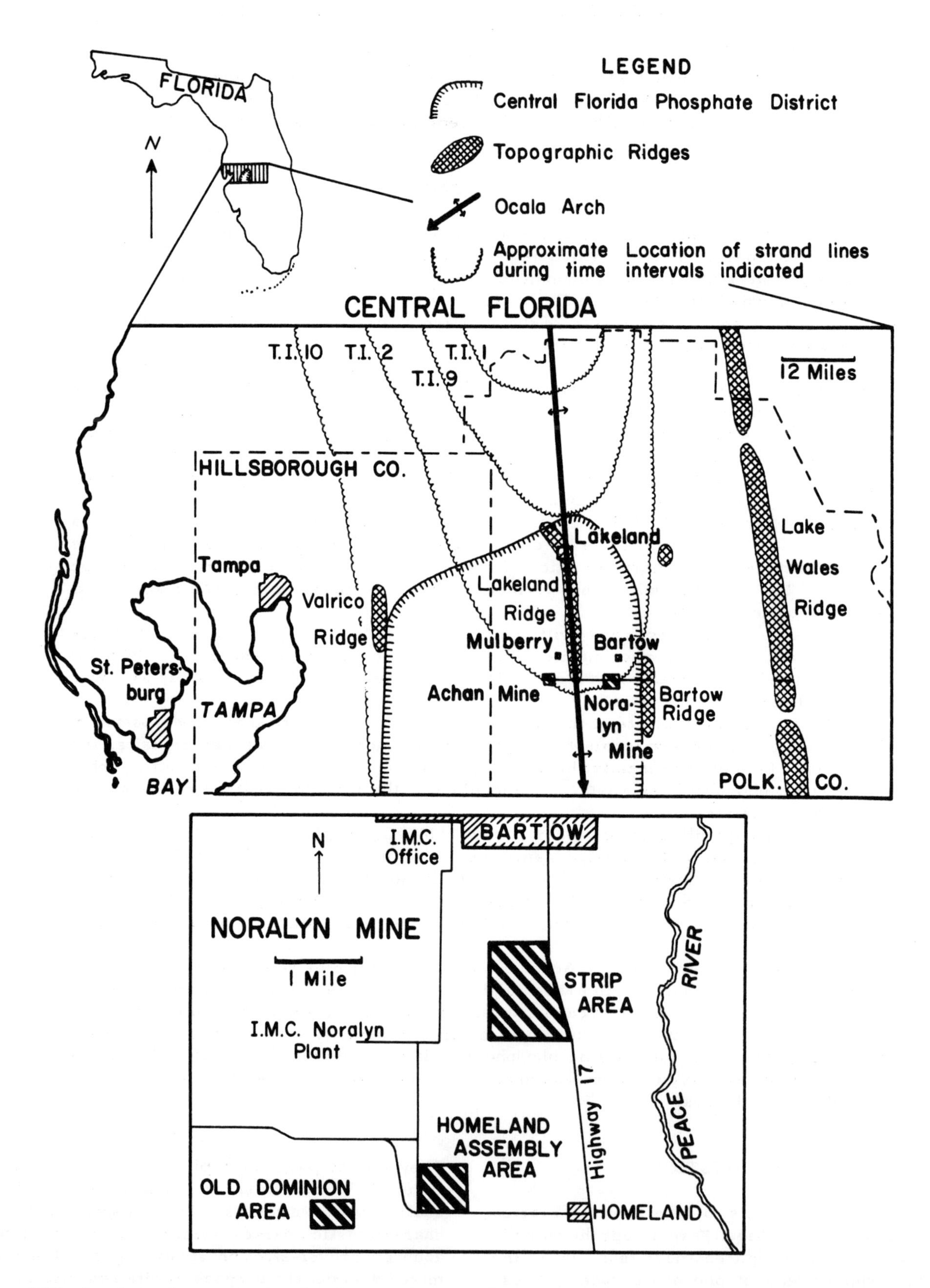

FIG. 1. Location maps for the Central Florida Phosphate District, Noralyn mine, and the Strip, Homeland Assembly, and Old Dominion areas.

erosion, redeposition, and weathering. These characteristics set the Florida deposits apart from all other known phosphate deposits of the world. The shallow-water environment of deposition of these unique deposits, therefore, is an interesting and challenging subject--especially to those of us who have attempted to discover new deposits of similar origin. This paper describes the stratigraphy and sedimentary environments of deposition of phosphorite in the Noralyn and Achan mines of International Minerals & Chemical Corporation. Work in other parts of the district suggests that the results of work in the mines is applicable to the entire Central Florida Phosphate District.

The Noralyn and Achan mines were selected for study because well exposed faces were readily accessible and the mines are widely separated in the district (Fig. 1). The study is by no means exhaustive, and similar studies should be conducted in the numerous other active mines throughout the district. The Noralyn mine was the subject of a Ph. D. dissertation by the coauthor (Riggs, 1967).

The writers are indebted to International Minerals & Chemical Corporation for permission to publish this paper.

GENERAL STRATIGRAPHY

A revised stratigraphic section for the Central Florida Phosphate District has been proposed by Riggs (1967) and is generalized in Figure 2. Although not formally published as yet, this section is described herein because it has been the basis of the stratigraphic work in the Noralyn and Achan mines.

The name "Hawthorn Group" is used by Riggs (1967) to include all phosphatic sediments of Central Florida. It contains the phosphatic dolomites and phosphorite sediments of Miocene age, as well as the reworked phosphorites of Pliocene to Recent age. Riggs (1967) has resurrected the name "Arcadia" to refer to the proposed Arcadia Formation, which is the phosphatic dolomite or limestone previously known as the Hawthorn Formation in the district. The formation is the "bedrock" of the district and constitutes the greatest part of the Hawthorn Group. The Tampa Formation is also regarded as part of the Hawthorn Group since it is phosphatic, and may be (in part) a time-equivalent of the Arcadia Formation.

The Bone Valley Formation, as proposed by Riggs (1967), is defined as the principal phosphorite unit overlying the Arcadia Formation, and containing predominantly terrigenous constituents (quartz and clay) and more than 10% phosphorite. This unit is composed of two phosphorite members which have been defined as the Noralyn and Homeland Members by Riggs (1967). The stratigraphically lower unit, the Noralyn Member, is Miocene in age and contains a marine vertebrate and invertebrate fauna. This phosphorite is widespread and is the principal source of phosphate in the district. The Noralyn Member is commonly overlain by the Homeland Member, a name also proposed by Riggs (1967). The Homeland Member is mostly a phosphorite which has been reworked from the underlying Noralyn Member. These sediments are highly cross-bedded, sinuous, fluvial deposits which contain abundant cut-and-fill structures and are composed of interbedded channel gravels, sands, and clays. Quartz granules and pebbles are common and almost diagnostic of this member. The geologic age of these deposits ranges from Early Pliocene through the Pleistocene and into the Recent as indicated by abundant articulated and indigenous land vertebrate fauna.

STRATIGRAPHY OF THE NORALYN MINE

A generalized stratigraphic section exposed in the mine faces of the Noralyn mine is shown in Figure 3, as compiled from many measured sections in the mine. The stratigraphic Noralyn and Homeland Members of the Bone Valley Formation are easily recognized, mappable, lithologic units occurring throughout the mine area. Each of the members is composed of numerous distinct sediment units, so designated by number and letters, and which are less continuous but are definitely recognizable and mappable within the mine area. Lateral variations in these sediment units account for most of the variation in the quantity and quality of the phosphate recovered in the mining operations. Detailed mapping of the mine faces and careful study of the sediment units have formed the basis of our knowledge of phosphatic sediments and their environments of deposition. In the following pages, the general sediment units will be described and their environments interpreted.

Arcadia Formation

The Arcadia Formation underlies the Bone Valley Formation everywhere in the Noralyn mine. The rock is typically a very pale orange, poorly sorted, massive, fossiliferous, clayey, sandy, phosphatic dolomite. The unit contains abundant sand-filled burrows and molds of mollusks.

Probably one of the most striking features of the Arcadia Formation in the Noralyn mine is the nature of the upper contact with the Bone Valley Formation. The upper surface, well exposed by the mining operations, has been mapped in the pits as well as from drill-hole data. It is characterized by a series of discontinuous topographic ridges or banks about 2 to 3 miles long, 1 mile

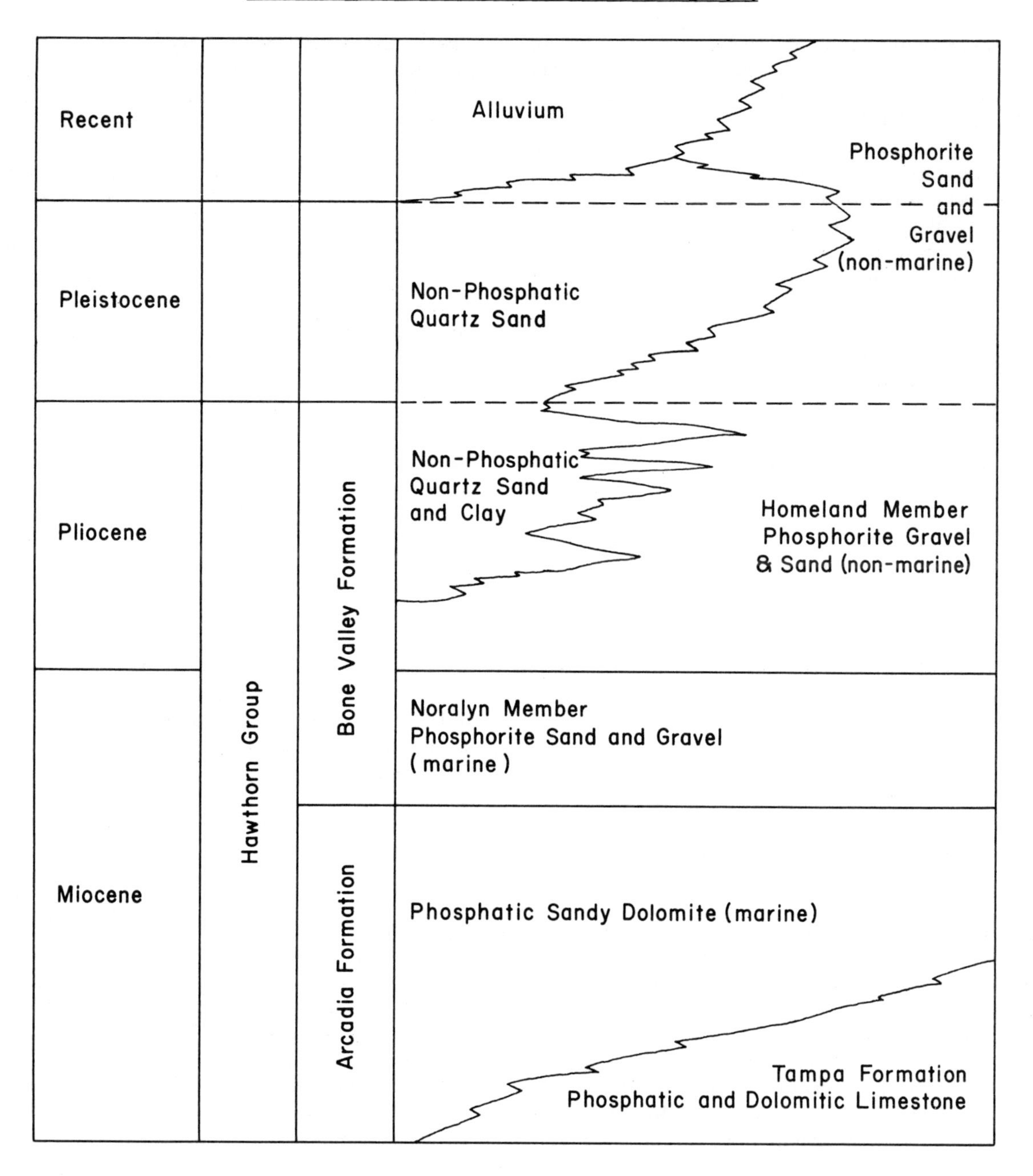

FIG. 2. Proposed stratigraphic section, Central Florida Phosphate District.

wide, and about 20 to 30 feet high. Mapping in the pits has revealed smaller mound structures, hundreds of feet long, tens of feet wide, and 3 to 10 feet high. These smaller structures occur nearly everywhere and cause the upper contact of the Arcadia Formation to "roll" in the length of a single mine cut. The orientation of the long axis of structures of both scales is north-northwest.

Detailed study of the Arcadia-Bone Valley contact has indicated that the banks and mounds are primary sedimentary features rather than features of tectonic origin or ground water solution. On the tops of the large banks, the contact with the overlying Bone Valley Formation is sharp, whereas in the low areas between banks the contact is gradational. Tops of the highs are irregular and contain solution pits, scoured surfaces, and pelecypod borings. These features suggest that the high areas were shoals where scour and sediment by-pass predominated, whereas continuous deposition took place in the adjacent lows. These structural highs are similar to banks and mounds on the modern ocean shelf such as occur in Florida Bay in South Florida.

The banks and mounds are persistent features of the Arcadia Formation and many of the sediments in the Bone Valley Formation can be related to these features. They undoubtedly per-

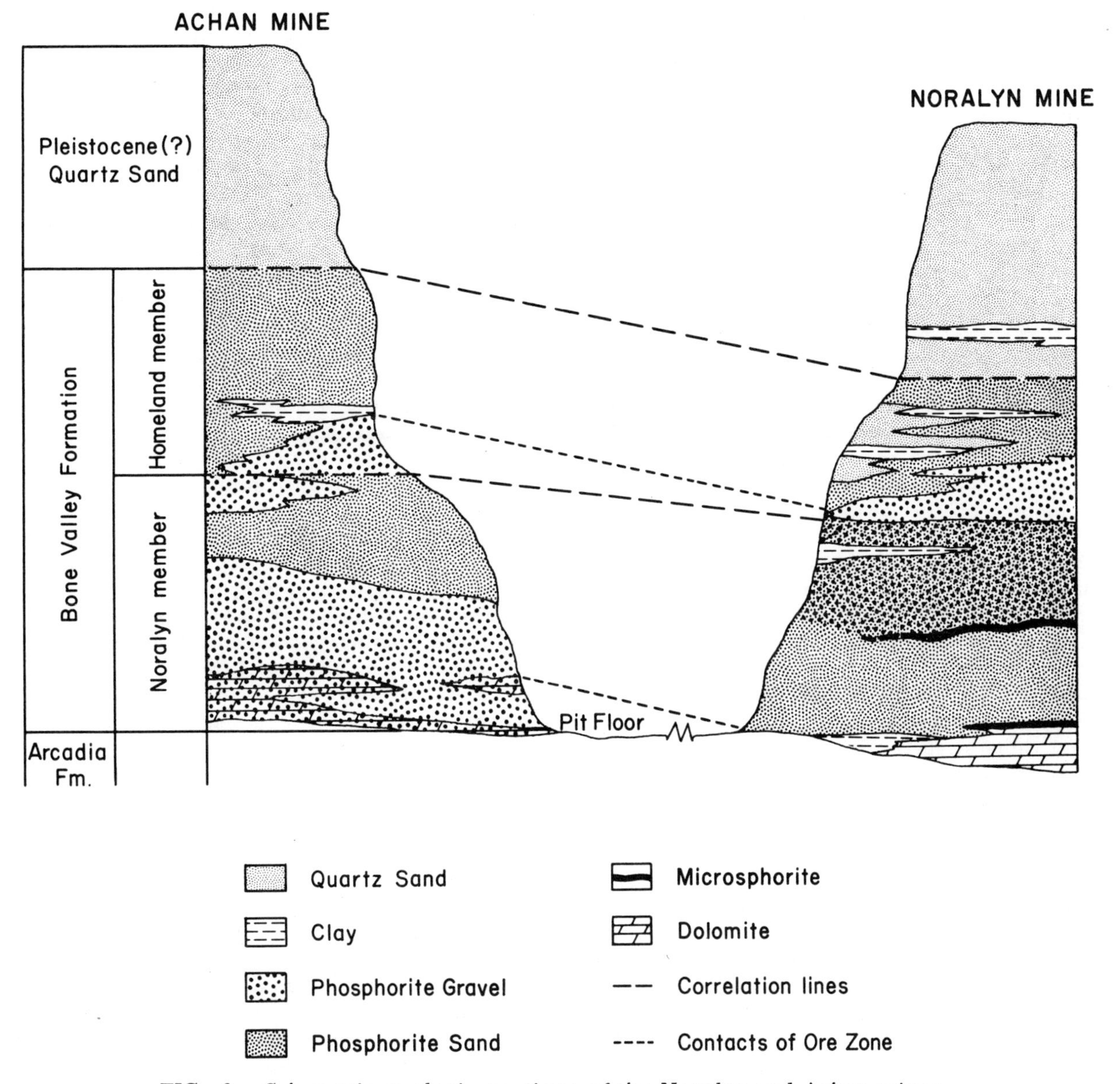

FIG. 3. Schematic geologic sections of the Noralyn and Achan mines.

sisted throughout deposition of the Bone Valley Formation; however, the only topographic expression on the present-day surface appears to be the very general location of the modern drainage pattern in the inherited lows between the banks.

Bone Valley Formation

Noralyn Member

The Noralyn Member of the Bone Valley Formation is comprised of a number of distinct sediment units which have been mapped over a wide area in the mines. These units are almost all phosphorites and contain the minable phosphate in the mine.

At the base of the Noralyn Member is a brown, sandy, phosphatic, dolomitic clay, which is typically an attapulgite-rich clay. This clay bed commonly is a few feet thick and is thickest in the low areas between the banks of the Arcadia Formation, pinching out onto the sides of the banks. It thins or is completely absent over the banks. The clay contains little phosphate, and is commonly referred to as "bed clay" in the mines.

Resting on the lower clay unit, or the Arcadia Formation, is a thin, discontinuous bed of microsphorite. This sediment is composed of clay-sized phosphorite which occurs as an indurated bed in which the microsphorite cements or binds coarser quartz and phosphorite grains. The term "microsphorite" is a contraction of "microcrystalline phosphorite." It was proposed by

Riggs and Freas (1965) and is described in detail by Riggs (1967). Two beds of microsphorite have been recognized in the Noralyn mine. The lowermost bed, which overlies either the brown basal clay or the Arcadia Formation, is not well developed and is discontinuous. The upper microsphorite bed, described later in this paper, is a thicker and more continuous unit.

The lower microsphorite bed, where it rests on the dolomite of the Arcadia Formation, has a gradational contact which appears to be the result of replacement of carbonate by phosphorite. The bed is normally thickest (3 to 4 inches) where it rests on the carbonate on the crests of the small mounds of the Arcadia Formation. The upper surface of the microsphorite bed is a smooth and hard scoured surface which has been bored by pelecypods. Off the mounds, the bed thins and pinches out. Locally, the phosphorite gravel beds occur lapping off the lower microsphorite bed.

Overlying the basal units of the Noralyn Member is a thick sequence of phosphorite sand which comprises most of the Bone Valley Formation in the Noralyn mine and is the main source for the silt- and sand-sized phosphate of the mine. The unit is composed of a well sorted quartz phosphorite sand which is up to 20 feet thick between the carbonate banks of the Arcadia Formation and thins slightly onto the banks. This unit contains three subunits which are characterized by the stratification. The interbank areas contain a homogeneous to mottled subunit and a subunit with good horizontal and cross-stratification. These two subunits grade into the third subunit on top of the banks, a horizontally laminated and bedded clayey quartz phosphorite sand. Short, thin, discontinuous clay laminae define the horizontal stratification in this subunit and are diagnostic of the unit. Clay beds up to 1.5 feet thick occur locally within the subunit and are normally restricted to the areas over the main banks of the Arcadia Formation. Plant impressions and articulated sirenian bones commonly occur in these clay beds.

This thick sequence of phosphorite sand is very widespread and is missing only where channels of the overlying Homeland Member have eroded the unit away. Phosphorite grains are mostly sand size, well rounded, brightly polished, and comprise up to 50 percent of the sediment. Fragments of marine vertebrates such as sirenian ribs, sharks' teeth, and ray plates are common. Clay molds of marine mollusks occur in the more clayey beds, clear evidence of the marine origin of these beds.

The best development of microsphorite directly overlies the thick phosphorite sand unit in the Noralyn Member. The microsphorite bed is very pale orange to moderate brown, partially indurated, fossiliferous, laminated, burrowed, and bored. The microsphorite is similar to the lower microsphorite bed, but ranges up to 2 feet thick. The bed is restricted to the tops of the still persistent topographic depositional area. Locally, it is broken up and its fragments disseminated in the overlying unit. The broken edges of the bed commonly have coatings of additional microsphorite which round off the truncated edges. The upper surface of the microsphorite bed is commonly smooth, hard, and bored. Burrows have been found which extend through the bed into the softer phosphorite below. These tube-shaped burrows are generally coated with thin, concentric laminations of microsphorite and filled with phosphorite grains, clay, and quartz sand.

Overlying the microsphorite bed is a sequence of very poorly sorted, clay- and gravel-rich quartz phosphorite sand. This unit, ranging up to 10 feet, contains abundant phosphorite gravel and constitutes a major source of "pebble" phosphate in the mine. Several subunits have been recognized and mapped over parts of the mine. Gravel-size phosphorite is more abundant near the base of the unit. Phosphorite sand and phosphorite clay beds occur upward in the unit, commonly suggesting graded bedding. Pavement of phosphorite gravel, 2 to 3 inches thick and up to hundreds of feet wide in the mine faces, interrupts the apparent graded sequences and cut-and-fill structures and truncates the normal stratigraphic sequence. Both marine vertebrate and invertebrate fossils are abundant in the unit. As in the lower phosphorite unit, thin clay beds with plant impressions occur locally, especially on the topographic banks.

Homeland Member

The Homeland Member is the upper phosphorite member of the Bone Valley Formation. It consists of a sequence of local stream channel deposits characterized by cut-and-fill structure, high angle cross-stratification, mixed types of phosphorite grains, and reworked marine fossils. Consequently, the individual lithic units are short, discontinuous, and complexly interbedded sediments which do not recur in sequential order. The Homeland Member contains a series of very well sorted, interbedded, and cross-bedded phosphorite gravels, sands, and clays. The most common interbedded sediment units include well sorted and well bedded phosphorite gravels and sands; poorly sorted homogeneous clayey, gravelly, quartz, phosphorite sands; well bedded and well sorted very clean quartz sands, massive gravelly, clayey, quartz sands and sandy clays; and very thinly laminated green clays.

These units are prolific sources of horse teeth and articulated skeletons of horses, camels, mastodons, mammoths, rhinoceros, birds, alligators, snakes, turtles, as well as sirenians, whales, porpoise, sharks, fish, and rays. Large quartz pebbles are also diagnostic in the sand and gravel units.

The Homeland Member has been mapped through successive cuts in the mines and suggests a dendritic pattern. This pattern is often reflected on the surface by the Recent stream system. The various units interfinger laterally, forming a complex sequence with little lateral continuity.

STRATIGRAPHY OF THE ACHAN MINE

The stratigraphy of the Achan mine, located about 12 miles west of the Noralyn mine, is very similar to the Noralyn mine. This mine is principally a source of pebble phosphate because phosphorite gravel is predominant over phosphorite sand. Bluish-green clay is a common constituent of the phosphorite, and the phosphorite grains are principally black. The darker color of the phosphorite may be accounted for partly by less intense weathering in the Achan area than in Noralyn. Still, the Bone Valley Formation at Achan is the main phosphorite unit which consists of two easily recognized, mappable members as at Noralyn.

Arcadia Formation

The Arcadia Formation is rarely exposed in the mining operation in this mine because the lower part of the Bone Valley Formation is commonly calcareous, contains low grade phosphate, and is not, therefore, excavated. In drill holes the Arcadia Formation has been penetrated and, in general, it is similar to the phosphatic dolomite of the Noralyn mine. The banks, which have been mapped at Noralyn, were not recognized in the pits at the time that this study was done at Achan. Structure contour maps of the upper surface of the Arcadia Formation, drawn from drill hole data, indicate that the banks do exist. Smaller scale topographic highs or mounds are, however, commonly exposed in the mining operation.

Lenses of carbonate are common near the top contact of the Arcadia Formation. These have been mapped in successive mine cuts and found to be large lens-shaped masses similar to the mounds at Noralyn. Below the carbonate lenses or mounds, clayey phosphorite sands may occur so that the lower contact of the Bone Valley Formation might be obscure. These lenses are referred to by the miners as "lime horses," because they are ordinarily not mined and cause abrupt rises in the floor of the pit.

Bone Valley Formation

Noralyn Member

The basal unit of the Noralyn Member of the Bone Valley Formation in the Achan mine is a thick pale bluish-green clayey and gravelly phosphorite sand which is one of the principal sources of phosphate, especially pebble. Phosphorite grains, commonly black, dark olive green, or brown, comprise from 25 to 75 percent of the sediment and 20 to 60 percent of the grains are coarser than sand size. The phosphorite grains are imbedded in a soft matrix of clay which is composed of a mixture of attapulgite, montmorillonite, and dolomite.

The unit is poorly stratified but several distinct sedimentary features are characteristic. Some beds contain a vertical sizing of phosphorite intraclasts suggestive of graded bedding. Coarser grained intraclasts occur at the base of such beds with successively finer grained intraclasts toward the top. These beds are commonly repeated several times. Other beds contain "gravel pavements" or thin beds consisting of a single line of large phosphorite intraclasts. Some of these beds persist hundreds of feet along the mine face.

In general, this unit is thickest in the lows between carbonate mounds. The pit bottom normally follows the lower contact of this unit so that it traces the tops of the mounds. On the highs, the basal unit of the Bone Valley Formation is only 2 feet thick, whereas in the lows, it reaches 10 feet thick. In all cases, the contact with the carbonate units below is gradational over several inches.

Poorly preserved fossils have been found in the unit. The fossils, mostly mollusks, still retain part of the original calcite shells and have been tentatively identified as equivalent to the Chipola and, therefore, of Middle Miocene age.

Overlying the basal gravelly phosphorite sand unit in the Achan mine is a bluish-green phosphorite clayey sand. This unit generally contains less than 10 percent fine- to medium-grained sand-size black phosphorite intraclasts. The unit is normally poorly stratified to very mottled with rare and poorly defined long planar cross-bedding. The abundant clay mottles are the most characteristic sedimentary feature of the unit. Sand-filled burrows are common and a few molds of mollusks similar to those of the lower unit have been observed. Plant fragments and rootlet impressions are also common. This unit is gradational and apparently conformable with the lower unit.

The uppermost unit of the Noralyn Member is a pale green to very pale orange, clayey, gravelly, phosphorite sand similar to the lowermost unit. It is thinner (up to 4 feet) and not everywhere present; locally it has been removed where the channel deposits of the Homeland Member have eroded into the older units of the Noralyn Member. The unit is poorly stratified and contains sedimentary features similar to the lower gravel unit. The lighter color of the unit is probably its most distinguishing characteristic. Pale yellow-brown and orange colors are more common than in the lower unit.

Homeland Member

The sequence of fluvially reworked phosphorite gravel and sand is well developed in the Achan mine. The phosphorite gravels, in places, constitute an important part of the phosphate recovered in the operations. The unit is similar to that in the Noralyn mine, but is not as altered, and consequently, easier to recognize in the mine.

The basal unit of the Homeland Member is a black, clayey, sandy, phosphorite gravel containing 60 to 90 percent phosphorite intraclasts. It occurs in cut-and-fill structures which constitute sinuous channel deposits. The gravels thin laterally and grade into clayey, quartz, phosphatic sands and interbedded clay and quartz sand.

The phosphorite gravel beds are well stratified and contain complex, long, moderately steep, planar cross-bedding. Ripple marks occur along the upper surfaces of the clay laminae in the finer grained parts of the unit. Gravel pavements and clay lithoclasts are common. The unit contains a prolific land vertebrate fossil assemblage.

The phosphorite gravel and sand unit grades up into yellowish gray or white sandy phosphatic clay and clayey quartz sand. The quartz sand unit is poorly cross-bedded and attains a thickness of 25 feet. It is widespread and comprises the main part of the overburden in the mine. No fossils have been observed in the unit, but its gradational contact with the underlying phosphorite gravel unit suggests that it is part of that succession of sedimentation.

ENVIRONMENTS OF PHOSPHORITE DEPOSITION

Sedimentation of Phosphorite

Coarse-grained phosphorite particles in the Florida phosphate deposits have presented a challenge to the sedimentologist. Study of the phosphorite grains by the writers has indicated that they are all composed of several minerals and are, therefore, rock fragments. The grains contain clay minerals, quartz grains, calcite or dolomite, small phosphorite grains--all of which are cemented by microcrystalline phosphorite. The larger phosphorite grains, which contain more nonphosphatic components than smaller grains, may be almost entirely composed of carbonate-fluorapatite.

The most common shape of sand-sized phosphorite grains is discoidal or ellipsoidal. The smaller grains are more spherical. All the phosphorite grains are well rounded. Even cobble-size phosphorite has well rounded edges. The surface of the grain is commonly smooth or brightly polished. Inclusions of quartz grains are beveled off where they protrude from the surface of the phosphorite grain. The larger phosphorite pebbles or cobbles are commonly bored by pelecypods.

A small part of the phosphorite in the Noralyn Member does not occur as grains visible to the unaided eye. This phosphorite forms a bed, up to 2 feet thick, of microcrystalline phosphorite in a matrix which cements or binds quartz, carbonate, and phosphorite grains. The bed contains an intricate network of burrows described previously, and the upper surface of the bed is commonly hard, well polished, and bored. This phosphorite is mineralogically and chemically similar to the discrete phosphorite particles which comprise most of the phosphorite in the sediment. The fine-grained phosphorite bed appears to be an in situ deposit; that is, it was formed in its present position as opposed to the grains which have obviously been moved.

The two types of phosphorite observed in the Noralyn Member comprise broad classes of phosphorite which are defined by textural characteristics, but which also have genetic implications. The fine-grained phosphorite bed is termed orthochemical phosphorite because it is an in place deposit which apparently formed by direct precipitation of phosphate or replacement of carbonate. The term microsphorite has been proposed for orthochemical phosphorite as a contraction of "microcrystalline phosphorite," which is analogous to "micrite" in carbonates (Riggs and Freas, 1965).

The clastic phosphorite grains are termed allochemical phosphorite since they are authigenic particles which were formed elsewhere in the basin during the same sedimentary cycle and subsequently transported as clastic particles. The allochemical phosphorite has not necessarily been transported a long distance, but it has been subjected to some degree of movement which would account for the external grain characteristics. Allochemical phosphorite can be composed of several kinds of particles or allochems. The majority of allochems are intraclasts (penecontemporaneous fragments broken from orthochemical beds of microsphorite), but fossil skeletal material, such as bones and teeth, coprolites and fecal pellets are common. Oolites are also allochems, but these have not been observed in the Florida deposits as they have in numerous other phosphate deposits of the world.

In addition to the types of phosphorite in the Noralyn Member, a third class of phosphorite occurs in the Homeland Member. This unit, as described previously, contains phosphorite reworked from the underlying Noralyn Member. The phosphorite particles, which range up to pebble size, are fragments of older rocks and, therefore, are termed lithochemical phosphorite. It is not always easy to identify an isolated phosphorite grain as lithochemical unless the entire sediment and its stratigraphic relationships are studied. However, the reworked phosphorite is characterized by its heterogenous mixture of all types of phosphorite grains with a variety of colors and degrees of alteration. In addition, the grains are commonly coated with a black stain although the interior of the grain may be light colored. The stratigraphic and paleontologic characteristics described previously are, however, the most conclusive evidence of the reworked origin of this phosphorite.

The occurrence of two types of phosphorite in the Noralyn Member suggests that sedimentation of phosphorite is indeed complex and involves mechanical as well as chemical processes. Most of the sediments generally have a strong bimodal size distribution. Phosphorite intraclasts are commonly coarser grained and poorer sorted than terrigenous constituents. Phosphorite sediments containing abundant granule- and pebble-sized intraclasts are commonly very clayey. These characteristics suggest that mechanical processes alone cannot account for the sediment types. Phosphorite particles must have been formed within the environment and moved enough to produce rounding and some degree of sorting. Movement was not, however, generally sufficient to produce well sorted phosphorite sediments, thus resulting in bimodally distributed size characteristics.

Examination of the phosphorite intraclasts and the microsphorite reveals many similarities in mineralogy, chemical composition, and textural relationships. Both are composed of a mixture of carbonate-fluorapatite, quartz, clay minerals, carbonate, and phosphorite grains. In general, the microsphorite contains less P_2O_5 than the intraclasts, which might be due to the effect of weathering on different size particles. Textural similarities of microsphorite and intraclasts are more obvious in the larger intraclasts. These large particles have many of the characteristic features of microsphorite, such as parts of burrows, borings, laminations, and included phosphorite grains. Furthermore, the microsphorite unit in the Noralyn mine is in places disrupted so that matching hard fragments are tilted and disjointed in the plane of the bed. The overlying unit also contains abundant large intraclasts near its base. These stratigraphic characteristics and mineralogical and textural similarities suggest that at least the coarse-grained phosphorite intraclasts were derived from an indurated phosphorite substrate such as the microsphorite beds in the Noralyn mine. The finer grained phosphorite intraclasts could be fragments derived from thin microsphorite laminae or they might be fragments of thick beds which have been abraded.

Sedimentation of phosphorite in the district is a two-stage process involving initial deposition of phosphorite as a phosphatic mud or possibly replacement of a carbonate mud. The phosphatic substrate was partially indurated and subsequently disrupted so as to produce abundant fragments or clasts. These fragments were transported, abraded, and sorted depending on the mechanical energy of the environment. The original environment of deposition of phosphorite is, therefore, more appropriately represented by the microsphorite bed.

Microsphorite beds have been found by the writers only in the eastern part of the district. Reports of microsphorite-type material have been made by miners who worked in the northeastern part of the district. Although a systematic search for microsphorite has not been made in all the pits of the district, it is thought that the material is confined to the eastern half of the district. The microsphorite observed in the Noralyn mine may be deposits on the fringes of a large area of microsphorite deposition to the north.

The microsphorite bed in the Noralyn mine appears to have been a soft sediment or mud at the time of deposition as indicated by burrows and internal molds of mollusks. The animals that produced the burrows could have lived only on a soft bottom. Internal molds suggest that the substrate was soft enough to flow inside the shells. This soft, muddy sediment accumulated in shallow water on the broad, flat tops of banks. The microsphorite bed nearly always occurs over the banks, which suggests that these shoal areas were favorable sites of phosphate precipitation. The microsphorite beds generally contain thin, hard pavements with smooth, polished, bored surfaces. These surfaces are interpreted to be surfaces of by-pass and could have been formed during periods of nondeposition and induration of the substrate or possibly by momentary exposure during flood tides. In either case, the amount of induration necessary to produce these features is not great.

Phosphate deposition might have occurred in areas of very quiet shallow water during times when climatic conditions caused upwelling along the east coast. In these areas, very little terrigenous detritus accumulated, phosphatic muds were deposited, and phosphate replaced any carbonate exposed on the bottom. The resulting sediment was a thin sheet of phosphorite. Conditions favoring phosphate deposition appear to have fluctuated so that there were periods when no deposition occurred or when there was momentary emergence. During these periods, the bottom became partially indurated and the bottom was disrupted forming intraclasts by minor storms. These were swept away by the same storms. This process may have occurred numerous times at a single favorable location so that these areas could have supplied an enormous quantity of intraclasts. The size of the resulting intraclasts would be a function of the thickness of the indurated layer, the strength of the disrupting currents, and the subsequent abrasion. The typical discoidal shape of the intraclasts could be expected from a thin fragmented sheet of phosphorite. The intraclasts, therefore, were carried as clastic grains by currents which sorted them to some degree. By this process, the particles would have come to rest in an environment quite unlike the environment of primary phosphate formation.

Paleogeography

The paleogeography of the Florida peninsula during the time of major phosphate deposition is summarized in Figure 4. The predominant positive structural element in the Miocene was the Ocala arch, which had a profound influence on the character of sedimentation during that period of time. The largest concentrations of phosphorite in the Hawthorn Group occur in a wide belt surrounding the Ocala arch as indicated by data from well cuttings examined from wells throughout the State. This arch appears, therefore, to have been a major control of paleogeography and phos-

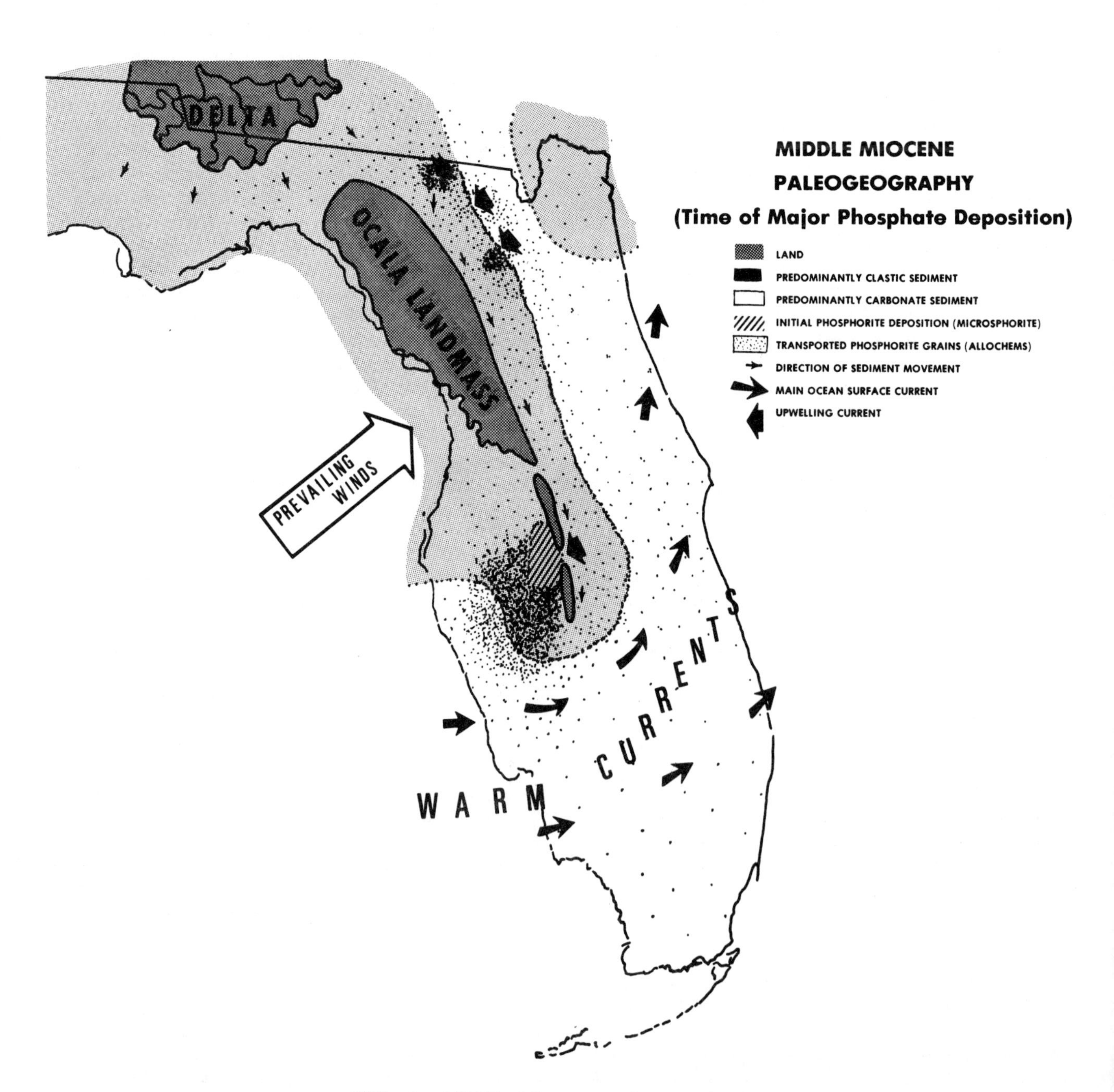

FIG. 4. Middle Miocene paleogeography.

phate deposition.

The Miocene was apparently the time of major phosphate deposition, as indicated by paleontological correlation of the marine phosphorite beds in North and Central Florida. During this time, the peninsula of Florida was a shallow shelf, dropping off abruptly to deeper water on the east and sloping gently westward into the Gulf of Mexico. The Ocala arch was a slowly emerging, low-lying landmass, probably a series of islands, swamps, and bays. In extreme northwestern Florida and Georgia, a delta extended out from the main landmass of the Piedmont. Most terrigenous detritus was deposited in the region of this delta, although some was carried out onto the ancient Florida bank by longshore currents sweeping the east coast of the Ocala landmass. These terrigenous sediments produced clastic facies surrounding the Ocala arch. Carbonate sediments were deposited farther offshore and were diluted with considerable terrigenous detritus in northern Florida. In southern Florida, carbonate sedimentation was extensive and less terrigenous detritus reached this region to contaminate the warm-water fossiliferous limestones and dolomitic limestone. These deposits were similar to those of the present-day Bahama Banks. In West-central Florida, clay and carbonate were deposited in the relatively quiet waters on the west side of the Ocala landmass.

There is no direct evidence of current directions during Miocene time. However, a hypothetical model of current movement is proposed which is based on distribution of sediment type. The warm-water deposits of southern Florida suggest that this region was open to the warm Gulf of Mexico. Currents flowing out of the Gulf would have crossed this bank and turned northward much like the modern Gulf Stream. Countercurrents, moving southward along the east coast of the Ocala landmass, would have transported terrigenous detritus from the north.

Phosphorite was deposited close to the Ocala landmass in the central and northern part of the peninsula. Upwelling could have occurred in these areas if the prevailing winds were from the south or southwest. The winds, blowing in the same direction as the northerly flow of the ancient Gulf Stream, could have pushed water offshore, thereby creating a condition for divergent upwelling of cool, deep water.

In Central Florida, a large spit consisting of a series of barrier islands extended almost north-south through Polk and Hardee counties. These islands were built by longshore currents which moved sand from the Ocala landmass. It is proposed that primary phosphorite sedimentation occurred behind this barrier island chain in what was normally a relatively low energy environment. As the phosphate-rich waters shelved, they flowed into the broad, shallow bank areas behind the islands.

Phosphorite intraclasts were formed in this area by the almost continuous fragmentation of partially indurated phosphorite substrate. These intraclasts were transported southwestward and southward from Polk County. The coarse-grained intraclasts (phosphorite gravel deposits) are prevalent west of the Ocala arch in Polk County. These coarse-grained intraclasts could have been swept by storm currents into an environment which was ordinarily one of relatively low energy. Southward, into Hardee, Manatee, and DeSoto counties, the amount of phosphorite gravel decreases rapidly. The concentration of phosphorite intraclasts in the sediments of this region is related to the topographic structures of the Arcadia Formation.

East of the Ocala arch, the phosphorite contains a lower percentage of phosphorite gravel than west of the arch. Southward, the phosphorite becomes fine grained and, at the same time, the overall phosphorite content decreases. Thus, a pattern of phosphorite sedimentation emerges in which primary microsphorite deposition took place largely in central Polk County and supplied phosphorite intraclasts which were mechanically transported away from that area. Coarser grained intraclasts travelled the least distance and accumulated on the tops or flanks of structural highs (shoals) where the fines were winnowed out. Fine-grained intraclasts were transported farther, becoming more diluted, and now make up the bulk of deposits such as those in South Florida. Small amounts of intraclasts were carried great distances and are now widely disseminated throughout the Hawthorn Group in Florida.

Toward the end of Miocene time, phosphorite deposition appears to have subsided or ceased. Carbonate sedimentation took place over much of the eastern part of northern Florida and southern Florida. The Ocala landmass appears to have been more emergent by the end of Miocene time.

In the Pliocene, the Ocala landmass extended into South Florida and its eastern shoreline essentially bisected the southern Florida peninsula. Quartz sand deposits formed along the eastern coast and carbonate deposits were laid down over the southern tip of Florida. Fluvial, estuarine, and lagoonal sediments were deposited in the Central and South Florida phosphate districts. Phosphorite from the underlying marine sediments was eroded, transported, and redeposited in these environments. In this way, the extensive deposits of reworked phosphorite were formed. The reworking continued through the Pliocene into the Recent.

REFERENCES

Freas, D. H. (1967) Exploration for Florida phosphate deposits: unpublished paper presented to the seminar on "Sources of Mineral Raw Materials for the Fertilizer Industry in Asia & the Far East," sponsored by ECAFE, Bankok, December 1967, 31 pp.

Riggs, S. R. (1967) Phosphorite stratigraphy, sedimentation, and petrology of the Noralyn mine, Central Florida phosphate district: Univ. Montana, unpublished Ph. D. thesis, 267 pp.

__________ and Freas, D. H. (1965) Stratigraphy and sedimentation of phosphorite in the Central Florida phosphate district: preprint, Amer. Inst. Min. Engrs., No. 65H84, 13 pp.

ENVIRONMENTAL FACTORS CONTROLLING OYSTER SHELL DEPOSITS, TEXAS COAST

Alan J. Scott
Department of Geological Sciences
The University of Texas at Austin
Austin, Texas

ABSTRACT

In recent years the demand for cement has increased tremendously along the Gulf Coast. This increase in usage is related to the urbanization and industrialization of the coastal plain and the acceleration of highway and other types of construction.

Several cement plants are in operation along the Texas coast, notably in the Houston metropolitan area and at Corpus Christi. Oyster shell forms the basic raw material for these plants. Shell is, in fact, the only significant source of lime along the coast and inland for a distance of over 100 miles. As such, it is an important mineral resource and is being actively exploited by several companies.

Two species of molluscs form shell deposits of economic importance in this area. Rangia cuneata (Gray), a brackish water clam, is widely used in Louisiana, especially as a source of road metal. The common edible oyster Crassostrea virginica Gmelin forms reefs in many bays along the northwestern Gulf Coast and is the most important lime producer in Texas.

The development of large oyster reefs is affected by a complex interrelationship of environmental and geological factors. Crassostrea is a sedentary form which filters suspended micro-organisms and organic material from sea water. It therefore requires a firm, stable substrate for the larvae or spat to settle and develop, and sufficiently strong currents to carry in nutrients and remove waste products. These factors strongly influence not only the location of, but also the geometry of reef masses. Oysters have a relatively broad range of tolerance to changes in salinity and temperature. Reefs are most frequently developed in areas of lowered salinity. This is due to two factors: first, oysters are able to thrive in brackish water and secondly, their common predators, especially oyster drills and starfish, are less common in these areas.

Dredging of living oyster reefs in the Texas bays is at present limited by law. Therefore, it is desirable to develop exploration methods for buried reefs. In the past, exploration programs have been based upon surveys with steel probes. The State of Texas also has completed a sonoprobe survey of some of the bays during the early developmental stages of acoustical profiling apparatus.

A more effective exploration program could be developed based upon a knowledge of the environmental tolerances of oysters, the geologic and depositional history of the area, and an analysis of cores to determine paleoenvironmental conditions. Large areas could be eliminated from consideration and favorable areas studied in more detail by applying these techniques.

INTRODUCTION

The urban industrial areas of Texas have been in a period of unprecedented growth during the past thirty years. Construction projects in the cities, coupled with expansion of the highway program, has led to a great increase in the demand for aggregates, cement and other materials. Several metropolitan areas in Texas, for example, Dallas, Fort Worth, San Antonio and Austin, are located on Cretaceous carbonate terranes. These limestones provide a convenient, inexhaustible and inexpensive source of aggregate and raw material for cement.

The situation is quite different in metropolitan areas along the Texas coast. The coastal plain, dominated by terrigenous clays and sands, extends inland for a distance of at least 100 miles. The only significant local sources of lime for the Houston, Beaumont and Corpus Christi metropolitan areas are shell deposits from the adjacent bays.

Crassostrea virginica Gmelin, more commonly known by gourmets as the eastern edible oyster, form extensive reefs in the brackish coastal bays. Ironically the carbonate shell of the oyster is more important economically than the meat of this delicious little beast. Pollution of many parts of the Texas coast has limited the taking of live oysters for human consumption.

Oyster shell has on the other hand become the basis for a multimillion dollar industry. It is the essential raw material for a variety of processes in the chemical industry. It is used in the manufacture of cement, lime base drilling muds, and in the extraction of magnesium from sea water. It is also widely used as concrete aggregate and for road metal.

Production and transportation costs are relatively low and lime from other sources is not competitive with shell in coastal areas. Kerr (1968) has recently discussed the economic factors of this industry.

The shell dredging industry is presently being threatened from different quarters. The bays currently being dredged are also used for other purposes by various industries and private citizens. Long the basis for an important commercial fishing industry, the bays are now extensively used for recreation and sport fishing. Such groups have been increasingly vocal about the effect of dredging operations on adjacent areas in the bays.

The other serious problem facing the shell industry is the rapid depletion of reserves. Shell is currently being dredged at a rate several times faster than it is being replenished by biological activity. At the present rate of consumption, reserves of only a few tens of years are currently known. The factors that influence the distribution of shell deposits are reviewed herein. Exploration programs based on these principles may result in the discovery of older, buried deposits of economic significance and increase the known shell reserves.

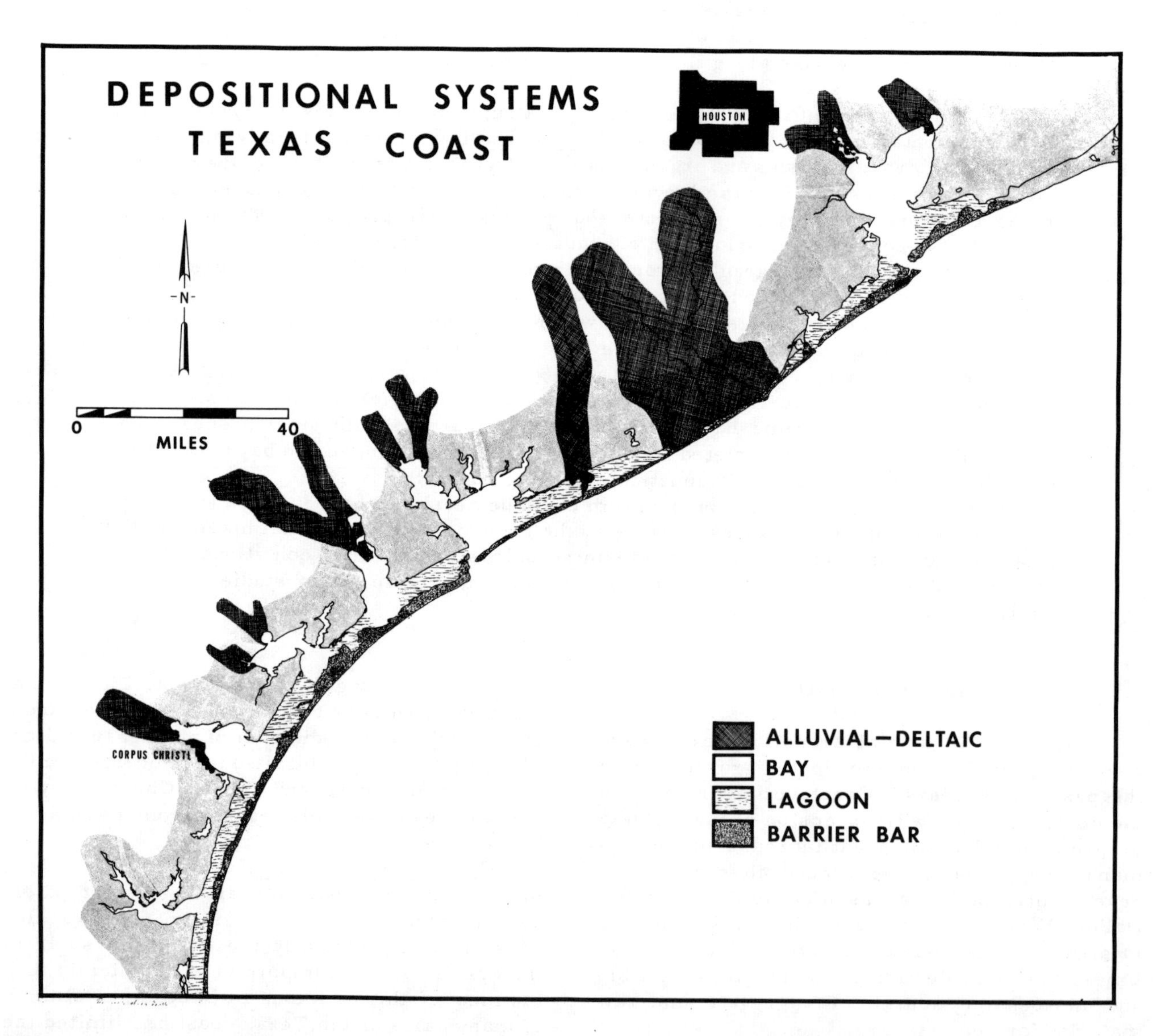

FIG. 1. Recent depositional systems, central Texas coast.

GULF COASTAL SHELL DEPOSITS

The late Quaternary history that led to the development of the present features of the Texas coast has been thoroughly reviewed by LeBlanc and Hodgson (1959) and Bernard and LeBlanc (1965) and will not be discussed in detail.

The Texas coast can be subdivided into several distinct depositional systems. These systems called environmental complexes by Hayes and Scott (1964a) are genetically related associations of depositional environments. As such they have a characteristic suite of sediments and can be mapped (Fig. 1). The occurrence of shell deposits can be related to these depositional systems as shown in table 1.

Two major types of shell deposits occur in association with the chain of barrier islands along the Texas coast. Shell content on Gulf barrier beaches varies considerably but is especially high in the Coastal Bend region south of Corpus Christi and north of Port Mansfield. Beach samples containing over 50 percent shell are common and have given rise to the place names "Big Shell" and "Little Shell" by local fishermen. Origin of these shell concentrations and their relationship to longshore drift have been discussed by Lohse (1955), Hayes (1965) and Watson (1968). The Coastal Bend region, although relatively inaccessible, is the only area with beach shell in sufficient concentration to be of economic importance. This region lies within the confines of the Padre Island National Seashore which precludes the possibility of commercial exploitation.

Tidal inlets or passes cut through the barrier

Table 1. Shell deposits in marginal marine depositional systems, northwestern Gulf of Mexico.

SHELL CONCENTRATIONS NORTHWESTERN GULF

DEPOSITS	DEPOSITIONAL SYSTEMS			
	BARRIER BAR	LAGOON	BAY	DELTA
SHELL BEACHES	X			
TIDAL CHANNEL LAGS	X	X		
SHELL BERMS		X	X	
OYSTER REEFS		X	X	
RANGIA BEDS				X
SERPULID WORM REEFS		X	X	

islands and permit exchange of waters from the lagoons and bays behind the barriers with the open Gulf. The strong currents in these channels concentrate shell debris as lag deposits. These deposits are associated with adjacent lagoonal, as well as barrier island depositional systems. The deposits vary in thickness but may be up to three feet thick. Examination of aerial photographs and old navigational charts shows numerous abandoned tidal channels. Also active tidal channels migrate along the coast due to the influence of longshore currents. Thus major portions of the barrier island chain are underlain by shell beds formed as channel lags.

It is doubtful that these lags will ever be exploited commercially. The shell is of relatively poor quality. Many of the shells are small, thin, and easily broken and would form a poor aggregate. They are commonly encrusted with fouling organisms and are extensively bored by clionid sponges. The sponge borings are filled with clay and fine quartz sand which cannot be easily removed. Introduction of large quantities of quartz sand to lime kilns used in the manufacture of cement would be highly undesirable.

Furthermore, the lag deposits are located in narrow tidal channels, several of which are major shipping lanes or are overlain by 20 to 30 feet of barrier island sands. Removal of this overburden for a thin shell bed of relatively poor quality would be, at present, uneconomical.

Large bodies of water frequently called 'bays" are located behind the chain of barrier islands along the Texas coast. LeBlanc and Hodgson (1959) and others have subdivided these water bodies into two major categories on the basis of genesis and geometry. Coastal water masses whose long axes are parallel to the strand are herein called lagoons. They are genetically related to the development of the offshore barrier islands. The other type of coastal water body is commonly elongated perpendicular to the strand line. These water bodies are herein referred to as bays. They are always associated with river valleys and represent the lower reaches of estuaries drowned during the sea level rise resulting from the melting of late Wisconsin glaciers. As seen in table 1, the discrimination between lagoons and bays is significant in considering the distribution and quality of shell deposits.

Shell berms are developed along the margins of both lagoons and bays. These berms or storm ridges are seldom over four feet in height, may be 40 or more feet wide and may extend for several miles. They are prominent features and can easily be mapped on aerial photographs. Berms are best developed adjacent to areas of live oyster reefs, or in areas associated with spoil from channel dredging. They also occur most commonly along northwestern shores of bays due to prevailing winds or southern shores due to the effect of "northers." Compound berms formed by accretion by several storms are formed in some areas, e. g., Indian Point adjacent to Corpus Christi Bay (Scott, 1964, p. 43).

A series of storm berms may contain a total of several thousand cubic yards of shell. Several problems limit the exploitation of berm deposits. The berms are located along the shorelines and serve as breakwaters for storm waves. As such they should be retained in certain areas. Berms are generally perched on soft clay-rich marsh deposits which will not support heavy equipment. The volume is too small to recommend the use of large dredges.

At present the only shell deposits on the Texas coast commercially exploited on a large scale are oyster reefs. Although reefs develop in both lagoons and bays, the largest and most important reefs are located in the bays. Oyster reefs are developed in all major Texas bays north of Corpus Christi. Shell production from these bays for the years 1966-67, 1963-64 and 1960-61 are summarized in Figure 2; more detailed figures are given by Kerr (1968). These production figures reflect proximity to market, as well as relative abundance of oyster reefs. For example, an average of over 8 million cubic yards of shell was produced annually from the Galveston-Trinity Bay area. This area contains well developed reefs and is most convenient to the Houston market. In the same period less than three million cubic yards were dredged per year from San Antonio Bay and there was no commercial exploitation of shell in Copano Bay. Both contain extensive reefs which were not worked in preference to more convenient deposits. Considering current dredging procedures and equipment, San Antonio and Copano bays represent the areas containing the largest shell reserves remaining on the Texas coast. These areas are discussed in greater detail in subsequent sections of this paper.

Oyster reefs are not developed in Baffin Bay located 30 miles south of Corpus Christi, Texas. The average annual rainfall in this area is less than 30 inches and no permanently flowing rivers discharge into Baffin Bay. At present, there is no natural pass through Padre Island in the vicinity of Baffin Bay. These factors and the high rate of evaporation result in salinities higher than those preferred by oysters. Serpulid worms have formed small reefs in this bay. These worms secrete calcium carbonate tubes which individually resemble drinking straws. The tubes are tightly cemented together forming wave resistant areas. The distribution, geometry and internal structure of these reefs have been reported on by Andrews (1964). It is interesting to note that there have been no reported occurrences of living serpulid worms from Baffin Bay although

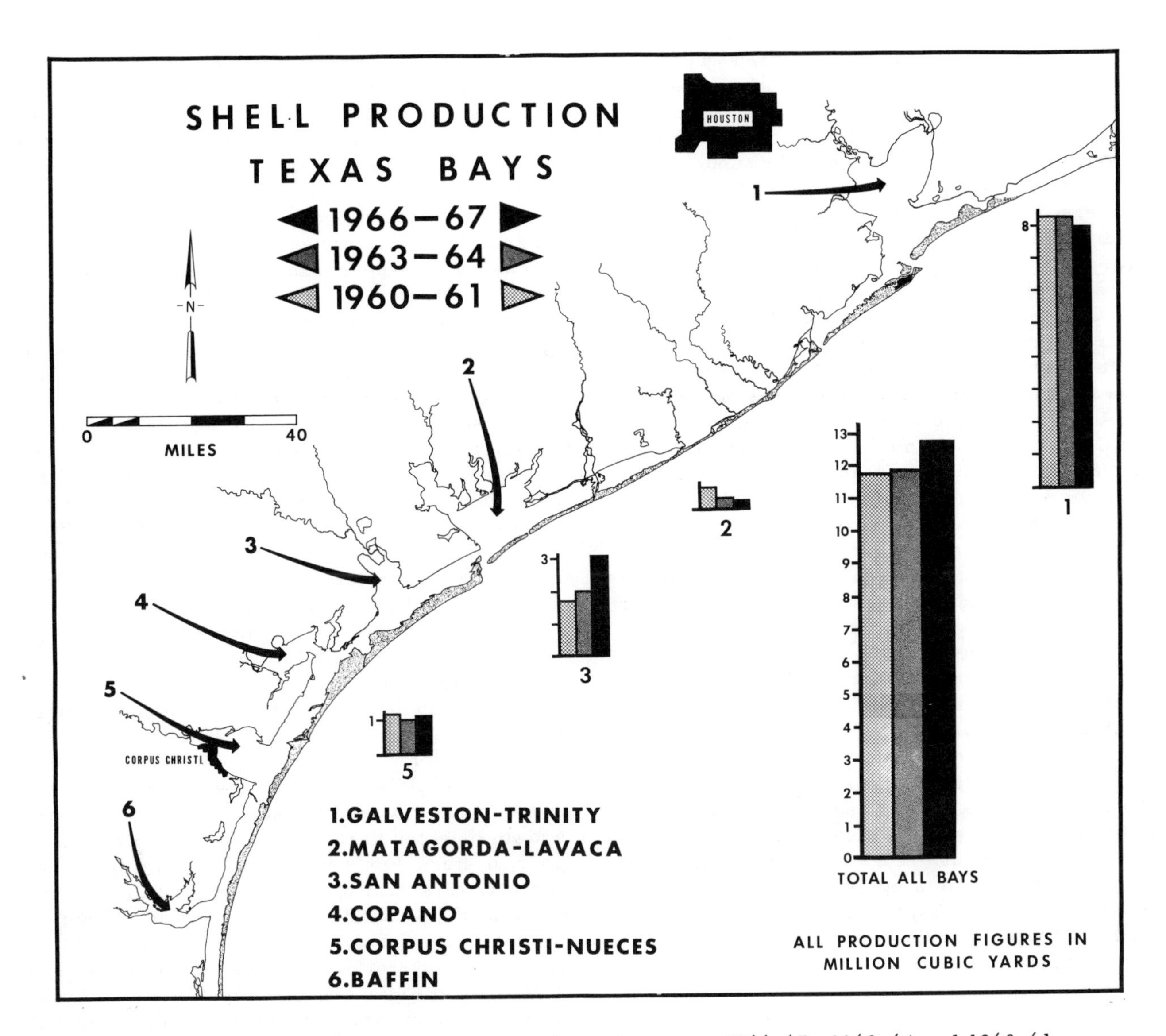

FIG. 2. Shell production from Texas bays for years 1966-67, 1963-64 and 1960-61.

they are present in sufficient numbers to be a serious fouling organism in other areas near Corpus Christi.

Several factors limit the value of serpulid reefs as potential lime sources. The reefs are relatively small in comparison to oyster reefs, the dead worm tubes fill with fine quartz sand which is removed with great difficulty, the tubes are tightly cemented to each other, and the serpulid reefs are not located convenient to a market. In short, they are currently more of a menace to small fishing boats than an economically important carbonate deposit.

Rangia shells are widely used as a source of lime and as road metal in southern Louisiana. This clam prefers waters of lower salinity than oysters. Rangia shells are currently not commonly utilized in Texas for this purpose, although the clam is abundant near rivers discharging into the various Texas bays. At present Rangia deposits are not competitive with oyster reefs, but they are a potential lime source for future exploitation in deltaic areas of the Texas coast.

Current dredging methods limit the exploitation of shell deposits to depths of about 40 feet below sea level and in relatively sheltered areas. Thus concentrations of shell occurring at greater depths in the bays and lagoons or in the open Gulf are not considered commercially important under present circumstances.

ECOLOGY OF OYSTERS

Due to their importance as a food source, literally hundreds of papers have been written on the ecology of oysters since the turn of the century. The factors that affect their abundance, rates of growth and mortality have been reported on at great length. It is beyond the scope of this report to discuss all aspects of oyster ecology. Readers interested in the biology of oysters may pursue the subject in the monographic study of Crassostrea virginica recently completed by Galtsoff (1964).

Considering the value of oyster shell as a natural resource in its own right, it is interesting that few studies have been completed in recent years on factors influencing the rates of development, geometry and location of oyster reefs. An excellent source of information on these topics are the United States Bureau of Fisheries oyster surveys, including those by Cary (1906), Galtsoff (1931), Grave (1905), Moore (1899, 1907, 1913a, 1913b), and Moore and Danglade (1915). These reports present considerable information on environmental conditions associated with major reefs in many coastal areas.

Galtsoff (1964, p. 397) lists the most important environmental factors affecting oyster populations (table 2). The significance of these factors in controlling the development and geometry of oyster reefs is discussed below. Two areas on the central Texas coast, San Antonio Bay and Copano Bay, are cited as case examples. Both areas have well developed oyster reefs although commercial dredge operations have not been as extensive as in other areas, for example, Galveston-Trinity bays and Nueces Bay. As such they represent the two most important areas for future exploitation.

Table 2. Major factors affecting oyster populations (after Galtsoff, 1964).

FACTORS AFFECTING OYSTER POPULATIONS

1. SUBSTRATE CHARACTERISTICS
2. WATER MOVEMENTS
3. SALINITY
4. TEMPERATURE
5. FOOD
6. RATE OF SEDIMENTATION
7. DISEASES AND PARASITES
8. COMMENSALS AND COMPETITORS
9. PREDATORS
10. POLLUTION

Substrate Characteristics

Oysters are sessile, benthonic organisms. They possess a relatively short lived planktonic larval stage. After settling of the spat the oyster remains attached and immobile throughout the remainder of its life. The nature of the substrate is therefore extremely critical for oyster development.

Two major types of substrates are extremely unfavorable for the establishment of large oyster populations. These are soft mud and shifting sandy substrates. Oysters growing on soft muds slowly sink into the sediment and are suffocated. Oyster spat and small individuals are buried by sands that are in motion at the sediment-water interface with similar results.

Oysters will grow equally well on hard rocky substrates, stable sands, and stiff muds that are capable of supporting their weight. One of the most favored substrates for spat settlement is the shell of other oysters, living or dead. This is one of the primary reasons that oysters form reefs rather than occurring as individuals widely spaced throughout a favorable habitat.

The shape of oyster shells commonly reflects substrate characteristics. Many shells conform to the shape of the substrate to which they are cemented. Clumps of oysters growing on firm sandy substrates have broad flat individuals with a relatively low length to width ratio. In contrast oysters that grow on muds often are extremely narrow and elongate with the axis of growth oriented vertically. Unlike many pelecypods, oysters do not have siphons. Water carrying nutrients and oxygen is taken in along the margin of the shell. If this margin is covered by sediment the oyster soon expires. Therefore, if the oyster can add new shell material at a rate equal to or greater than the rate of subsidence it survives. On soft substrates growth is concentrated along a narrow axis resulting in an elongate shell.

Surficial sediment characteristics of the central Texas bays has been studied by Shepard and Moore (1955 and 1960) and many other workers. The generalized distribution of major sedimentary facies in the vicinity of San Antonio and Copano bays is shown in Figures 3A and 3B.

The bay center is essentially flat with water depths averaging 8 to 12 feet, except where reefs rise above the adjacent bay floor. The bay margins slope gently toward the bay center. At a depth of about four feet this slope steepens. The resulting shape can be compared with a broad soup plate with a gently sloping rim and a flat bottom separated by a steeper slope. The dominant sediment type in the bay center is soft muds and sandy sediments are characteristic of the bay margin facies.

The shape of oyster clumps and even reefs

is also strongly influenced by the firmness of the substrate. Oyster clumps have been observed in soft muddy areas in Redfish Bay near Aransas Pass, Texas, that were only one foot in diameter and extended more than 2-1/2 feet beneath the sediment-water interface. Probing near reefs associated with soft bay center muds has revealed that many of the reefs have sharp, nearly vertical walls. A probe taken on the edge of a reef may penetrate 10 or more feet of shell. A second probe made only 30 feet away may not penetrate any shell beds. Similar observations have been reported by Norris (1953) on buried reefs probed in San Antonio Bay.

The plan view of reefs in bay centers is generally narrow elongate bodies. The long axis of the reef tends to be oriented at right angles to currents.

The sharp lateral facies change across the reef is due to two important factors. The muds adjacent to the reefs are not as favorable for the establishment of oyster spat as the shell substrate. The sediment underlying the reef may initially have been capable of supporting a thin layer of oyster shell. As the reef grew, the increased weight caused subsidence of the shell

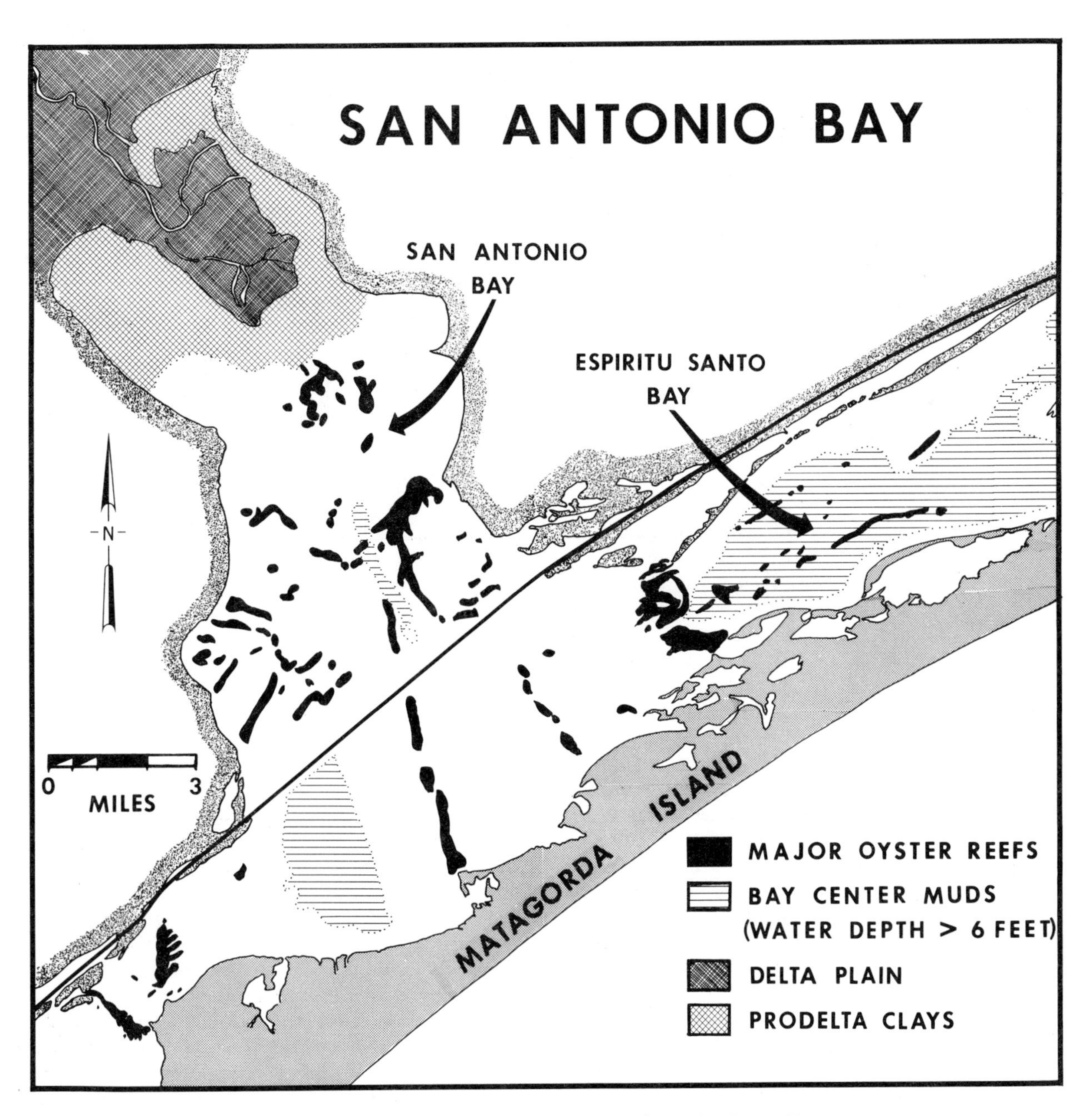

FIG. 3A. Sedimentary facies, San Antonio Bay, Texas.

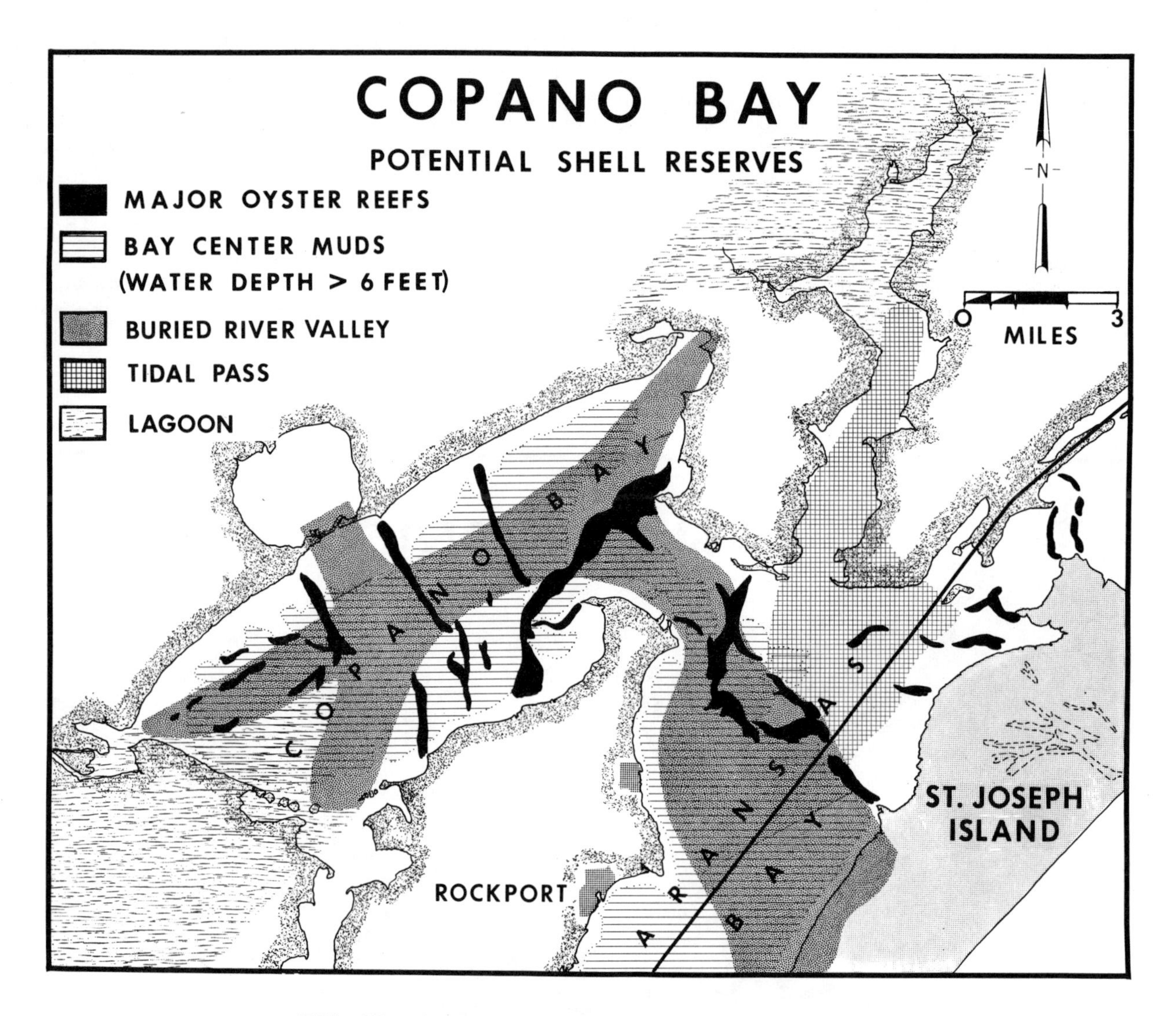

FIG. 3B. Sedimentary facies, Copano Bay, Texas.

resulting in a ridge-like reef with sharp walls. In contrast reefs growing along sandy bay margins form broad flats which develop into thin, laterally persistent shell beds.

Two rivers, the Aransas and the Mission, flow into Copano Bay. These rivers have relatively small drainage basins and carry relatively small sediment loads. The Mission River flows into Mission Bay, a small enclave on the western shore of Copano Bay, before actually passing into the larger water mass. Mission Bay thus acts as a settling basin for Mission River sediments. Therefore, no large delta is forming at the head end of Copano Bay (Fig. 3B).

The Guadalupe River that discharges into San Antonio Bay is a major river that traverses the entire width of the coastal plain. Its sediment load is sufficient to form a large bayhead delta (Fig. 3A). This delta is fringed by delta front sands and prodelta clays. Neither of these lithotopes is favorable for the establishment or maintenance of oysters.

Water Movements

Water circulation is essential for the establishment and maintenance of oyster populations. Oysters feed on planktonic organisms suspended in the water column. Since they are sessile they are dependent upon water currents to bring food as well as oxygen to them. The suspended material is trapped on sheets of mucous on the oysters' gills; this mucous is then moved through the digestive tract by ciliary action. After the nutrients have been digested, mucous containing large quantities of inorganic materials such as clay particles is expelled in the form of long strands. Currents sufficiently strong to carry these strands away from the oyster are necessary for development of reefs or the material settles and eventually suffocates the oyster colony. Galtsoff (1964, p. 400) states that the ideal condition for oyster populations is a steady nonturbulent flow of water, strong enough to remove waste materials and to provide food and water. He also points out the importance of water currents in the dispersal of oyster larvae.

Circulation in estuarine environments, such as the Texas bays, is extremely complex. Currents are caused by the interaction of tidal exchange, river discharge, winds and density differences due to variations in temperature, salinity and suspended matter. Minor shifts in wind direction or river discharge can cause major changes in circulation patterns in shallow bodies of water. A limited number of measurements of currents and wind velocities was made in the vicinity of San Antonio and Copano bays as a part of Project 51 of the American Petroleum Institute. These data were summarized by Shepard and Moore (1955, pp. 1481-1493). Parker (1959, p. 2112), utilizing this information, the distribution of sediments, surface isohalines and isotherms, published a generalized map of circulation patterns for the bays of the central Texas coast. This map is the basis for Figures 4A and 4B.

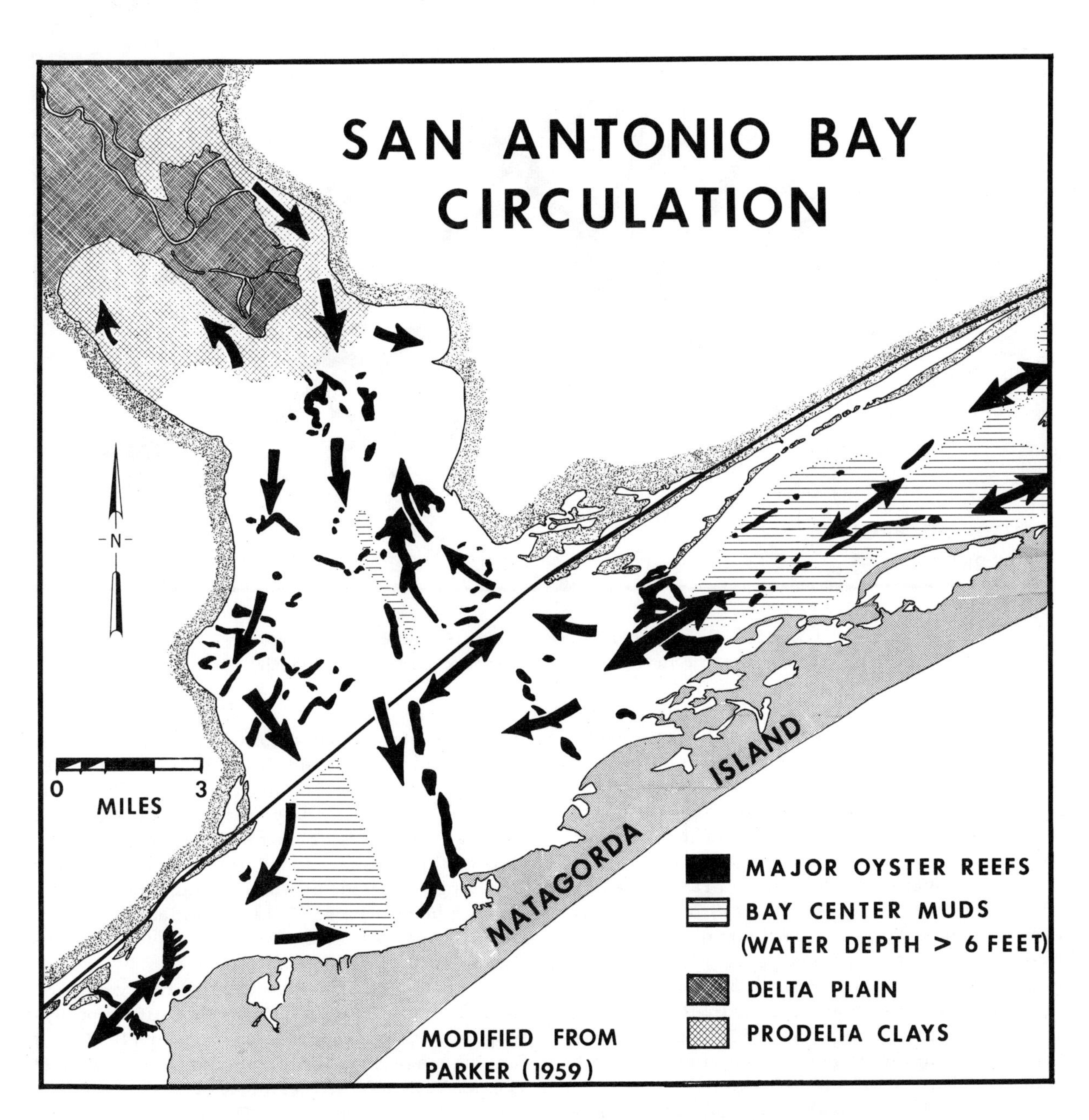

FIG. 4A. Generalized circulation, San Antonio Bay, Texas. Size of arrow indicates comparative current strength.

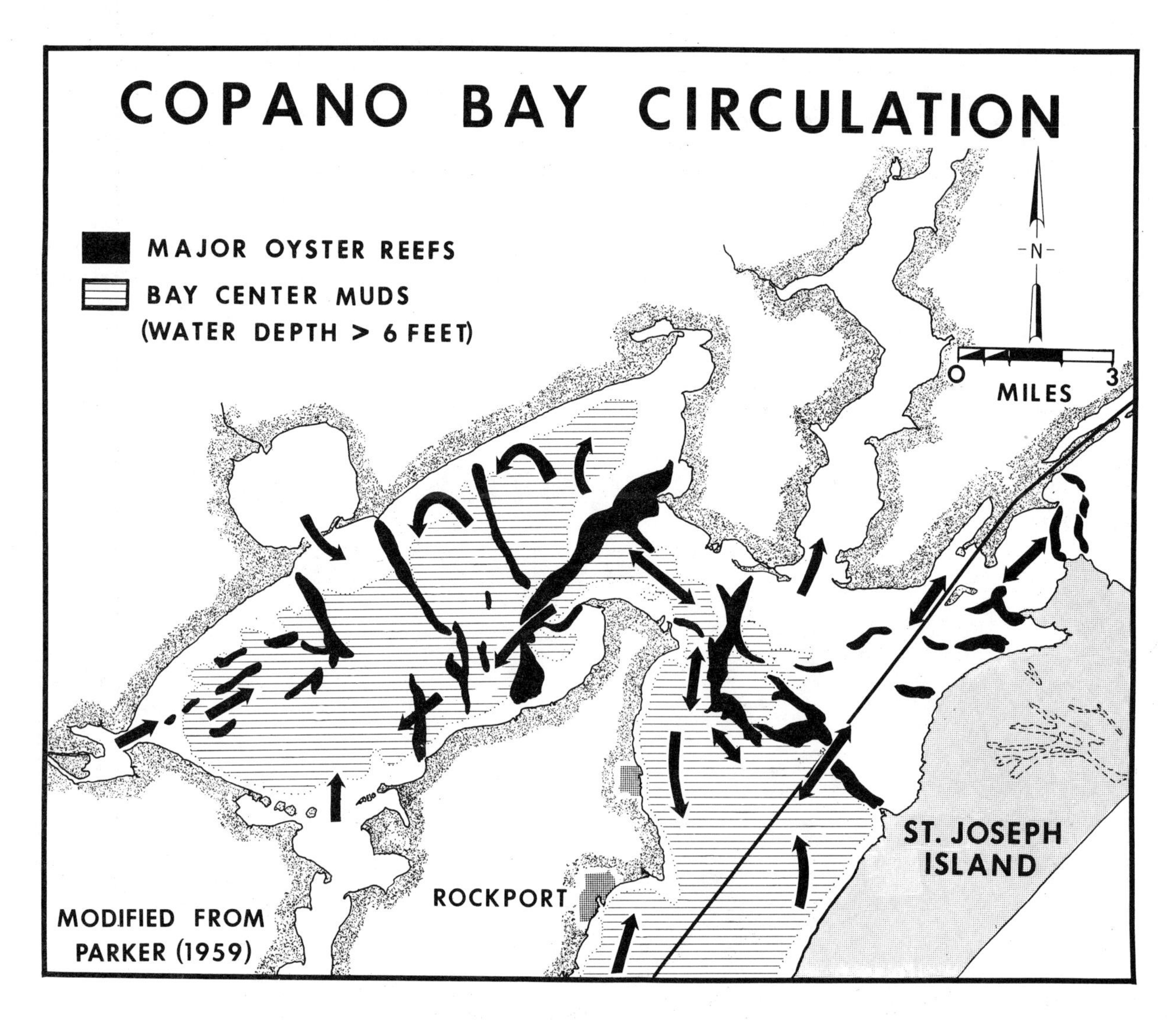

FIG. 4B. Generalized circulation, Copano Bay, Texas. Size of arrow indicates comparative current strength.

A detailed study of the physical oceanography of the Texas bays has not been undertaken to date. The generalized patterns based upon deductions and qualitative field observations are sufficient to emphasize a very strong correlation between water circulation patterns and the location, size and shape of oyster reefs. In general, it can be said that reefs are largest in areas of relatively strong currents where the greatest volume of food bearing waters are passed over the oysters. These larger reefs also tend to be elongate with their long axis oriented perpendicular to the major currents. Examples of elongate reefs oriented across currents are obvious in both Copano and San Antonio bays.

Galtsoff (1964, p. 404) points out that oysters are euryhaline and are able to live in sea water of a wide range in salinity. They prefer brackish waters with salinities from 30 o/00 to about 5 o/00 but can survive sudden changes in salinity of short duration, such as those caused by flood waters diluting the saline bay waters. This is accomplished by merely closing their valves and isolating themselves from the unfavorable water. Prolonged periods of flooding can totally decimate oyster populations. This is caused not only by the lowered salinity but also by the sudden increase in the rate of sedimentation.

Prolonged exposure to salinities greater than 32 o/100 also has an unfavorable effect on oyster populations as reported by Parker (1955). Such an increase in salinity occurred in the central Texas bays as the result of a six-year drought from 1948 to 1953. The fauna of the oyster reef community was altered with Ostrea equestris, a species preferring higher salinities, gradually replacing Crassostrea virginica. The rate of shell production is much less in O. equestris than the larger species C. virginica.

Increased salinity has another deleterious effect on oyster populations. Many oyster predators prefer waters of higher salinities than is optimum for C. virginica. Therefore, higher population densities and consequently greater rates of shell production are maintained in waters of salinities from approximately 15 to 25 o/00. Salinity measurements in Texas bays have been made by many workers over the forty years. Galtsoff (1931) published the first isohaline maps of the Copano-San Antonio bay region based upon measurements made in the winter of 1926. Portions of these maps are presented in Figures 5A and 5B.

These figures show a strong salinity gradient ranging from less than 4 o/00 near the mouth of the Guadalupe River to over 26 o/00 in Espiritu Santo Bay near the tidal inlet Pass Cavallo. This gradient is due to the dilution of the saline waters of the bay by the river. Copano Bay also shows a salinity gradient, as does Aransas Bay, the lagoon south of Rockport. This gradient is not as sharp nor does it encompass such a great

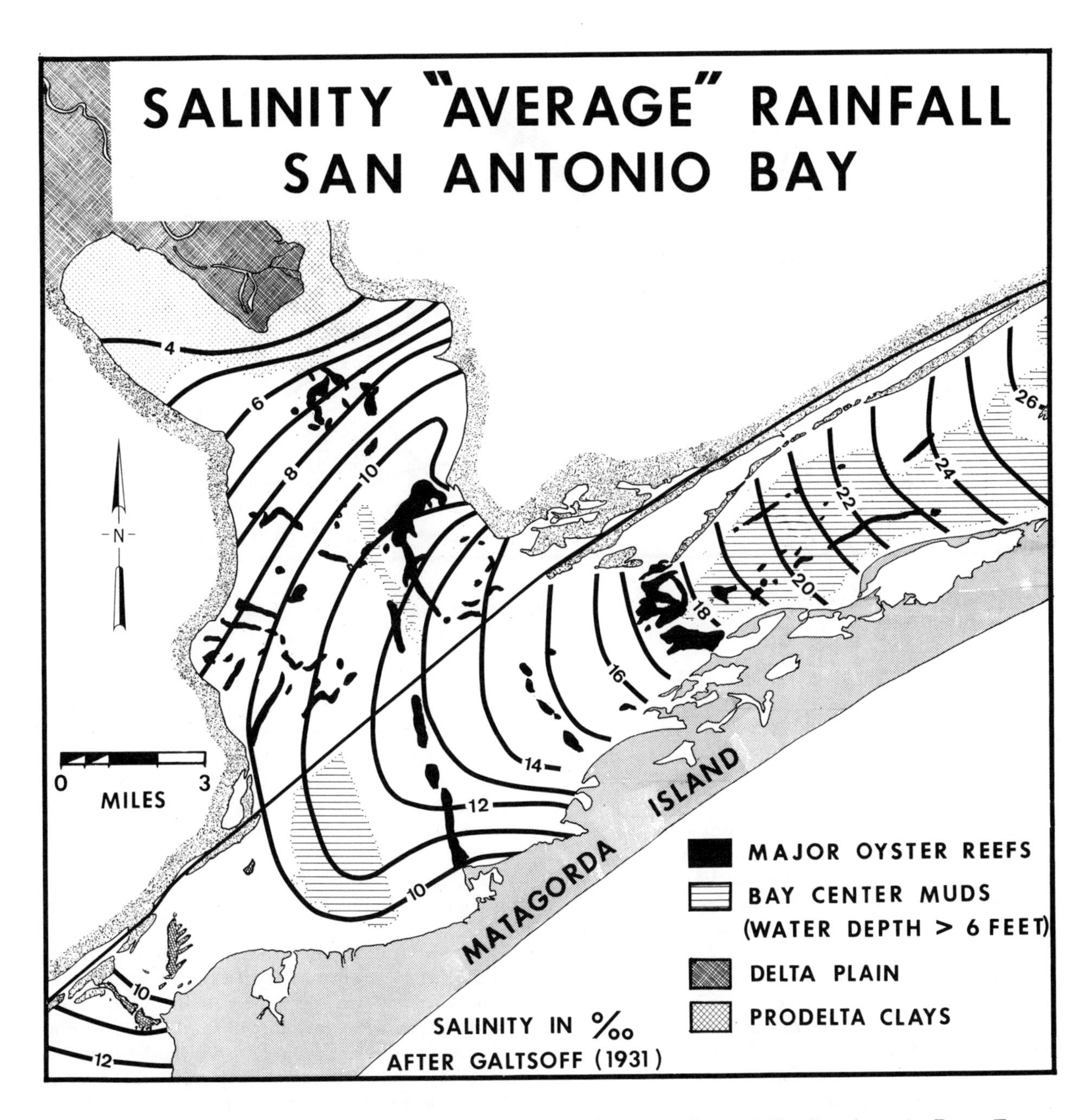

FIG. 5A. Generalized isohalines during period of "average" rainfall, San Antonio Bay, Texas.

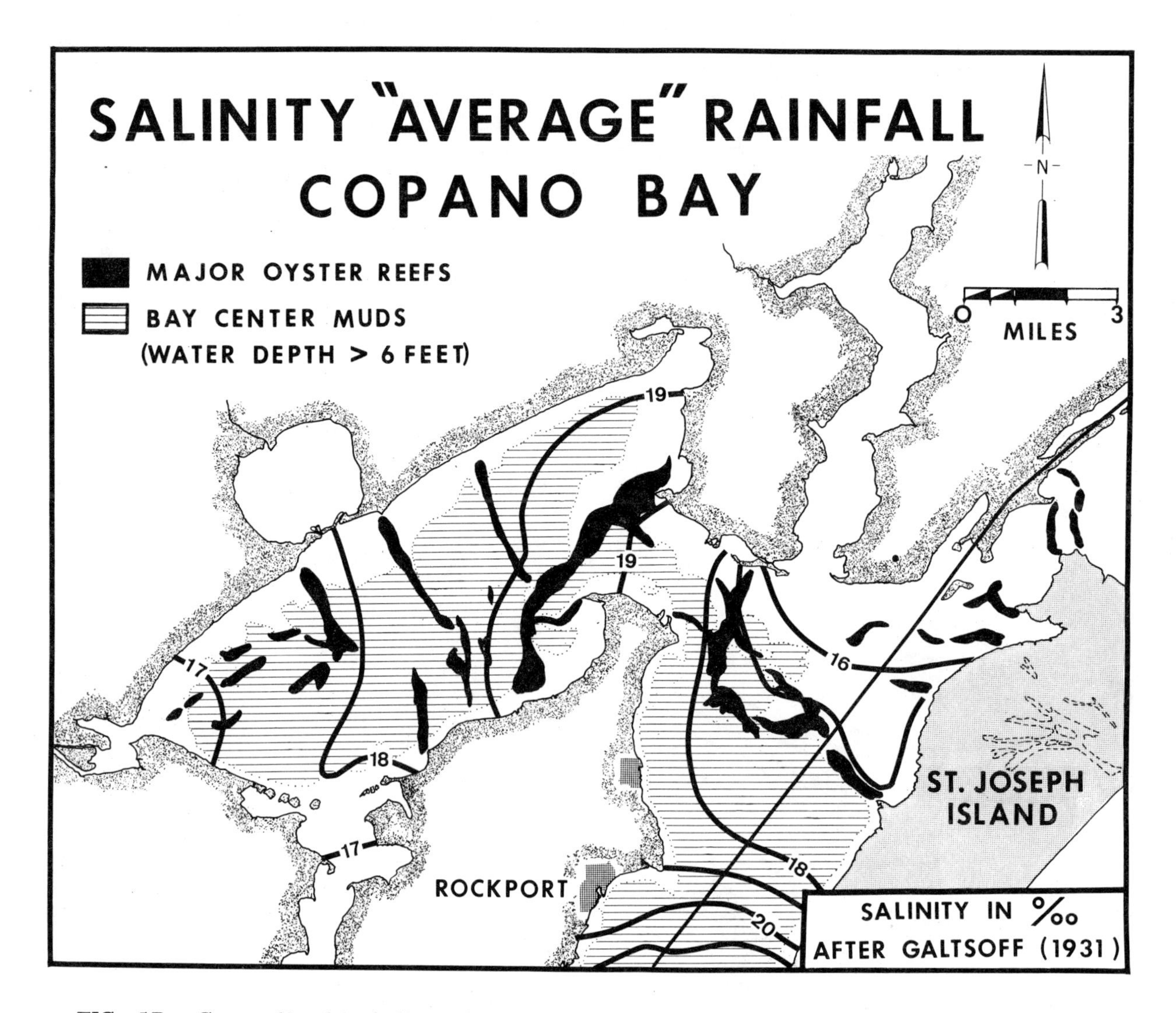

FIG. 5B. Generalized isohalines during period of "average" rainfall, Copano Bay, Texas.

range of salinities due to the lower discharge levels of Aransas and Mission rivers. Oyster populations, therefore, do not vary along the axis of Copano Bay as they do in San Antonio Bay, but remain high throughout.

Collier and Hedgepeth (1950) and Parker (1959) have described salinity changes associated with major floods. Copano and San Antonio bays were virtually fresh after a flood in June, 1957, as reported by Parker. Similar freshings are not uncommon. Baffin Bay, south of Corpus Christi, is normally hypersaline. Heavy rains, totaling over 30 inches, were associated with Hurricane Beulah in September, 1967. Behrens (1967) had been monitoring the salinity of Baffin Bay for several months prior to the storm. He reported that fresh water completely filled the bay for weeks after the storm. Continued heavy rains in the area kept salinities at a low level (about 15 o/00) for over six months.

Phleger and Lankford (1957), Parker (1955, 1959) and others have reported on salinities in San Antonio and Aransas bays at different stages in the prolonged drought from 1948 to 1953. During this drought, salinities in the bays rose well above 36 o/00 and remained there for several months. The changes in the composition of the oyster reef community and other invertebrate populations have been documented by Parker (1955).

From the above discussion it can be seen that salinity fluctuations sufficient to cause drastic changes in oyster populations and even faunal assemblages are common in the bays of the central Texas coast. Such changes appear to be of relatively short duration geologically speaking, seldom being longer than a few years. Patterns similar to those shown in Figures 5A and 5B have probably been prevalent for at least the past thousand years based upon the known distribution and volume of shell reefs dominated by Crassostrea virginica.

Temperature

Diurnal and seasonal water temperature fluctuations closely parallel air temperatures in the shallow waters along the Texas coast. Oysters are eurythermal organisms and are seldom exposed to lethal temperatures in the Texas bays. Exceptions occur when oysters are exposed subaerially during extreme low tides caused by strong winds. If this occurs during the summer the sun can cause the temperature of the body fluids of the oysters to exceed the upper limit of tolerance.

Food

Oysters feed on microplankton and other suspended organic matter. They are not selective in their diet and feed upon whatever larvae, phytoplankton, and other organisms that pass through their mantle margins. Galtsoff (1964, p. 409) points out that extremely high concentrations of phytoplankton have a harmful effect on oyster populations. The ideal condition for the feeding of oysters is in water free of pollution and suspended sediment with low concentrations of small diatoms and dinoflagellates flowing across the bottom in a nonturbulent flow.

Phytoplankton production and species composition varies seasonally; however, productivity is sufficient to sustain large oyster populations in most areas that are otherwise suitable.

Rate of Sedimentation

This factor is of great importance to sessile epibenthic organisms such as oysters. Oysters as previously mentioned live on the sediment-water interface and have no means of extricating themselves should they sink into soft sediments or be buried by newly deposited sediment. Even a layer of clay one millimeter thick deposited after a flood or storm can prove disastrous to oyster spat.

The rate of sedimentation in the center of Texas bays is low and these are, therefore, quite favorable to oysters in this respect. Unfortunately the soft muds limit the distribution of oyster populations to the crests of elongate transverse reefs. This pattern of reefs alternating with mud having few oysters in Copano Bay (Fig. 3B) is in part due to differences in sedimentation rates. The reef crests are covered with less than a foot of water at high tide. Currents strong enough to remove the mucous strands and other wastes are developed as the wind drives water across these reefs. Water in the interreef areas averages about eight feet in depth and currents, especially near the sediment-water interface, are weak. These interreef areas, therefore, form excellent settling basins for the sediment-rich mucous strands. This slimy coating on the sediment interface inhibits the establishment of oyster spat and suffocates young oysters.

Absence of exposed oyster reefs in the prodelta facies in San Antonio Bay (Fig. 3A) is also due to the high rate of sedimentation characteristic of this environment. Studies of the Guadalupe Delta have shown that progradation of the delta has buried several large oyster reefs.

Diseases and Parasites

A variety of diseases and parasites infect oysters and can completely decimate local populations. Such epidemics are not uncommon but rarely persist for prolonged periods of time. Most oyster shell deposits are formed over a period of several hundreds or in many cases thousands of years. Therefore, short term effects of oyster diseases are difficult to evaluate and are probably geologically insignificant.

Commensals and Competitors

Oysters compete with all other suspension feeders; however, food is seldom the factor that limits the distribution of oysters. More important is the "Lebensraum" factor. Oyster spat compete with other larvae for space to settle and develop. At higher salinities, or in areas otherwise unfavorable for Crassostrea, fouling organisms become a serious threat to oysters. They feed on oyster spat, occupy favorable settling areas, grow over shell margins and suffocate them with feces and pseudofeces.

Several species bore into living or dead oyster shells. Among the most common of these boring organisms are the clionid sponges. Oyster shells in some areas are completely riddled by a complex network of borings. Such borings greatly weaken the shell and decrease the value of the shells.

Predators

Oysters are eaten by a variety of other organisms such as oyster drills, crabs, starfish, as well as man. The following example illustrates the complexity of predator-prey relationships and the effect of other environmental factors. Many of the oysters' most voracious predators, such as the drills and starfish, prefer salinities greater than 30 o/00. Crassostrea virginica can tolerate and even thrive at these higher salinities; however, the increased rate of predatation limits the

most important reefs currently being dredged for shell are located in this zone.

The volume of shell decreases away from this ridge in both directions. Series of transverse reefs commonly grow across the central portion of the bays. These reefs are from several hundred feet to over four miles in length and average only a few hundred feet wide. They may be 20 or more feet thick and commonly have steep walls.

The head end of the bay is characterized by a decrease in exposed oyster reefs. The high rate of sedimentation in this area, due to the progradation of bayhead deltas, is unfavorable for oysters. Probes and cores in this area and on the deltaic plain reveal the presence of buried oyster reefs. These reefs may be transverse reefs or longitudinal reefs associated with old channels, but more commonly they represent pancake reefs.

Charts of the Texas lagoons indicate the presence of many oyster reefs. Several of these are sufficiently large to form extensive shell deposits. In general, the lagoonal reefs are located in higher salinities, are exposed to more oyster predators and have a lower rate of growth. They are also associated with sandy sediments.

FUTURE PROSPECTS OF THE SHELL DREDGING INDUSTRY

At present rates of consumption, shell deposits currently being exploited in bays convenient to Houston and Corpus Christi will be exhausted in a very few years. Dredging operations will then undoubtedly move into other bays such as Copano and San Antonio. The greater distances from deposit to plant will add to production costs, but shell will still be competitive with cement or lime transported overland to the Houston market.

The shifting of dredging operations to other less convenient bays will only postpone the problem for a few more years. What will happen after these reserves are also depleted? Exploration for other shell deposits should be concentrated on three potential areas.

Present dredging methods limit operations to a depth of less than 40 feet below sea level. Thus, shell can be processed only 30 feet below the sediment interface if the dredge is floating in 10 feet of water. It has been suggested that simple modification of the dredges to permit deeper cuts would solve the problem of the vanishing reserves. I doubt that this will be the case. The bays are not wells of shell. They are partially filled river valleys. Thus deeper dredge cuts would eventually penetrate alluvial or nonmarine sediments. Acoustical profiling methods are currently available to determine accurately the thickness of Recent sediments overlying the Pleistocene surface. Cores taken in conjunction with this program would delineate the thickness of estuarine facies capable of supporting oysters. The areas of thickest sediments filling the old river valleys are outlined on Figures 9A and 9B. Exploration for deeper deposits should be concentrated in these areas.

A second area for future exploration is in Pleistocene depositional systems. Bernard and LeBlanc (1965) and others have pointed out the remarkable similarity between the distribution of facies associated with the Ingleside Barrier and the present coastal physiographic elements. The Ingleside represents sediments deposited during the late Wisconsin at a time when sea level was approximately the same as it is now. A system of barrier islands and lagoons developed that are analogous to the present features.

Pleistocene depositional environments are easily mapped on aerial photographs due to the contrast in floras (Hayes and Scott, 1964b). Live oaks predominate on the well drained sands of the barrier island. Their dark tones are easily distinguished from the lighter mesquites that grow on the poorly drained lagoonal clays. Pleistocene deltaic plains are characterized by the presence of cultivated fields. Figures 10A and 10B show the areal distribution of Pleistocene Ingleside facies in the vicinity of San Antonio and Copano bays. Note the presence of a Pleistocene tidal pass near the mouth of Copano Bay. This north-northeast trending channel is presently occupied by St. Charles Bay and has the same alignment as the modern tidal passes Cedar Bayou and Aransas Pass.

One of the most promising areas for shell deposits is in the Ingleside lagoonal facies. As pointed out previously, reefs are common in similar areas today. This land is poorly drained and is generally not under cultivation.

A third area for future exploration is the coastal region near the Brazos and Colorado rivers. Bays are associated with all of the major rivers that cross the coastal plain of Texas (Fig. 1) with but three exceptions: the Rio Grande, Colorado and Brazos rivers. These rivers carry the heaviest sediment loads and have succeeded in filling their valleys by deltaic progradation. It is reasonable to assume, especially in the case of the Colorado and Brazos rivers, that their bays contained major oyster reefs prior to their being filled.

Since deltaic and alluvial facies obscure details of the underlying bay facies, a major exploration program would be necessary to determine the extent of buried shell deposits. Such a program should involve shallow seismic profiling, coring and detailed analysis of aerial photographs.

Potentially, shell deposits of the magnitude of those in Galveston, Trinity and Nueces bays underlie the alluvial and deltaic plains of the Colo-

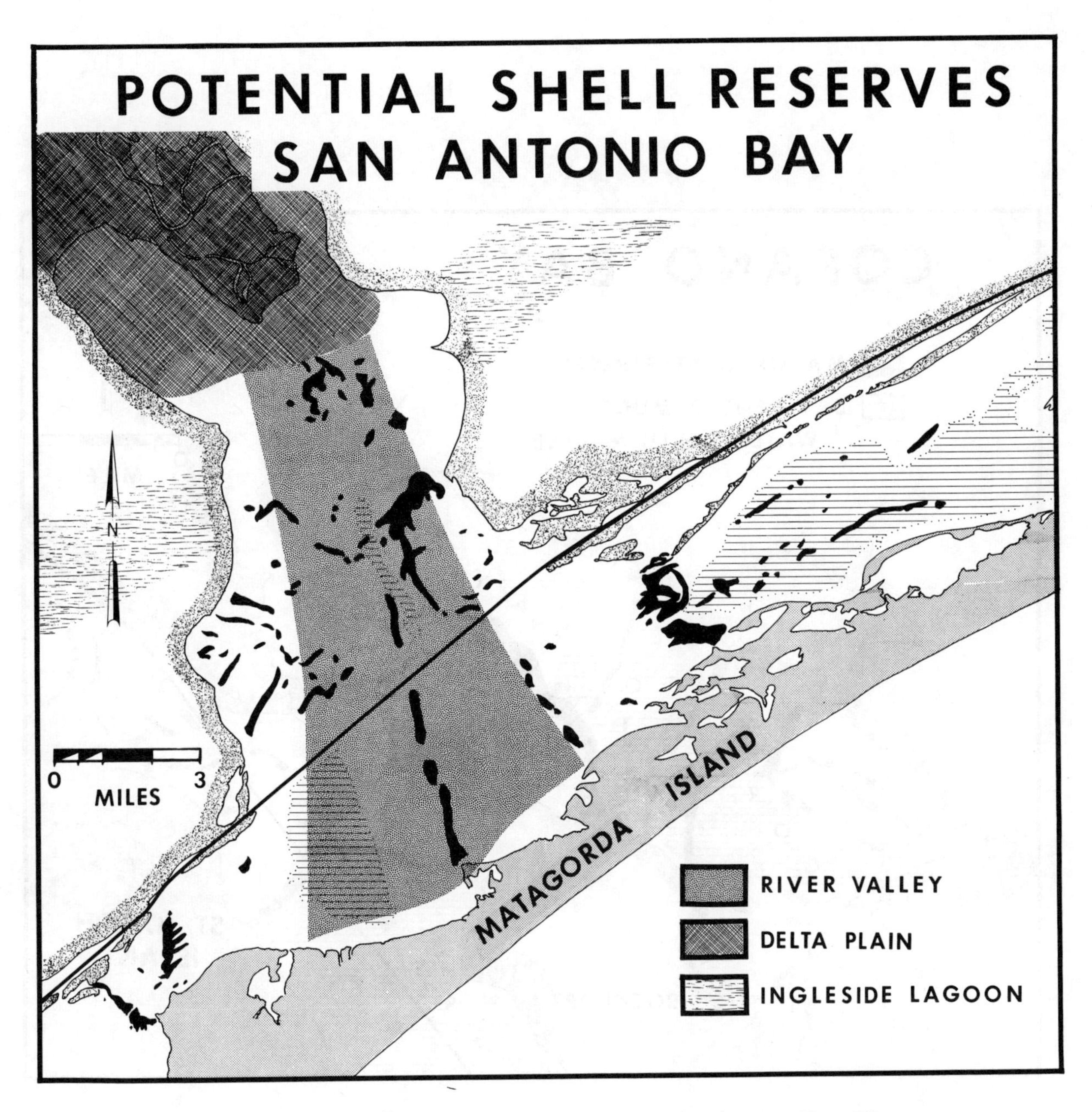

FIG. 9A. Areas of potential shell deposits, San Antonio Bay, Texas.

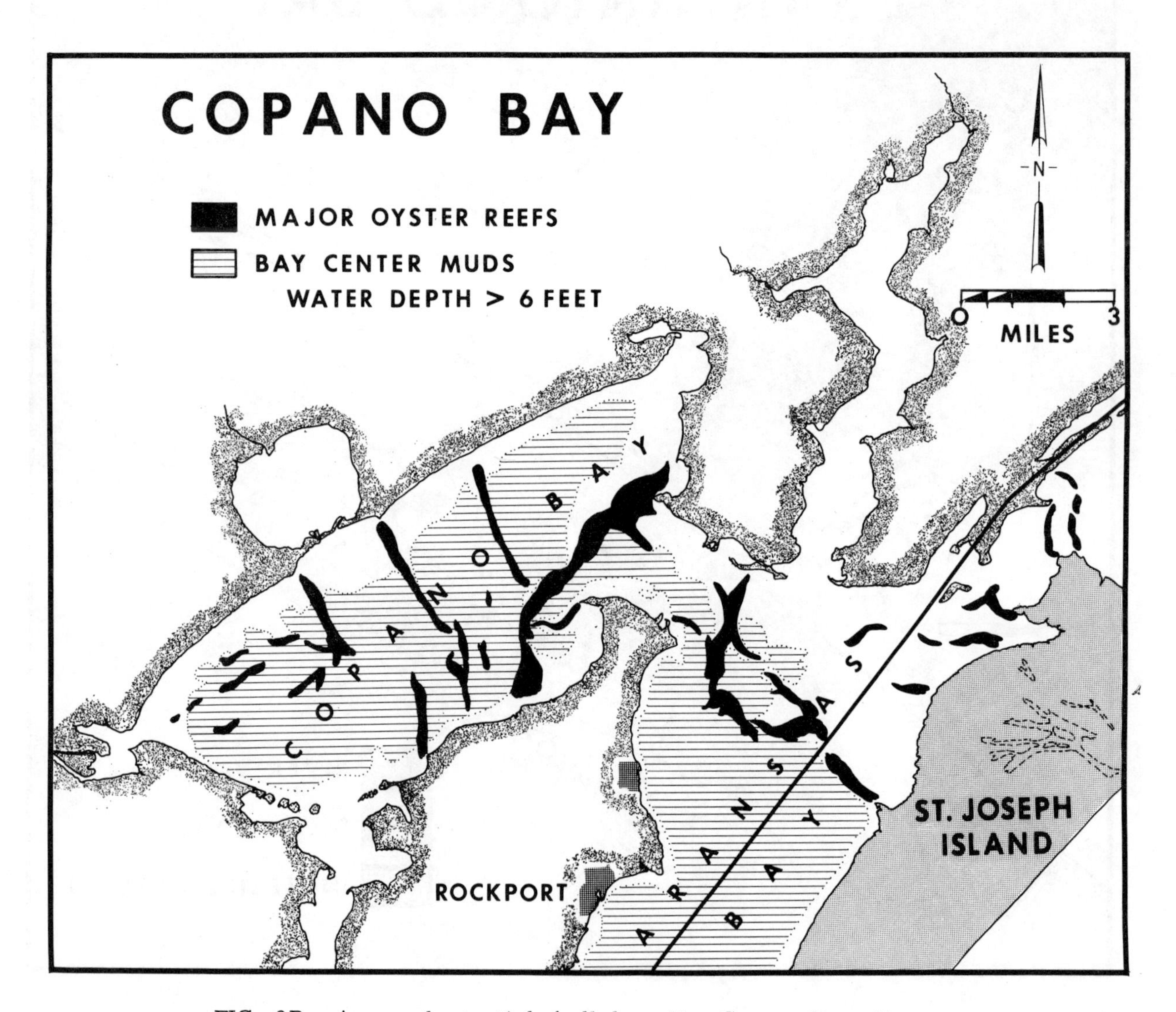

FIG. 9B. Areas of potential shell deposits, Copano Bay, Texas.

rado and Brazos rivers. This large complex lies only 40 to 50 miles south of the Houston metropolitan market. Costs of exploiting these deposits would be much higher than reefs currently being mined due to increased costs of land acquisition, transportation and production methods.

The mere presence of shell or any other mineral deposit does not necessarily mean that it is economically feasible or even desirable to develop it. Other political, social and economic factors play an important role. Oyster shell is an important natural resource along the entire Gulf Coast. An exploration program for additional reserves based upon a sound scientific basis should be undertaken to expand the rapidly diminishing reserves. Expense of this program would be repaid many times over by the discovery of additional shell deposits.

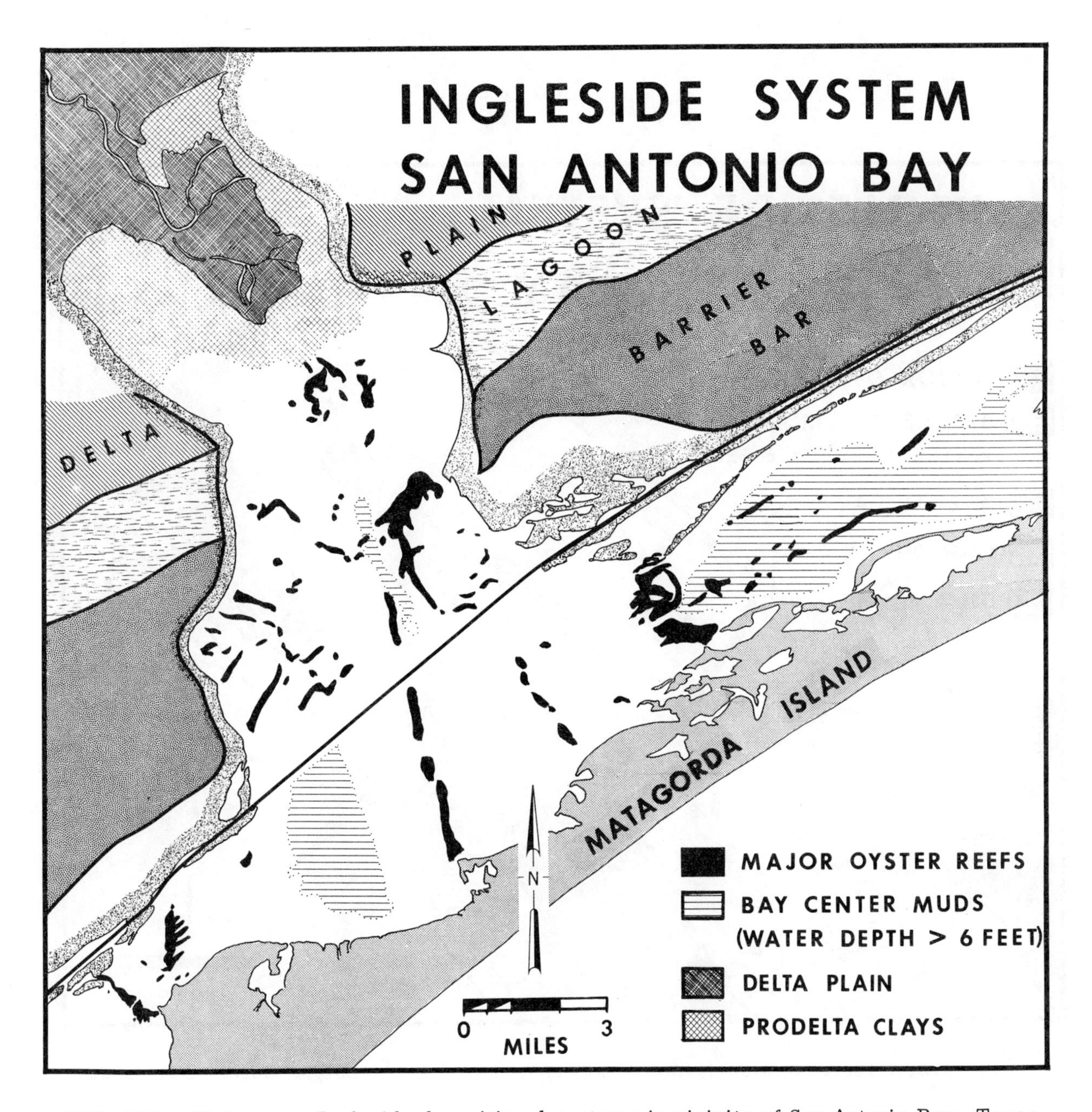

FIG. 10A. Pleistocene Ingleside depositional systems in vicinity of San Antonio Bay, Texas.

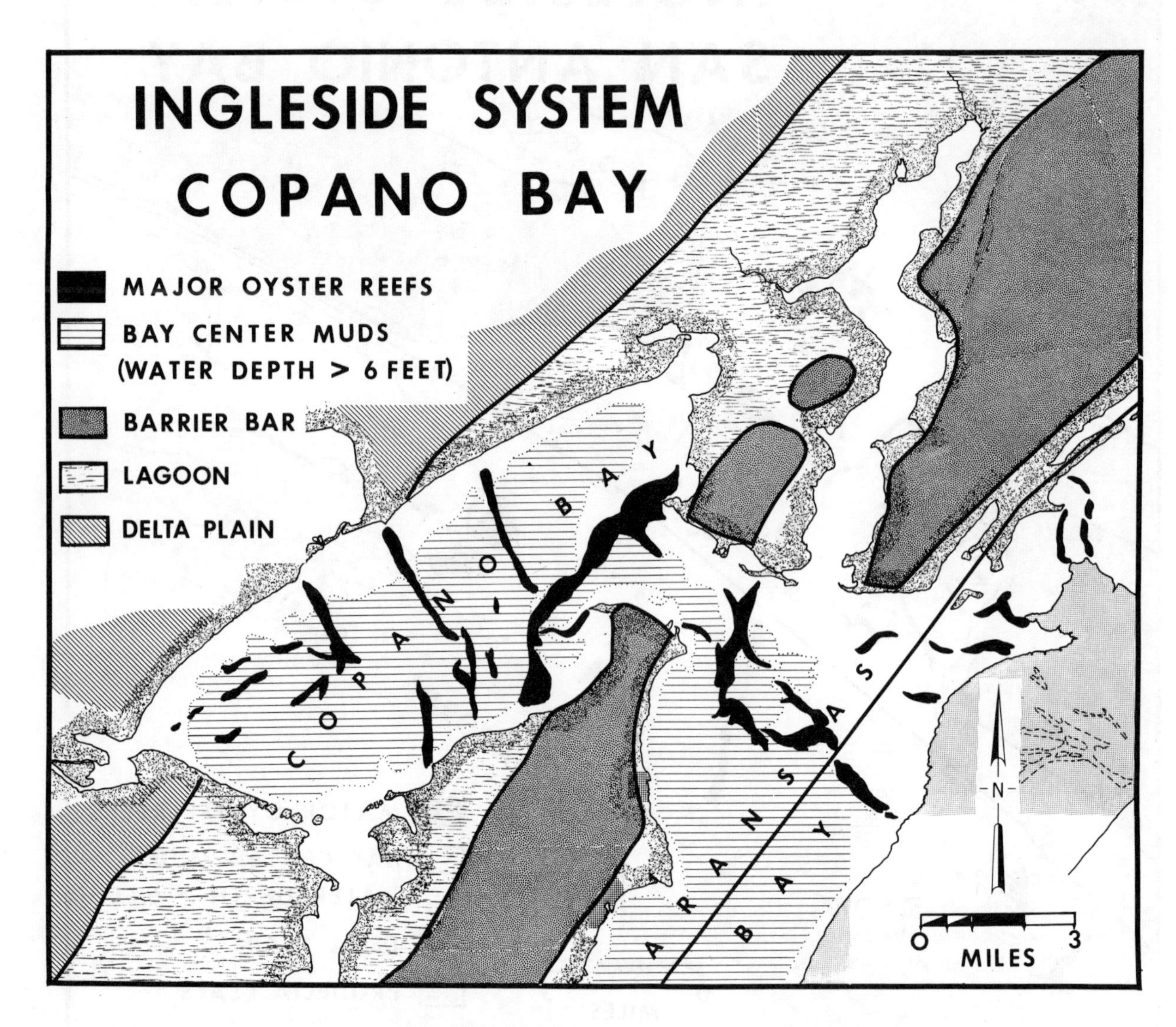

FIG. 10B. Pleistocene Ingleside depositional systems in vicinity of Copano Bay, Texas.

REFERENCES

Andrews, P. B. (1964) Serpulid reefs, Baffin Bay, southeast Texas, in Depositional environments, south-central Texas coast--Gulf Coast Assoc. Geol. Socs., Ann. Mtg., Field Trip Guidebook: Gulf Coast Assoc. Geol. Socs., Austin, Texas, pp. 102-120.

Behrens, E. W. (1967) Hurricane effect on a hypersaline bay system--a case study (abst.): Program of UNESCO, International Symposium on Coastal Lagoons, Mexico, D. F., Nov. 1967.

Bernard, H. A., and LeBlanc, R. J. (1965) Résumé of the Quaternary geology of the northwestern Gulf of Mexico province, in The Quaternary of the United States: Princeton Univ. Press, Princeton, New Jersey, pp. 137-185.

Butler, P. A. (1954) Summary of our knowledge of the oyster in the Gulf of Mexico, in Gulf of Mexico, its origin, waters, and marine life: [U. S.] Fish and Wildlife Service, Fishery Bull. 89, vol. 55, pp. 479-489.

Cary, L. R. (1906) The conditions for oyster culture in the waters of the parishes of Vermilion and Iberia, Louisiana: Bull. Gulf Biologic Sta., vol. 4, 27 pp.

Collier, A., and Hedgpeth, J. W. (1950) An introduction to the hydrography of the tidal waters of Texas: Inst. Marine Sci., vol. 1, no. 2, pp. 120-194.

Galtsoff, P. S. (1931) Survey of oyster bottoms in Texas: Bureau Fisheries Inv. Rept. No. 6, 30 pp.

__________ (1964) The American oyster Crassostrea virginica Gmelin: [U. S.] Fish and Wildlife Service, Fishery Bull., vol. 64, 480 pp.

Grave, C. (1901) The oyster reefs of North Carolina, a geological and economic study: Johns Hopkins Univ. Circ. No. 151, 9 pp.

__________ (1905) Investigations for the promotion of the oyster industry of North Carolina: Rept. U. S. Fish Comm., 1903, pp. 247-341.

Hayes, M. O. (1965) Sedimentation on a semiarid, wave-dominated coast (South Texas); with emphasis on hurricane effects: Univ. Texas at Austin, Ph. D. dissertation, 350 pp.

__________, and Scott, A. J. (1964a) Environmental complexes South Texas coast: Gulf Coast Assoc. Geol. Socs. Trans., vol. 14, pp. 237-240.

__________, and __________ (1964b) Pleistocene Ingleside barrier complex, in Depositional environments of south-central Texas: Gulf Coast Assoc. Geol. Socs. Field Trip Guidebook, Oct. 30-31, 1964, pp. 63-64.

Hedgpeth, J. W. (1954) Bottom communities of the Gulf of Mexico, in Gulf of Mexico, its origin, waters, and marine life: [U. S.] Fish and Wildlife Service, Fishery Bull. 89, vol. 55, pp. 203-214.

Kerr, Alex (1968) The Texas reef shell industry: Univ. Texas at Austin, Bur. Business Research, Texas Industry Series No. 11, 80 pp.

LeBlanc, R. J., and Hodgson, W. D. (1959) Origin and development of the Texas shoreline: 2d Coastal Geography Conf., April 6-9, 1959, Louisiana State Univ., pp. 57-101.

Lohse, E. A. (1955) Dynamic geology of the modern coastal region, northwest Gulf of Mexico, in Finding ancient shorelines--a symposium: Soc. Econ. Paleontologists and Mineralogists Special Pub., No. 3, pp. 99-103.

Moore, H. F. (1899) Report on the oyster beds of Louisiana: U. S. Fish Comm., Rept. no. 24, pp. 45-100+.

__________ (1907) Survey of oyster bottoms in Matagorda Bay, Texas: U. S. Bur. Fish. Doc. 610, 86 pp.

__________ (1913a) Condition and extent of the natural oyster beds and barren bottoms of Mississippi Sound, Alabama: U. S. Bur. Fish. Doc. 769, 61 pp.

__________ (1913b) Condition and extent of the natural oyster beds and barren bottoms of Mississippi east of Biloxi: U. S. Bur. Fish. Doc. 774, 41 pp.

__________, and Danglade, E. (1915) Condition and extent of the natural oyster beds and barren bottoms of Lavaca Bay, Texas: U. S. Bur. Fish. Doc. 809, 45 pp.

Norris, R. M. (1953) Buried oyster reefs in some Texas bays: Jour. Paleontology, vol. 27, no. 4, pp. 569-576.

Parker, R. H. (1955) Changes in the invertebrate fauna apparently attributable to salinity changes in the bays of central Texas: Jour. Paleontology, vol. 29, no. 2, pp. 193-211.

__________ (1959) Macro-invertebrate assemblages of central Texas coastal bays and Laguna Madre: Bull. Amer. Assoc. Petr. Geol., vol. 43, no. 9, pp. 2100-2166.

Phleger, F. B., and Lankford, R. R. (1957) Seasonal occurrences of living benthonic foraminifera in some Texas bays: Cushman Found. Foram. Research Contr., vol. 8, pt. 3, pp. 93-105.

Price, W. A. (1947) Equilibrium of form and forces in tidal basins of coast of Texas and Louisiana: Bull. Amer. Assoc. Petr. Geol., vol. 31, pp. 1619-1663.

__________ (1954) Oyster reefs of the Gulf of Mexico, in Gulf of Mexico, its origin, waters, and marine life: [U. S.] Fish and Wildlife Service, Fishery Bull. 89, vol. 55, p. 491.

Scott, A. J. (1964) Indian Point, discussions of field trip stops, in Depositional environments of south-central Texas coast: Gulf Coast Assoc. Geol. Socs. Field Trip Guidebook, Oct. 30-31, 1964, pp. 43-52.

Shepard, F. P., and Moore, D. G. (1955) Central Texas coast sedimentation: characteristics of sedimentary environment, Recent history and diagenesis: Bull. Amer. Assoc. Petr. Geol., vol. 39, no. 8, pp. 1463-1593.

__________, and __________ (1960) Bays of central Texas coast, in Recent sediments, northwest Gulf of Mexico: Amer. Assoc. Petr. Geol., pp. 117-152.

Watson, R. L. (1968) Origin of shell beaches, Padre Island, Texas: Univ. Texas at Austin, Master's thesis, 121 pp.

GRAVEL DEPOSITS OF THE MINNEAPOLIS QUADRANGLE, MINNESOTA

R. K. Hogberg
Minnesota Geological Survey
The University of Minnesota
Minneapolis, Minnesota

ABSTRACT

The position of surficial glaciofluvial gravel deposits within the Minneapolis 15-minute quadrangle was controlled by the prominent St. Croix Moraine. Beginning approximately 13,000 years ago, the Grantsburg Sublobe, flowing from the southwest, overrode morainal topography left by the Superior Lobe (>13,700 - <40,000 B. P.), which had advanced into the Minneapolis lowland from the northeast and formed the St. Croix Moraine at its terminal position. During the retreat of Grantsburg ice, three types of glaciofluvial gravel deposits were formed--braided outwash fan, crevasse, and valley train.

Braided outwash-fan bodies were formed near the leading edge of Grantsburg recessional moraines and in topographic lows within or on the flanks of the buried St. Croix Moraine. Crevasse bodies were deposited within fractures developed by tensional forces in the stagnating ice at the crest or on lee slopes of the buried moraine. Gravel-bearing valley trains accumulated adjacent to a deep preglacial valley now occupied by the Minnesota River. Valley train bodies were formed by late Grantsburg meltwater rivers during the breaching of the St. Croix Moraine. In early post-glacial time most river terrace bodies were formed from the lagged gravels of eroded moraine.

Lithologic analyses of -5/8, +3/8 rock particles from the four types of deposits indicate that pre-Grantsburg regolith was a source for a substantial amount of the gravel. Events of late glacial and post-glacial time had little effect on the deposits.

INTRODUCTION

Gravel-bearing sands accumulated within certain restricted depositional environments during the Pleistocene Epoch. Knowledge of the present day position of these depositional environments, gained from studies of regional and local stratigraphic controls, can aid appreciably in the search for commercial gravels.

This approach was used in a study of the 15-minute Minneapolis quadrangle, within the Twin Cities metropolitan area of southeastern Minnesota (Fig. 1). Approximately 217 square miles of land area was mapped; the area includes most of the southwestern urban and suburban Twin Cities, in which about one-half million people live.

The glacial deposits, which contain the gravel bodies, were mapped on 7-1/2-minute topographic base maps. Upon completion of field mapping, detailed field investigations were carried out. This work included: (1) measurement of pebble orientations in till to determine ice-flow directions and (2) detailed examination of the stratigraphy of individual gravel bodies to learn their relative position, size, and sequence within the pile of glacial drift. Laboratory procedures consisted of clay mineral and lithologic analyses to determine the provenance mixture. An estimate of gravel quality was made from an analysis of pebble lithologies.

Regional and local geologic relationships will be discussed first. Later, I will describe the gravel-bearing sands within the limits of generalized stratigraphic models.

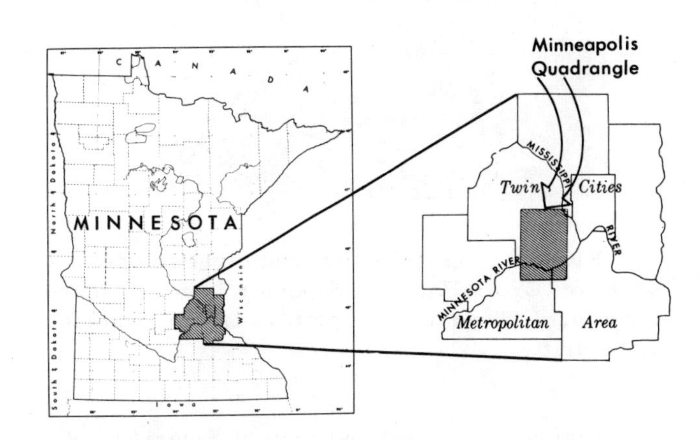

FIG. 1. Map showing location of Minneapolis quadrangle.

BEDROCK GEOLOGY

Glacial deposits ranging in thickness from 100 to 450 feet cover the Paleozoic bedrock surface of the Twin Cities metropolitan area. The direction of glacier ice flow was controlled partially by pre-glacial bedrock topography, which is dominated by valleys, now buried, that are 50 to 450 feet deep and 0.25 to 1.25 miles wide. One of these valley complexes--now beneath the Minnesota River valley--had a profound effect on the deposition of commercial gravels.

GLACIAL GEOLOGY

Wisconsin glaciation of the Pleistocene Epoch in the Twin Cities area is represented by exposed drift from two major ice advances--Superior Lobe and the Grantsburg Sublobe (Wright and Ruhe, 1965). Both Superior and Grantsburg deposits contain a considerable amount of rock material from earlier Pleistocene glaciation.

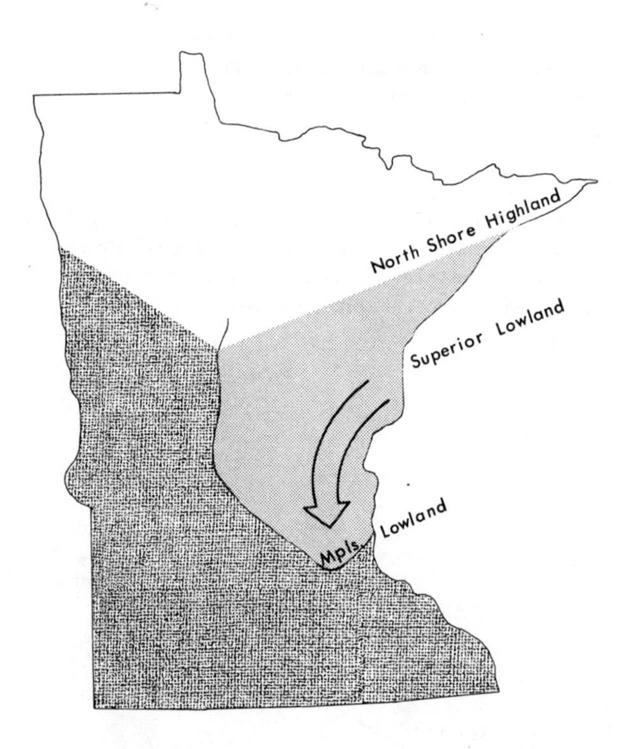

FIG. 2. Regional extent and direction of flow of the St. Croix Phase of Superior Lobe, Wisconsin Epoch. The dark gray pattern is pre-Superior Lobe drift.

Figure 2 is a regional map of Superior Lobe glaciation. The open arrow indicates direction of ice flow from the Superior lowland southwesterly into the Minneapolis lowland. Terminal position of the Superior Lobe is marked by the St. Croix Moraine, which within the Minneapolis quadrangle ranges in width from 5 to 10 miles and in height from 150 to 300 feet. Most of the Superior Lobe drift is very sandy, but clay- and silt-size particles make up more than 50 percent of some exposures. The St. Croix Moraine is breached in the northeastern corner of the quadrangle by the Mississippi River, and in the southern part by the early Minnesota River (Fig. 1).

Figure 3 is a regional map of the Des Moines Lobe glaciation, the last Wisconsin ice advance in Minnesota. The Grantsburg Sublobe ice sheet flowed northeasterly into the Minneapolis lowland, beginning about 12,700 B.P. (Wright and Ruhe, 1965). The surficial bodies, described in this paper, are confined to the Grantsburg drift.

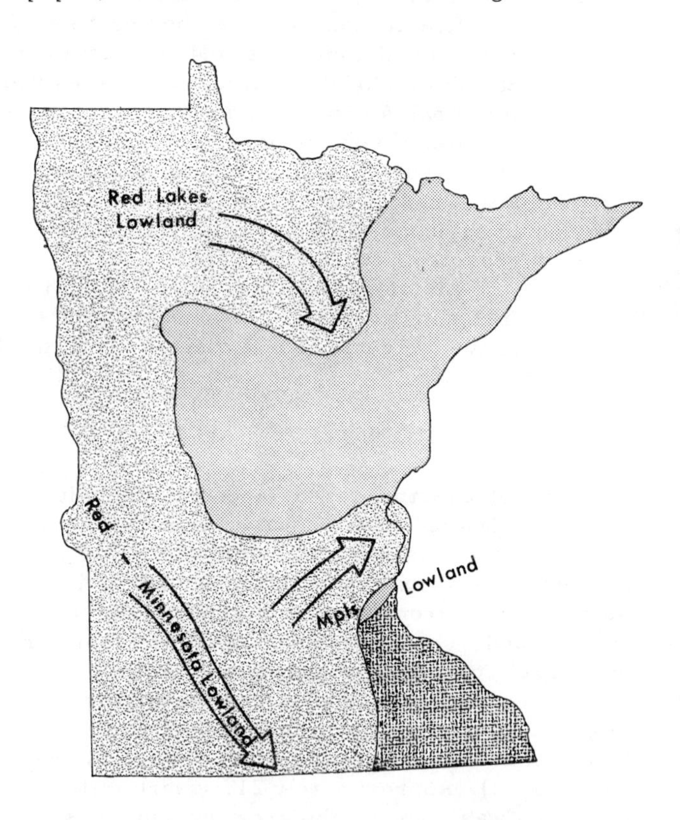

FIG. 3. Regional extent and direction of flow of the Grantsburg Sublobe of the Des Moines Lobe.

Grantsburg ice overrode deposits left by the Superior Lobe, which had withdrawn from the Minneapolis lowland approximately 1,000 years before (Wright and Ruhe, 1965). Erosion was minor during the short interstadial period between the two ice advances, as indicated dramatically by scattered surficial deposits of the earlier Superior Lobe drift. These Superior Lobe deposits were still standing as scattered, ice-defended, nunatak moraines when the Grantsburg ice sheet overrode the Minneapolis lowland.

Figure 4 is a generalized map of surficial deposits of the Minneapolis quadrangle. The map, compiled from field data, shows the location and extent of the units of Grantsburg drift and two small exposed areas of Superior Lobe drift (in black) that were sufficiently large to be shown at the map scale. The small solid arrows depict probable mean directions of Grantsburg ice flow. Near its terminal position in the southern part of the Minneapolis quadrangle, ice flowed toward a deep, preglacial, bedrock valley which underlies the present-day Minnesota River valley (Sloan and Austin, 1966). During retreat from its terminal position, Grantsburg ice formed several recessional moraines which are delineated on Figure 4 by a series of east- to southeast-trending exposures of till. The short, south to southwest-trending lines represent crevasse ridges--one of the types of gravel bodies--that are fillings of fractures developed in the stagnating ice.

Gravel bodies deposited by meltwaters from the retreating Grantsburg Sublobe ice can be fitted into one of four generalized stratigraphic models: (1) braided outwash fan, (2) crevasse, (3) valley train, and (4) lagged bodies. However, none of the gravel bodies fits perfectly into any one specific stratigraphic model.

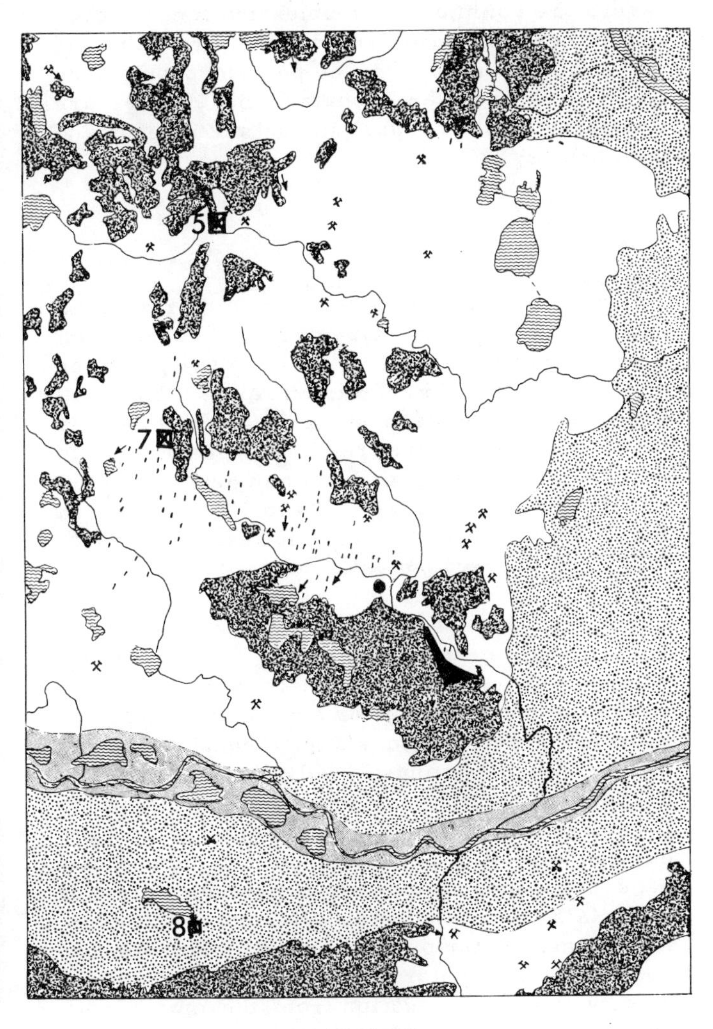

EXPLANATION

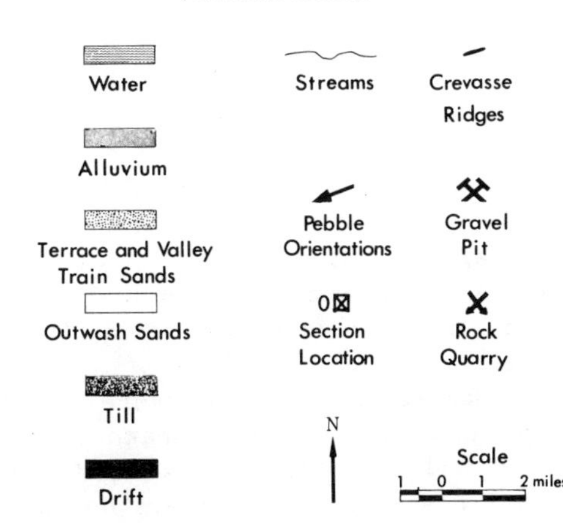

FIG. 4. Surficial geologic map of the Minneapolis quadrangle, Minnesota.

GLACIAL HISTORY OF THE MINNESOTA RIVER VALLEY

To observe the glacial history of two of the types of gravel bodies, let us look through the "window" provided by the Minnesota River valley (Fig. 1). During Grantsburg glaciation, meltwater rivers choked with ice blocks carried large volumes of gravel-size material that accumulated in lens-shaped bodies; bases of initial gravel bodies rested directly on the bedrock shoulder of the deep valley. These valley-train gravel bodies were formed while meltwater rivers were breaching the St. Croix Moraine downstream.

Major post-glacial events in the Minnesota River valley were: (1) deposition of an alluvial veneer upon the earlier valley train deposits, and (2) erosion of glacial River Warren--the river that drained the famous Lake Agassiz, which occupied a basin that included parts of northwestern Minnesota, northeastern North Dakota, and Saskatchewan, Manitoba, and Ontario, Canada (Matsch and Wright, 1967). Glacial River Warren eroded its channel down to bedrock and eventually migrated northward and downward into unconsolidated sediments that filled the deep preglacial bedrock valley mentioned earlier. During downcutting, coarse glacial materials remaining from the erosion of St. Croix Moraine were reworked by River Warren and redeposited on the bedrock surface as lagged gravel bodies. When drainage of Lake Agassiz shifted from south to north, the early Minnesota River--the underfit stream occupying the former River Warren channel--became an aggrading stream. Deltas formed within the channel when the volume of tributary stream flow became greater than that of the trunk stream. Thus, the channel was filled to the present flood plain level in post-River Warren time.

GRAVEL BODY MODELS

With this background on the general geologic setting, I will now describe depositional environments of three types of gravel bodies within limits of generalized stratigraphic models.

Braided Outwash Fan Bodies

Braided outwash fans are the most abundant and, therefore, the most important of the gravel bodies within the Minneapolis quadrangle. The bodies studied are located near the leading edge--ice contact--of the Grantsburg recessional moraines (Fig. 4) within topographic lows or on the flanks of the buried St. Croix Moraine. The gravel bodies are lens- to wedge-shaped, as much as 150 feet thick, 200 to 1,320 feet wide, and 300 feet to a mile long.

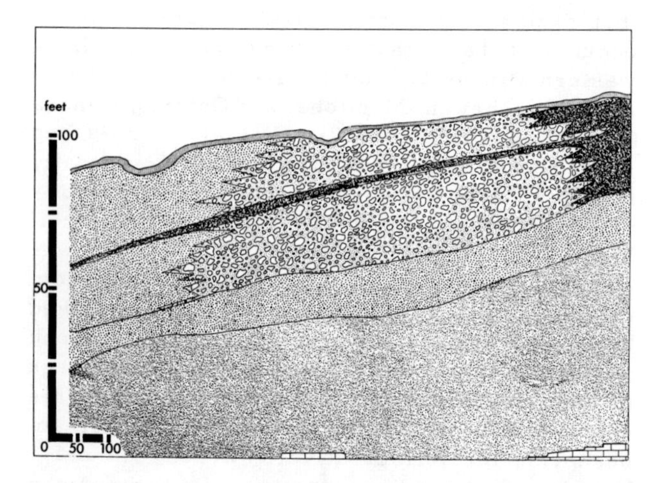

FIG. 5. Generalized stratigraphic section of a braided outwash-fan body. See Figure 6 for legend.

Figure 5 is a longitudinal section through a braided outwash fan--a model generalizing the field observations of this type of gravel body.

The stratified layering of the body shows that there were repetitive changes in the competence of meltwater streams. Most of the body is composed of sand-size particles in which pebble-rich layers are common and cobble-rich layers rare, except near the basal and upper parts. Pit-run materials average 25 to 40 percent pebble-size (0.187 inch) and larger particles. Pebbles are composed of about equal amounts of igneous and sedimentary rocks.

Crevasse Bodies

Figure 6 shows the topographic form of the second, or crevasse, type of gravel body, which formed near the crest or at lee slope positions above the buried St. Croix Moraine. This map shows an area of stagnant ice features, the most prominent of which are crevasse ridges and kettle lakes. Crevasse ridges range in height from 50 to 125 feet, in width from 200 to 500 feet, and in length from 500 feet to 1-1/2 miles. The nearly level tops of crevasse ridges were inherited from their deposition within ice walled drainageways, which in most cases drained ephemeral lakes. Each of the stratigraphic units was accordant through the ice-walled channels with the elevation of the upstream lake or other glacial deposit. In detail, this type of body (Fig. 7) consists mostly of layers of silty sands; pebbly sands occur near the base. Flow rates of the drainage stream were low; therefore, the waters for the most part carried fine-size particles. A thin layer of till caps the deposits, which indicates slump of materials from adjacent ice that stood at higher elevations. Later overriding by glacier ice accounts for the contortion of the upper sequence of silty sand. Information from the model indicates that this type of crevasse body has a very low gravel potential.

Valley Train Bodies

Figure 8 is a section through one of the valley train deposits in the western part of the Minnesota River valley (Fig. 1). This stratigraphic model also has a capping of till--residual material melted from a post-gravel advance of the Grantsburg ice sheet.

Three separate gravel lenses were deposited by point-bar migrations of the meandering ice-block choked meltwater rivers issuing from the front of the ice sheet. Valley train bodies are less stratified--had more uniform stream flow volumes--than braided outwash-fan bodies. The valley train deposits also contain coarser rock

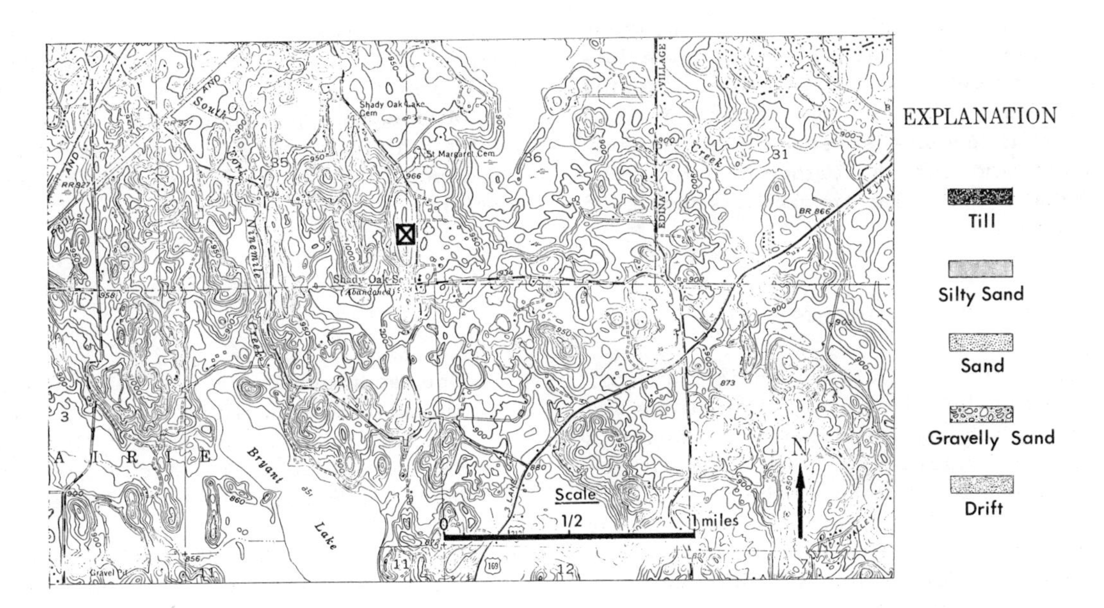

FIG. 6. Topographic map of stagnant ice terrane; ridges are ice crevasse features (taken from U. S. Geological Survey, Hopkins 7-1/2-minute quadrangle).

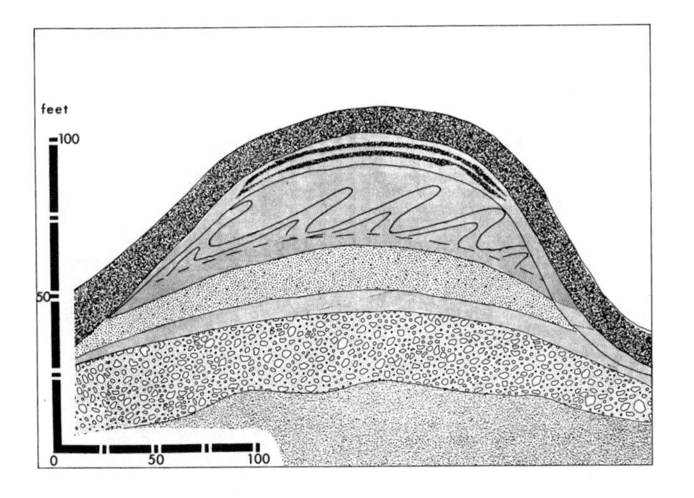

FIG. 7. Generalized stratigraphic section of a crevasse body. See Figure 6 for legend.

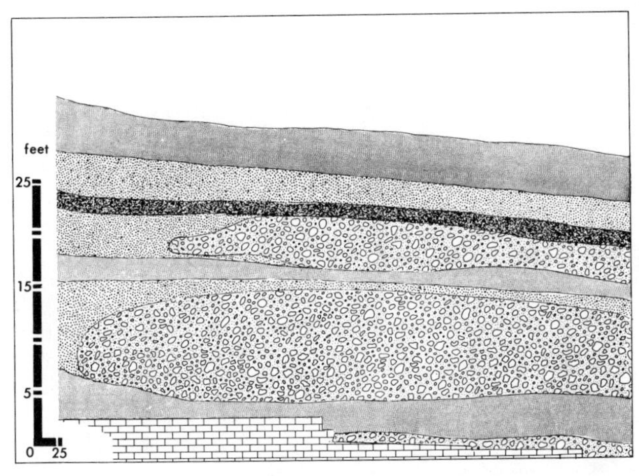

FIG. 8. Generalized stratigraphic section of a valley train body. See Figure 6 for legend.

particles than the braided outwash fans because meltwater rivers had coarser source materials and greater competence than meltwater streams of outwash fans. The average particle size of the valley train deposits decreases in an easterly direction outward from the former ice front. The model indicates that valley train bodies have good gravel potential.

CONCLUSIONS

In conclusion, surficial geologic mapping, supplemented by detailed field examinations and laboratory studies, indicates that gravel deposits in the Minneapolis area were formed by glaciofluvial processes associated with the recession of the Grantsburg Sublobe ice. Positions of the gravel bodies were controlled mostly by the nature of preadvance surface, which was dominated by the St. Croix Moraine. Gravel occurrences can be fitted arbitrarily into one of four stratigraphic models. Braided outwash-fan bodies are the most important and abundant; valley train and lagged bodies of the Minnesota River valley are less abundant, but nevertheless important deposits. Crevasse deposits are abundant in restricted areas and have some importance.

I would suggest from results of this work in the Minneapolis quadrangle that studies of both geologic history and glacial processes should be combined in the search for commercial glacial gravels.

REFERENCES

Matsch, C. L., and Wright, H. E., Jr. (1967) The southern outlet of Lake Agassiz, in Mayer-Oakes, William J., ed., Life, Land and water: Proceed. Conf. Environmental Studies of Glacial Lake Agassiz region, Univ. Manitoba Press, Winnipeg, Canada, pp. 121-140.

Sloan, R. E., and Austin, G. S. (1966) St. Paul Sheet: Minnesota Geol. Survey, Geol. Map of Minnesota.

Wright, H. E., Jr., and Ruhe, R. V. (1965) Glaciation of Minnesota and Iowa, in Wright, H. E., Jr. and Frey, D. G., eds., The Quaternary of the United States: Princeton Univ. Press, Princeton, N. J., pp. 29-41.

UTILIZATION OF DEPOSITIONAL MODELS IN EXPLORATION FOR NONMETALLIC MINERALS

J. H. McGowen
Bureau of Economic Geology
The University of Texas at Austin
Austin, Texas

ABSTRACT

Prospecting for specific clay, lignite, and carbonate deposits is facilitated by recognition that these mineral deposits are components of particular depositional systems. The use of three dimensional facies analysis of genetically related ancient sediments indicates areas favorable for mineral exploration. Two examples are considered--a terrigenous clastic sequence containing lignite and ceramic clay, and a carbonate sequence containing dolomite and high-purity limestone.

The Wilcox Group of Texas (Lower Eocene) is a thick terrigenous sequence that accumulated in fluvial and deltaic depositional environments and in laterally associated strandplain, barrier bar-lagoonal, and shelf environments. The lower Wilcox fluvial system consists of three distinct, contemporaneous facies. Elongate lignite deposits occur in two of these facies. Downdip from the fluvial facies relatively widespread lignites were deposited in marshes and swamps on the delta plain. Quality of lignite is different in these two depositional systems.

Ceramic clays with high kaolinite content are common to all three Wilcox fluvial facies. One of the facies is characterized by widespread occurrence of sand-kaolin. In the other two fluvial facies, clays with high kaolinite content occur commonly associated with channel sands.

The Edwards Formation (Lower Cretaceous) of Texas, a platform carbonate deposit, is comprised of facies analogous to some platform carbonates of the modern Bahamas. Four broad facies characterize the Edwards. These are from the seaward edge to platform center: (1) basinal facies, (2) reef-grainstone facies, (3) lagoonal facies, and (4) platform facies. In this model the reef-lagoonal couple controlled both initial carbonate deposition and subsequent diagenetic processes which produced two distinct dolomite facies.

INTRODUCTION

The purpose of the depositional model in mineral exploration is to relate mineral occurrence to a depositional framework with which its relationship to other facies becomes apparent. By recognizing similar facies relationships in other sedimentary sequences, areas where specific potential mineral deposits should occur can be predicted. A realistic approach to exploration for many industrial minerals demands a basic understanding of conditions, both primary and diagenetic, that controlled distribution of facies of which these deposits are a part.

Regional three-dimensional studies of the Wilcox Group (Eocene) of Texas, a thick sequence of terrigenous sediments (Fisher and McGowen, 1967), and the Edwards Formation (Lower Cretaceous) of Texas, composed predominantly of limestone and dolomite (Fisher and Rodda, 1967) have provided depositional and diagenetic models for the occurrence of kaolinite, lignite, dolomite, and chert. These models could be constructed only after recognition of a larger genetic unit, the depositional system, of which models of component genetic facies, recognized in terms of modern analogues, are a part. Recognition and delineation of depositional systems are fundamental.

Lignite and kaolinite deposits are component facies of specific depositional systems which comprise the Wilcox Group; diagenetic dolomite and chert facies are closely associated with specific depositional facies comprising the Edwards Formation. The following discussion concerns specifically fluvial, delta, and bay-lagoon systems of the Wilcox Group and a platform carbonate system of the Edwards Formation.

LOWER WILCOX DEPOSITIONAL SYSTEMS

Within the lower part of the Wilcox, Fisher and McGowen (1967) delineated seven principal systems (Fig. 1). Of special concern here are three systems interpreted as delta, fluvial, and lagoon. The dominant depositional element, a delta system comprised of mud, sand, and carbo-

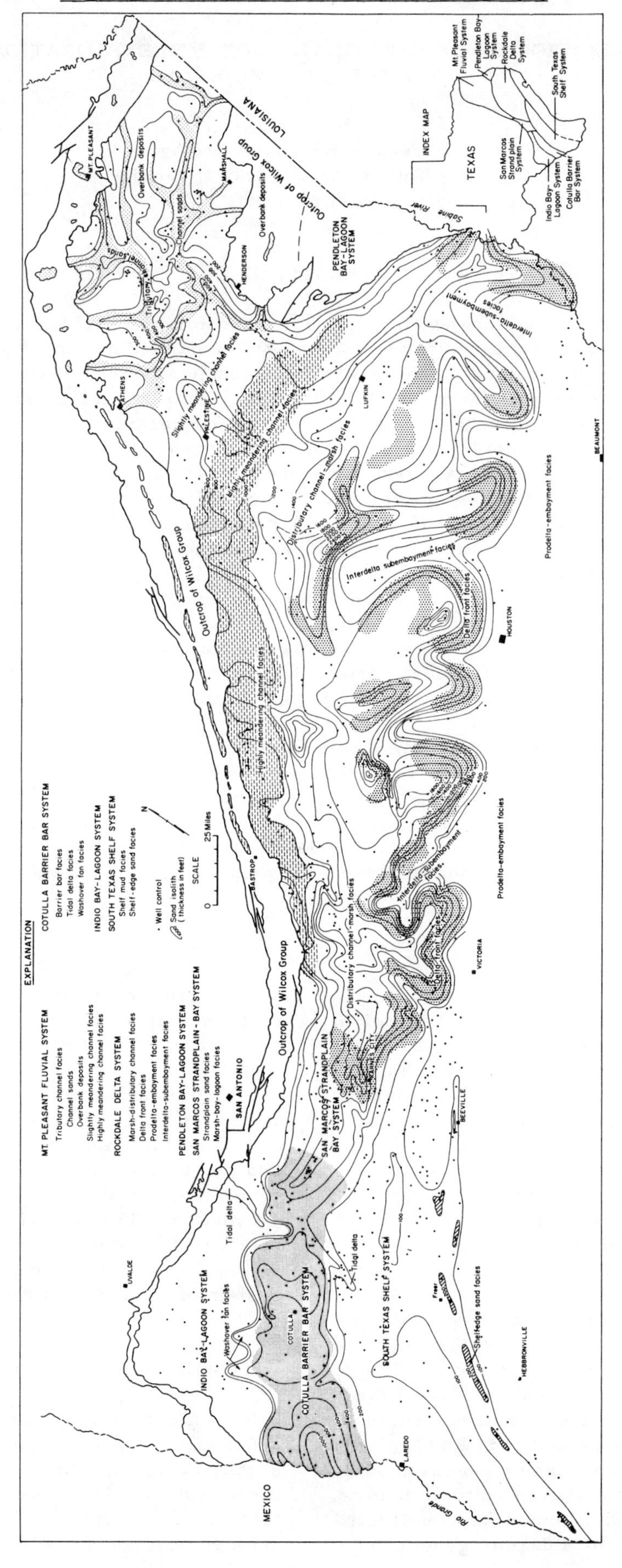

FIG. 1. Principal depositional systems of the lower Wilcox Group, Texas.

naceous deposits (chiefly lignite) of several delta progradations is up to 5,000 feet thick. It grades up paleoslope into a thinner fluvial system. Southwestward transport of sand from the delta system and deposition within embayments marginal to the delta resulted in distinct depositional systems lateral to the delta system. A strandplain developed just west of the delta system with barrier bar and complementary bay-lagoon systems occurring farther along strike to the west.

Three systems with which either kaolinite or lignite are associated are the fluvial system, delta system, and the bay-lagoon system. The fluvial system includes most of the Wilcox in the outcrop belt north of the Colorado River and over the northern half of the Sabine Uplift. Three facies within the fluvial system are interpreted as tributary, slightly meandering, and highly meandering channel facies, based on sand body geometry, sand isolith trend, sequence of sedimentary structures, relative abundance of channel and overbank sediment, and textural properties.

Wilcox Fluvial Systems: Kaolinite and Lignite

Kaolinite is a common associate of fluvial sandstone bodies, either as a lateral deposit genetically related to the fluvial sediment or as terrigenous grains, authigenic minerals, and replacement products contained within the sandstone. Feldspar and muscovite are commonly altered to kaolinite; Weaver (1960) stated that clay minerals in sandstones are more apt to be altered after deposition than clays in other rocks.

Fluvial sandstone, lignite, and kaolinite associations have been reported in Paleozoic, Mesozoic, and Tertiary rocks. In many of these associated facies, kaolinite content of the sandstone is high though laterally contemporaneous shales commonly have considerably lower kaolinite content. Depositional environments in which authigenic kaolinite occurs have been reported as fluvial (Potter and Siever, 1956) and fluviatile or fluviolacustrine (Glass, 1956) for Pennsylvanian sandstone bodies in Illinois; fluvial, deltaic, and marine for the kaolinite-bearing Cretaceous Tuscaloosa Group of northwest Alabama (Clark, 1964); and deltaic plain and associated environments for the Tuscaloosa of Georgia (Kesler, 1956). Similar fluvial-deltaic-lagoonal depositional environments have been interpreted for the kaolinitic lower Ione Formation (Eocene) of California (Pask and Turner, 1952), and Wilcox Group of Texas (Fisher et al., 1965; Fisher and McGowen, 1967; and Kohls, 1967).

All these reported kaolinitic deposits have fluvial components in common. Kaolinite in these facies is reported to have formed postdepositionally as an authigenic mineral or by replacement of feldspar and muscovite.

Wilcox Component Facies

Tributary channel facies.--In northeast Texas lower Wilcox sandstones occur in definite elongate bodies; isolith pattern shows an overall tributary plan (Fig. 2). Sand comprises about 40 percent, and mud and clay about 60 percent of the total facies. Individual sandstone bodies are up to 80 feet thick. Lower parts of the channel fill are conglomeratic and trough cross-bedded; these grade upward into wedge set cross-beds. These channels cut into underlying and laterally adjacent laminated muds and silts. Woody lignites are common within this facies, and occur mainly as elongate deposits that apparently accumulated in swamps developed in abandoned channel courses. Lignites of commercial thickness are uncommon in interchannel areas. Clays within and immediately adjacent to channel deposits are chiefly kaolinitic, whereas clays of nonchannel overbank deposits are mostly illite, chlorite, and montmorillonite (Fisher et al., 1965). Ceramic clay plants and abandoned lignite mines, as a consequence of channel-sand and kaolinite-lignite association, are located along the outcrop adjacent to these local Wilcox channel sand deposits.

Slightly meandering channel facies.--Down depositional slope, a second fluvial facies is encountered; sand isolith pattern is less dendritic than for the tributary facies (Fig. 2). This pattern suggests that channels of this part of the fluvial system were slightly more meandering than tributary channels. Sand makes up about 60 percent of the facies. Primary sedimentary structures and relationships of channel-fill sand, kaolinite, and lignite are similar within tributary and slightly meandering facies.

Highly meandering facies.--The most downdip fluvial facies is interpreted to be a highly meandering channel facies. This facies is the so-called sand-kaolin belt, or Simsboro Formation, of outcrop. Kaolinite occurs in genetically related deposits lateral to channel-fill sandstone, and within sandstones as mud clasts, authigenic minerals, and replacement products. Kaolinite in this facies is a result of both depositional and diagenetic conditions related to the primary depositional facies. Outcrop of this kaolin-rich facies extends from northern Freestone County southward to about central Bastrop County, and thickness ranges from 10 to about 400 feet. Sandstone comprises about 90 percent of the facies. Meandering channels laid down continuous sands in belts up to 30 miles wide. Sand isoliths are relatively uniform, reflecting lateral persistence of the units.

Sedimentary Structures

Types of sedimentary structures.--Interpretation based on vertical and lateral succession of

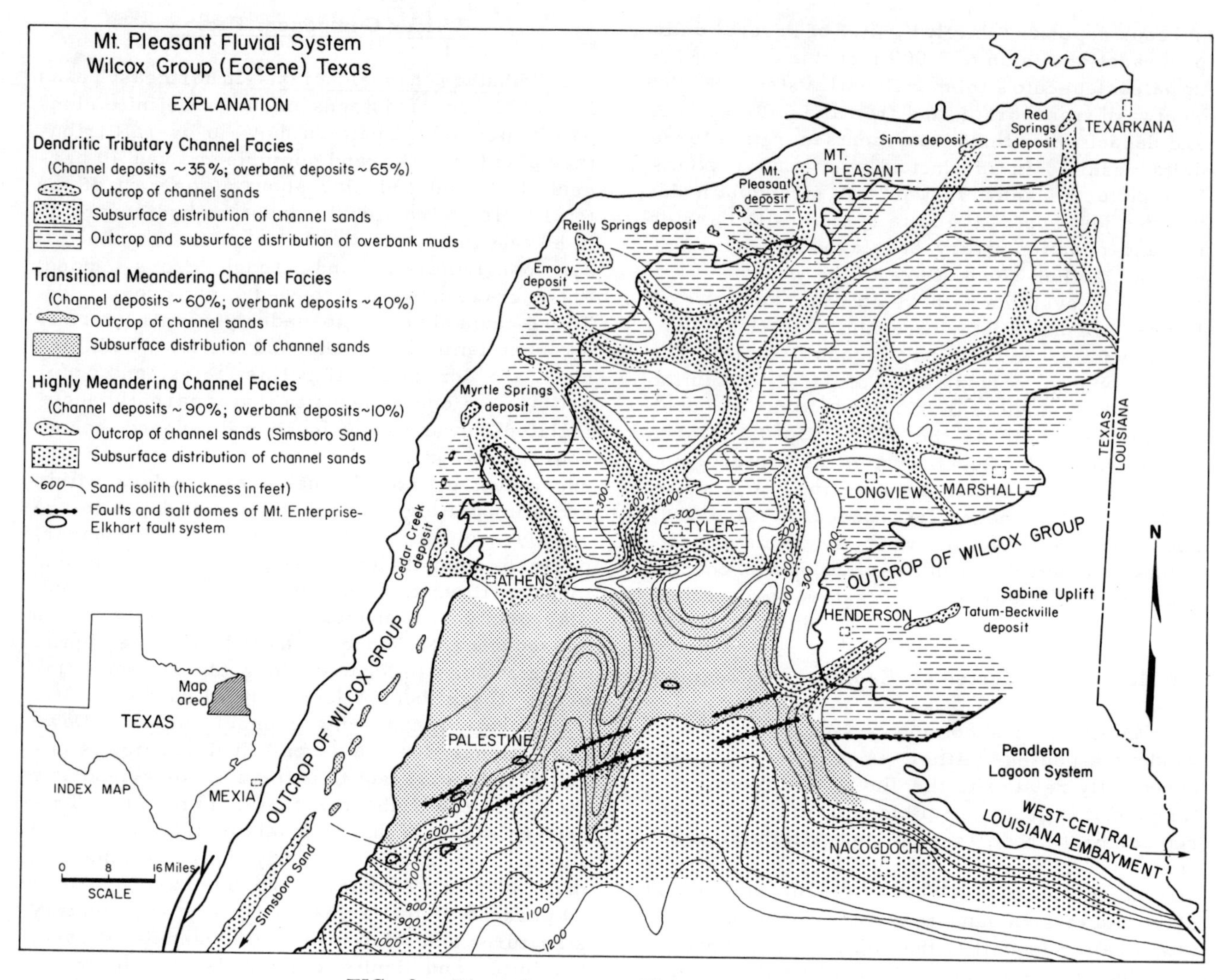

FIG. 2. Fluvial system, Wilcox Group, Texas.

sedimentary structures indicates that most of the preserved sandstone units represent channel lag and lower point bar depositional units (Fisher and McGowen, 1967) and only in the stratigraphically highest sandstone units are upper point bar and overbank sediments preserved. Characteristic sedimentary structures (Fig. 6) include: (1) large trough cross-beds, (2) tabular cross-beds, (3) wedge cross-sets, and (4) parallel laminated beds.

Trough cross-bedding is common in channel lag sediments. Commonly associated with large trough cross-bedded units are climbing ripples that begin near the base of a trough and migrate both laterally and upward to the flank of a trough where they grade into thin to very thin beds composed of parallel laminae.

Tabular cross-bedded units range in thickness from about 0.25 to 0.8 foot; locally within the tabular cross-bedded unit are shallow, trough cross-bedded units and small dunes that have the dune form preserved. These types of structures grade laterally into and overlie channel lag sediments. These stratification types are interpreted as lower point bar deposits.

Overlying tabular cross-beds are wedge sets or epsilon cross-beds (Allen, 1963) that range in thickness from 3 to 4 feet. These beds are interpreted to be upper point bar and, in the lower parts of the highly meandering channel facies, they are normally truncated by channel lag sediments and the sequence of structures 1, 2, 3 (above) is repeated.

Thin to thick beds composed of parallel laminae are present in the uppermost beds of the facies. This structural type either overlies and grades laterally into wedge sets or overlies massively bedded sandstone containing a few shallow trough cross-bedded units. Thin, carbonaceous, kaolinitic mudstone beds are locally interbedded with the parallel laminated structural types. This unit is interpreted as upper point bar and overbank sediment. Mudstone beds of this facies were the source of clay and mudstone granule to boulder clasts that locally comprise a significant part of the channel lag sediments. Upper point bar and overbank sediment is preserved only in the stratigraphically highest parts of the highly meandering facies (Simsboro Sand).

Sedimentary structures of the highly meandering fluvial facies are somewhat different from point bar structures that have been reported in fine grained sand of meandering channel systems (Harms et al., 1963; Frazier and Osanik, 1961; and Beutner et al., 1967). Sedimentary structures in point bars on the modern Amite River, East Feliciana and East Baton Rouge Parishes, Louisiana, which carries a coarse sediment load, are analogous to those of the Wilcox highly meandering channel facies.

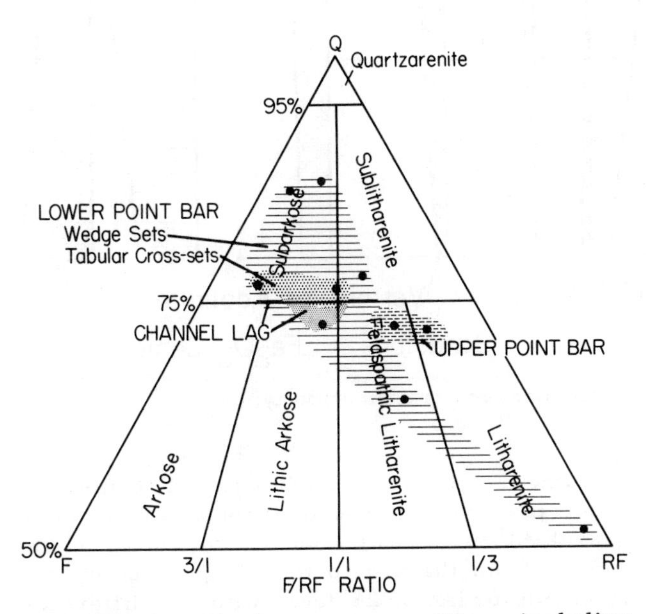

FIG. 3. Sandstone composition, including mud clasts.

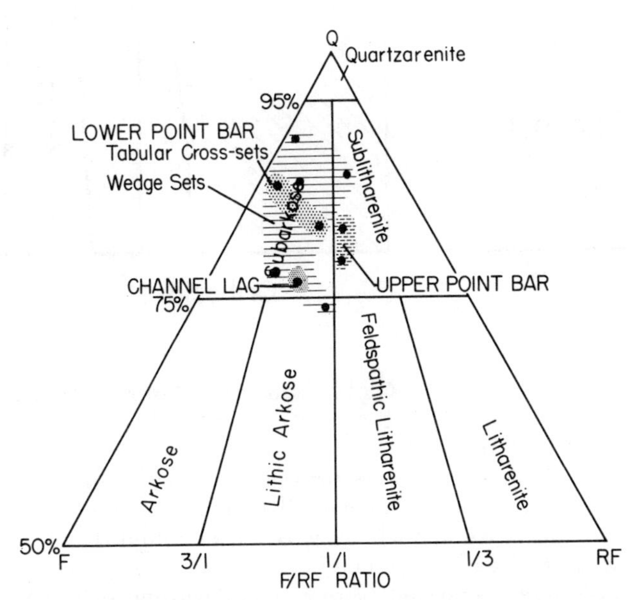

FIG. 4. Sandstone composition, exclusive of mud clasts.

Petrography.--A thin section study was made of each of the stratification types present in the highly meandering channel facies, using Folk's (1968) descriptive sandstone classification. Compositionally, sandstone of this facies ranges from subarkose to sedlithite, depending upon the quantity of mud or clay clasts. Figures 3 and 4 show variations in sandstone composition when mud clasts are counted as sedimentary rock fragments. Most of the mud clasts contain kaolinitic clay. Evidence from thin sections indicates that most of the kaolinite contained within this facies was formed within the sandstone body.

Average composition of sedimentary structure types shows some trends when plotted in order of stratigraphic occurrence (Fig. 5). Quartz and orthoclase are more or less constant throughout the facies. Muscovite increases as grain size decreases. Mud clast increase results from both sorting and preservation of the highest parts of the fluvial system. Vermicular kaolinite is a cement in upper point bar and overbank sediment; its upward increase is attributed to preservation of those units only in the uppermost parts of this channel facies, and to the fact that these units were subjected to intense leaching by meteoric water.

Postulated Kaolin Model

A reconstructed cross section through a channel near the end of fluvial activity for the highly meandering facies of the lower Wilcox is shown in Figure 6. Composition is indicated for each depositional unit. Channel profile and associated sedimentary structures relative to specific sections along the profile are illustrated. Migration was to the left; lateral erosion of older upper point bar and overbank sediment with production of mud clasts is shown on the left. As the channel migrated, only channel lag and lower point bar units were preserved.

Highest percent kaolinized feldspar and vermicular kaolinite in upper point bar and overbank units resulted from periodic flooding and drying of these units. Mud lenses deposited in oxbows, meander scrolls, and other floodplain environments were kaolinized by organic acid-rich meteoric water that percolated downward and laterally through sand of the floodplain and upper point

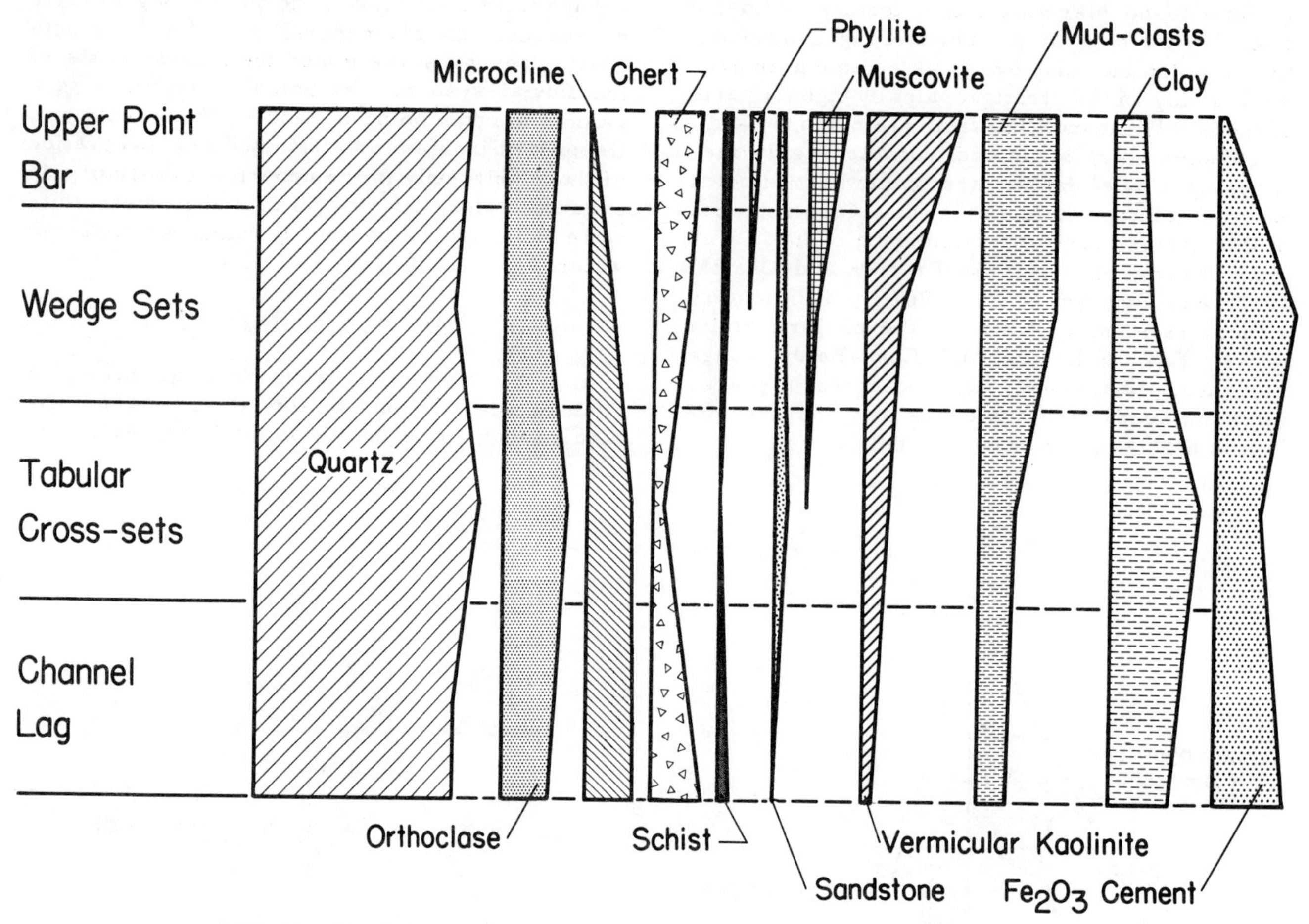

FIG. 5. Sandstone composition related to specific stratification types.

bar. Erosion of these lenses contributed kaolinitic mud clasts to channel fill sand.

Kaolinite associated with Wilcox fluvial sandstone bodies is not unique; similar occurrences have been reported from various regions in North America in sedimentary sequences ranging in age from Paleozoic through Tertiary. This association is empirical, and can be expected where water with high pH is brought into contact with clay minerals, feldspar, etc. in the presence of permeable media for removal of water and material in solution. Requisites of active leaching by acid water and permeable sediment through which the water could migrate are met in Wilcox fluvial sediments. Kaolinized mud and clay of tributary and slightly meandering facies probably contributed some of the kaolinitic mud clasts to the fluvial system.

The highly meandering facies (Simsboro Sand) characteristically shows an upward fining of grain size only when considered in its entirety. Individual depositional sequences within the unit do not always show such a trend. Analogous point bar sequences on the Amite River near Magnolia, Louisiana, likewise do not display an upward fining trend. Overall grain size for the Amite is somewhat coarser (cobble gravel to medium sand) than the Simsboro (pebble and granule gravel to fine sand). Maximum relief of the Amite, from thalweg to top of upper point bar, is on the order of 25 feet, with the difference between flood and low water stages being about 16 feet.

Because of the coarse grain size of the Amite, percolation of water through upper and lower point bars (that mass of sediment between flood and low water stage) is quite rapid. Floodplain of the Amite is covered by a dense growth of trees as a consequence of annual average rainfall of about 60 inches (Wisler and Brater, 1963).

Highest kaolinite present in the Simsboro occurs in upper point bar-overbank sediment. These sediment types, interpreted as being analogous to those of the Amite, were subjected to intense chemical activity by percolating, high pH, meteoric water.

In ancient highly meandering fluvial sandstones composed of coarse sand, the most favorable sites for exploration for authigenic kaolinite are upper point bar-overbank units, especially those representing the last (highest stratigraphic units) depositional events of a particular system.

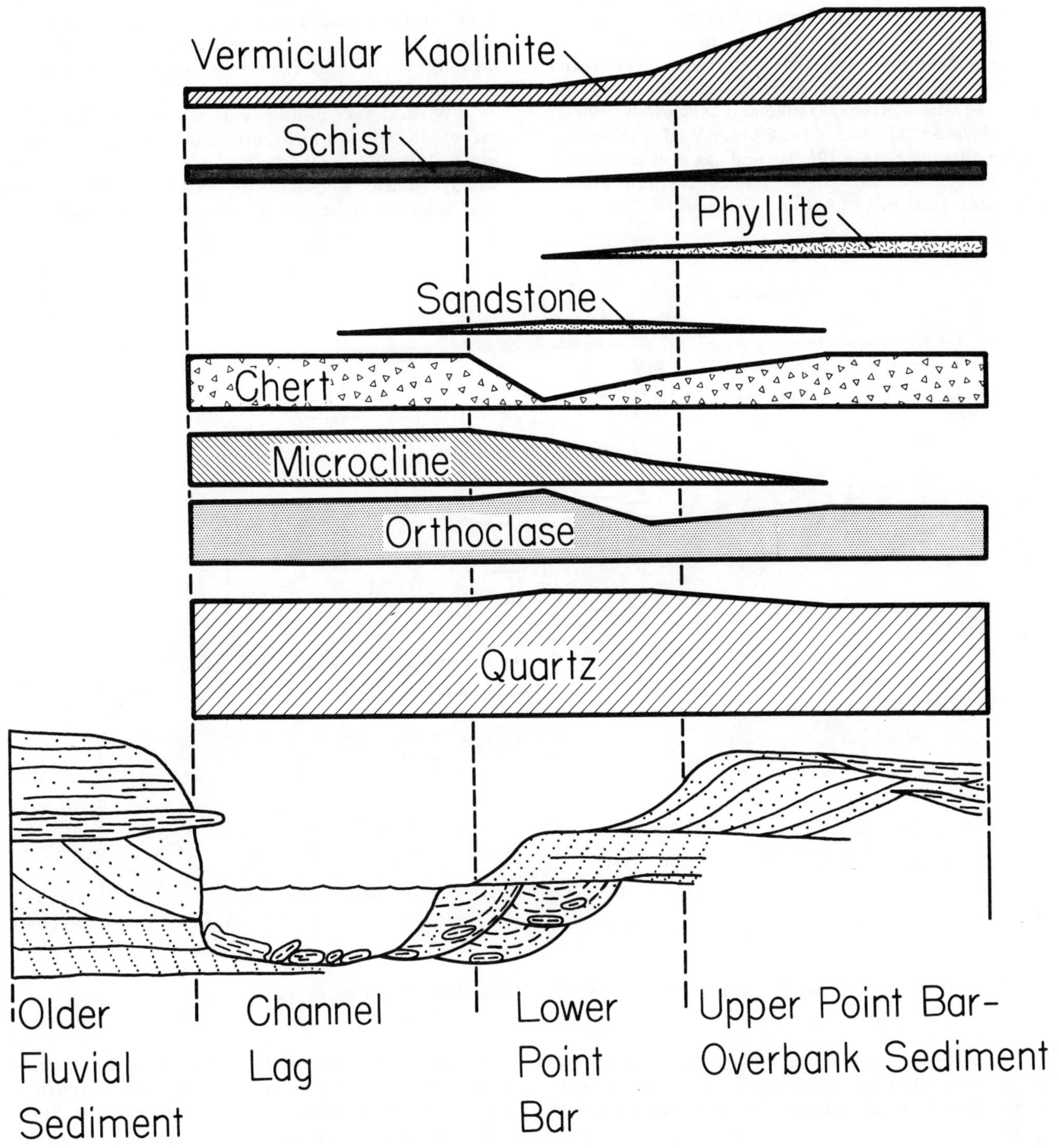

FIG. 6. Sand-kaolin fluvial model.

Wilcox Delta and Bay-Lagoon Systems: Lignites

Lignite is a component facies of fluvial, delta, and bay-lagoonal systems. The occurrence of coal in fluvial sediments has been reported in ancient rocks of various ages (Potter and Siever, 1956; Fisher and McGowen, 1967; and Brown, in press). Fluvial associated lignites are commonly thin and of limited lateral extent. Areally more extensive and thicker lignite and coal beds are a facies of delta plain sediments, which accumulated under marsh and/or swamp conditions. Ancient lignite and coal beds have been interpreted as interdistributary and transgressive (destructional phase) salt marsh-swamp deposits. Interdistributary coals have been reported in the Yoredale

Series of England (Moore, 1958); the Wilcox Group of Texas (Fisher and McGowen, 1967); and in Upper Pennsylvanian-Lower Permian rocks in north-central Texas (Brown, in press). The more widespread coal beds have been interpreted as part of the delta destructional phase of the Wilcox Delta System (Fisher and McGowen, 1967); fresh-water lake basin for Golden Valley (Eocene) sediments (Freas, 1962); and as interdeltaic-subembayment for Upper Pennsylvanian-Lower Permian coal beds (Brown, in press).

and interdistributary deposits which accumulated on delta plains; the massive sand facies represents delta front deposits laid down chiefly as distributary mouth bars and associated units; the downdip thick mud sequence is interpreted to be prodelta facies.

Within the delta system, the delta plain facies is the main lignite-bearing unit. Delta plain lignites are tabular and commonly non-woody; some are areally extensive but others are laterally restricted. Restricted deposits are

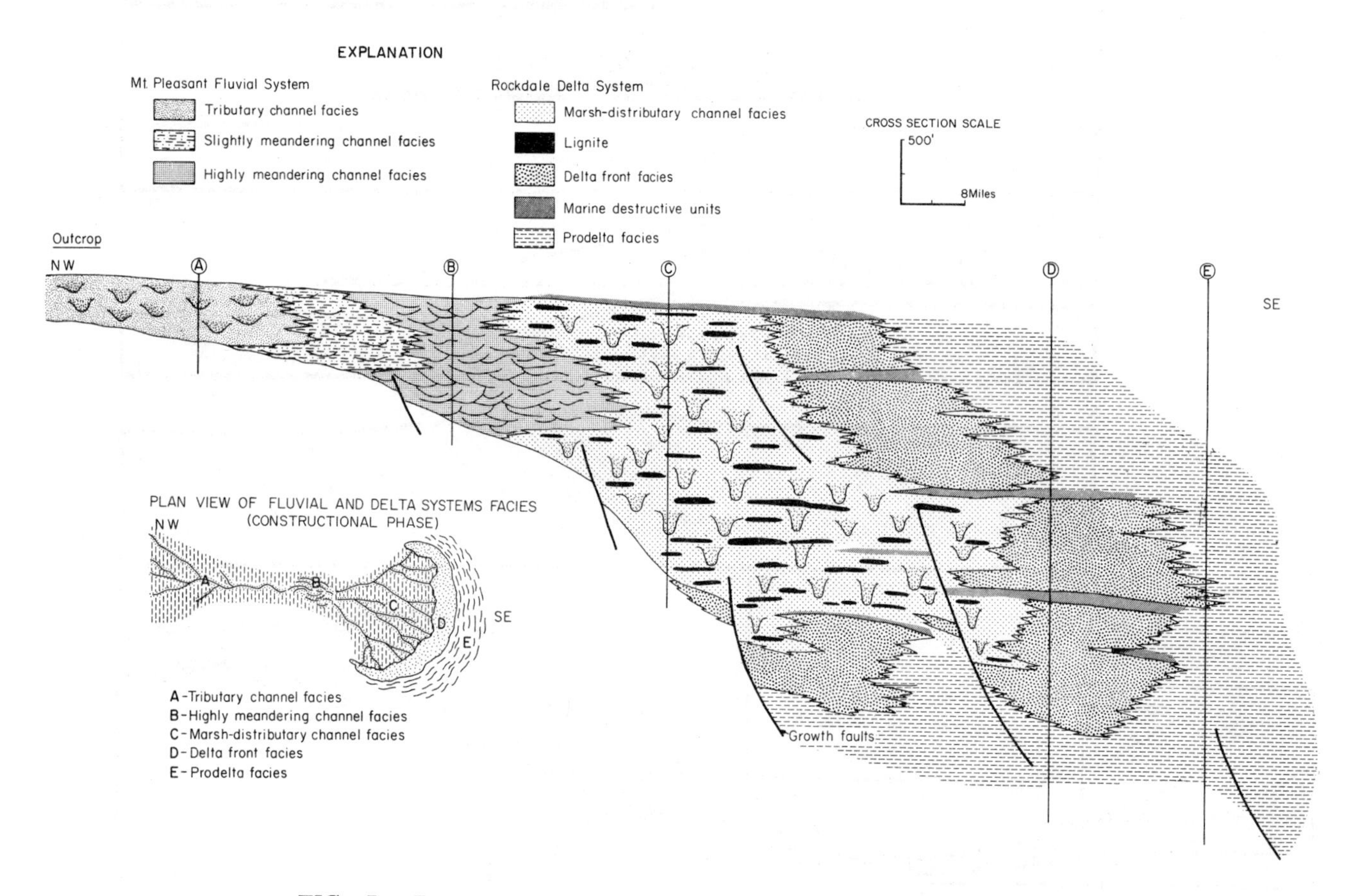

FIG. 7. Dip section through Wilcox fluvial and delta systems.

Quality of lignite and continuity of deposits were controlled by depositional conditions that were operative in specific Wilcox depositional systems.

Delta System

Downdip from fluvial facies, the lower Wilcox consists of sands alternating with muds and silts containing numerous lignite beds (Fig. 7). Farther downdip this alternating facies is replaced by thick massive sands, which in turn grade downdip and along strike to thick, uniform mud sequences. The alternating sand and mud-silt-lignite sequence is interpreted to represent distributary channel interpreted as interdistributary marsh deposits (Fisk, 1958; Frazier, 1967), and the more extensive beds probably developed on the delta plain following cessation of sedimentation and during compactional subsidence of the delta. Delta plain lignites, in addition to being the most widespread, also attain the greatest thickness of any Wilcox carbonaceous deposits. Within the delta plain facies, some lignite beds attain thicknesses of 40 feet or greater.

Bay-Lagoon System

Southwest of the delta system are barrier bar and lagoonal systems. Lignite occurs in the lagoonal fill, which is characterized by burrowed, massive mud; thinly bedded, laminated sand and mud; and mound structures composed of impure micrite and alternating mud and sand laminae. Lagoonal sediment becomes more sandy seaward and landward. Lignite beds occur in upper parts of the lagoon system and represent a final phase of lagoon fill.

Comparison of Wilcox Lignite and Modern Peat Deposition

Interpretation of the significance of Wilcox lignite quality is based upon comparison of Wilcox deposits with Mississippi delta marsh environments. Fisk (1958), Frazier (1967), and Kolb and van Lopik (1966) have discussed the physical control of peat accumulation in several areas on the active Mississippi delta and abandoned older deltas. They mapped marsh environments on the active delta and on most of the abandoned delta plain (Fig. 8). The largest single area of fresh-water marsh, because of frequent flooding, is on the active delta. In areas away from active delta-tion, marsh environments occur as well defined zones ranging from salt marsh through fresh-water marsh, and inland into swamp environments. Development of marsh on the active delta is a part of delta construction, while other marsh areas are chiefly a part of destructional processes. Widespread Wilcox lignites are interpreted as part of the delta destructional phase, and the laterally restricted tabular deposits as a constructional facies within interdistributary areas.

The St. Mary Parish marsh and peat deposits in south-central Louisiana (Fig. 9) studied by Lytle and Driskell (1954) are a modern analogue of some of the Wilcox lignites. St. Mary Parish marsh-swamp environment is more areally restricted than the more extensive Wilcox lignite facies. In general, marsh-peat deposits are higher in NaCl, sulfates, and organic matter, and lower in ignition loss, silica, and alumina than the swamp peats. Sodium chloride and sulfates are highest in marsh peats because of their proximity to the sea. During tropical storms, swamp vegetation is protected from

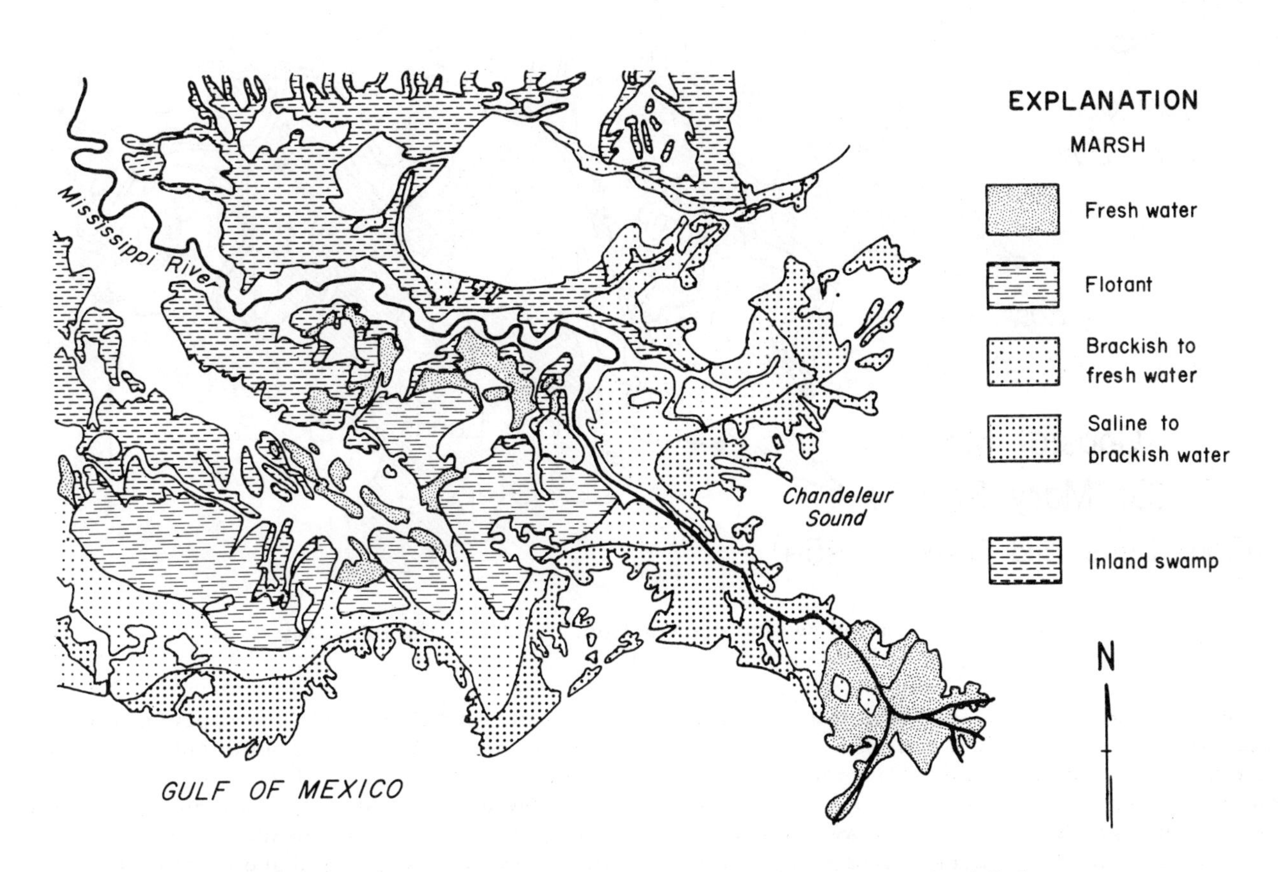

FIG. 8. Marsh and swamp distribution of the Mississippi Delta.

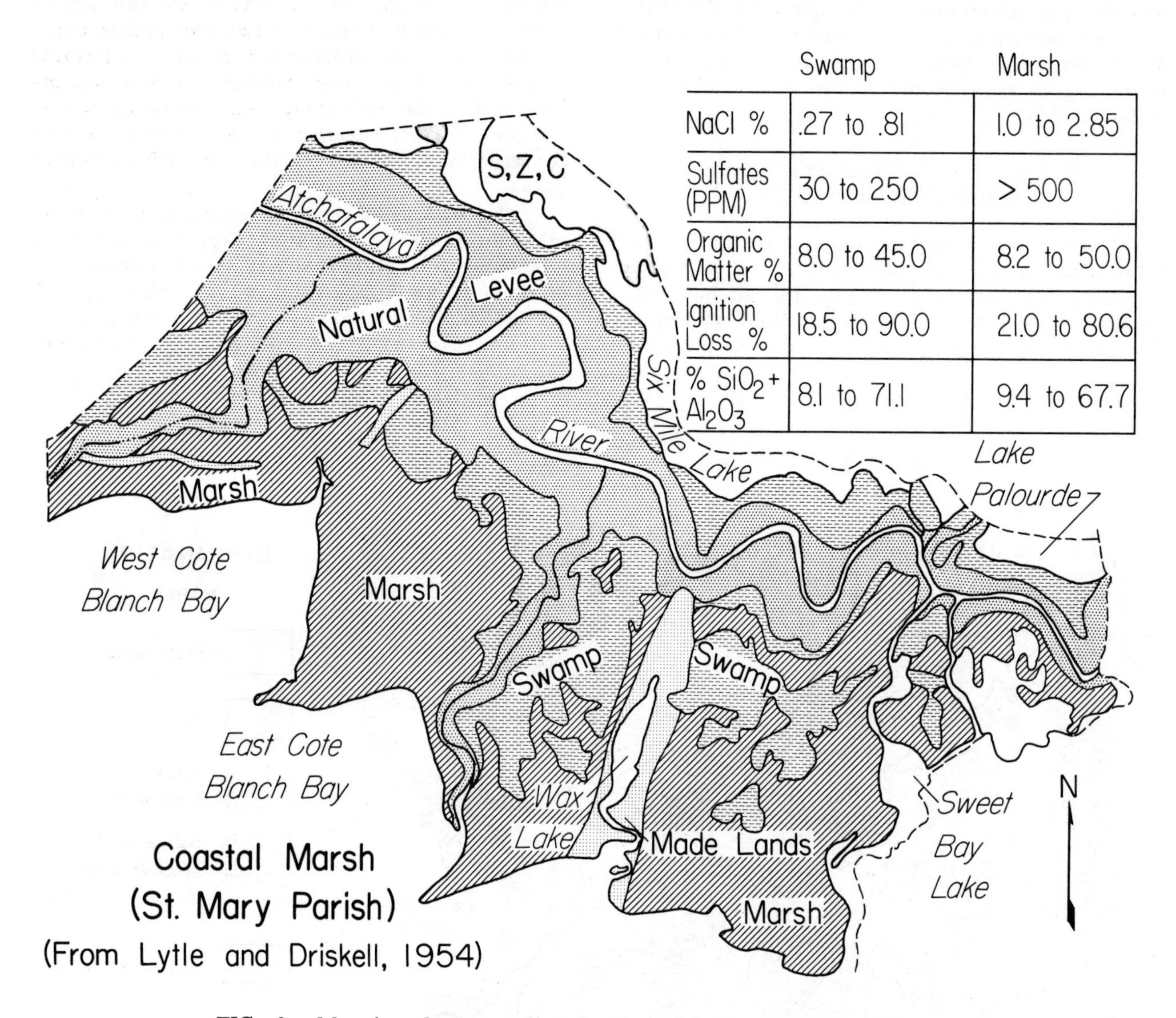

	Swamp	Marsh
NaCl %	.27 to .81	1.0 to 2.85
Sulfates (PPM)	30 to 250	> 500
Organic Matter %	8.0 to 45.0	8.2 to 50.0
Ignition Loss %	18.5 to 90.0	21.0 to 80.6
% SiO_2 + Al_2O_3	8.1 to 71.1	9.4 to 67.7

FIG. 9. Marsh and swamp distribution, St. Mary Parish, Louisiana.

saline waters by a head of fresh water derived from river flood water and run off from adjacent high areas. Contamination by terrigenous sediment is greatest in those parts of the swamp nearest the levees, and at the seaward edge of salt marshes where the area is being cut back by marine processes.

Changes in composition of Wilcox lignites along the outcrop are interpreted to indicate differences in depositional environments of these deposits and to reflect chemical and physical conditions analogous to those in the St. Mary Parish area. Fisher's (1963) study of Wilcox lignites in Texas shows that the Colorado River is the approximate boundary between higher quality lignite in the north and lower quality in the south, and this marks the approximate boundary between Wilcox delta plain and embayment-strandplain facies (Fig. 10).

Fluvial associated lignites average about 5 feet in thickness; they are few in number. Delta plain lignite beds average about 10 feet thick, and are more numerous than in other facies; the delta plain facies commonly contains as many as 40 lignite beds. Much thinner lignite beds, averaging about 1 to 2 feet thick, occur in lagoonal sediments; lagoonal lignites contain more clay or mud partings than either fluvial or delta plain deposits.

A comparison of fluvial, delta, and lagoonal lignite characteristics indicates that, in addition to being the thickest and most extensive deposits,

FIG. 10. Characteristics of fluvial, delta, and lagoonal lignites, Wilcox Group.

the delta plain lignites are the best quality. Fluvial associated lignite is intermediate in quality, and lagoonal deposits are the poorest quality.

High percent woody material and relatively low sulfur content reflect the fresh-water origin of fluvial associated lignites. Low ash content, at least for the thicker commercial grade deposits, suggests that contamination by terrigenous sediment was at a minimum in these swamps; high percent volatile material and intermediate specific gravity are a function of the high percentage of woody material. Restricted distribution of beds was controlled by accumulation in cut-off channel segments and possibly by partial removal by changes in stream courses.

In outcrop, delta plain lignite beds represent inland displacement of marsh-swamp environments as the delta system was foundering. The relatively high quality of lignite suggests that little terrigenous sediment was carried into the environment. Low sulfur content and presence of some woody material indicate that swamps, fresh water, and perhaps some brackish marshes coexisted during accumulation of these lignites. Relatively high specific gravity and low percent volatile matter is a function of either low percent woody material and high percent marsh plant material, or that the material accumulated as mucks. Delta plain lignites are thicker and areally more persistent than fluvial and lagoonal deposits because the compacting thick mud section of the delta system permitted marsh-swamp conditions to persist for longer periods of time than in areas of fluvial and lagoonal systems.

The generally poor quality of lignite in lagoonal sediments resulted from contamination by terrigenous sediment. High sulfur content suggests that lignite probably accumulated in salt marsh environments. Lignite beds are thin and discontinuous because of the slow rate of subsidence of sediment underlying the marsh relative to rates of organic and terrigenous sediment accumulation.

There appears to be close correlation between lignite quality, continuity of deposits, and depositional environments; the most favorable areas for exploration are delta plain facies. Coal and lignite deposits of this type are widely distributed in North America in Paleozoic, Mesozoic, and Tertiary rocks. A well known U. S. example is the Pennsylvanian deposits of the central interior region. In Russia, delta plain coals occur in the Donets Basin in Carboniferous through Cretaceous rocks (Koperina, 1958). Many Cretaceous coals and lignites of the western U. S. accumulated on delta plains.

The thickest beds and best quality lignite occurs in the delta plain facies of the Wilcox delta system. Location of delta plain lignites can be predicted after the fluvial-delta system relationships have been recognized. Control of the continuity of delta plain marsh-swamp environment is a balance between the rate of compactional subsidence and production of plant material. The mass of delta sediment controlled the site of compactional subsidence, directly controlled the duration of conditions favorable for a marsh-swamp environment, and controlled the ultimate thickness of lignite deposits. Proximity of marsh or swamp to either fluvial or marine environments determined the amount of terrigenous material contained within lignite, and to a large degree, the chemical quality, such as sulfur and sodium chloride content.

EDWARDS FORMATION: CARBONATE PLATFORM DEPOSITIONAL SYSTEM

In recent years the Edwards Formation (Lower Cretaceous) of Texas has been studied by Rodda et al. (1966), Fisher and Rodda (1967), Frost (1967), and Moore (1967). Most of the data for this section was taken from the 1967 paper of Fisher and Rodda, and the interpretations of both primary and diagenetic facies is credited to them.

The Edwards Formation has been interpreted to represent a platform carbonate depositional system, comparable in many respects to the Bahama Platform--that is, a platform essentially isolated from influence of terrigenous sediment (Fig. 11). A Lower Cretaceous platform designated the Comanche Platform was encircled by areas of shallow and deep oceanic water. Rudistid reefs developed along the platform edge, and enclosed by these reefs was a broad shallow area, somewhat analogous to the Bahama shelf lagoon; the lagoon was separated by a broad, slightly higher platform area into two smaller lagoonal areas. Prominent biohermal reefs that rimmed the platform edge restricted circulation across the shelf lagoon, and exerted tremendous influence on carbonate deposition on the platform. Reduced physical energy, or turbulence, in the area enclosed by reefs, coupled by changes in water chemistry, acted together to produce carbonate and evaporite sediments distinct from those which formed reefs and reef associated facies. A diagenetic dolomite facies was superimposed upon some of the primary carbonate facies and was, to a large degree, controlled by the reef-lagoon couple (Fisher and Rodda, 1967).

Primary Component Facies

Contemporaneous with deposition of platform carbonates were deeper water basinal sediments which accumulated in the North Texas--Tyler Basin and the ancestral Gulf Basin. Characteristics of basinal and platform carbonates are listed below (for details on these particular facies, see Fisher and Rodda, 1967).

Basinal Sediments

Rocks deposited in the shallow oceanic basin consist of an upper sequence of nodular micrite, biomicrite, and mudstone, and a lower sequence of mudstone and Gryphaea biomicrite. Basinal rocks of the deeper basin consist of micrite and biomicrite (Bishop, 1966).

Platform Sediments

Depositional facies at the outcrop include rudistid bioherm-biostromal facies, platform grainstone facies, and lagoonal facies; dolomites represent diagenetic facies.

Rudistid reefs, which formed the north edge of the platform, fringed the north part of the lagoon. A framework of rudistids with micrite interareas comprise the bioherms. Reef flank material normally consists of large shells, both whole and fragmented, in a micrite matrix; biosparite is minor. Interreef sediments consist of biosparite and micrite with varying amounts of allochems.

The platform facies consists of alternating, thin, evenly bedded biosparite and biomicrite, thick beds of cross-laminated biosparite, nodular micrite, and rudistid bioherms. Chert occurs throughout much of this facies.

The lagoonal facies presently exposed at the outcrop consists of thinly bedded and nodular micrite, and intensely burrowed micrite. Evaporites initially formed a significant part of the lagoonal fill, but only relatively small patches of gypsum were preserved. Collapse structures and solution features in the limestone above lagoonal evaporites are evidence of former existence of gypsum.

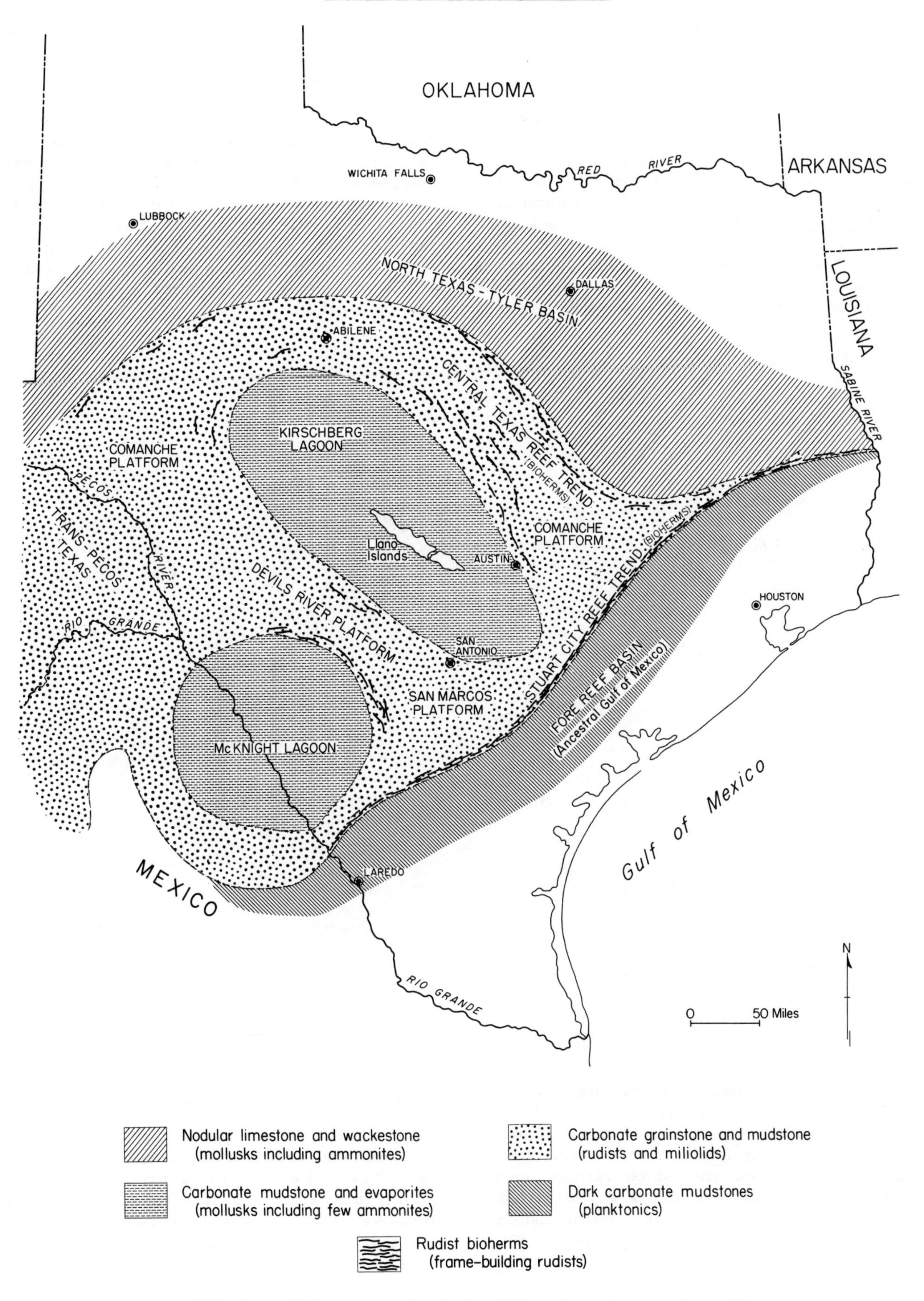

FIG. 11. Paleogeography during Edwards carbonate deposition.

Diagenetic Facies

Dolomite

Dolomite occurs predominantly marginal to the lagoonal facies (Fig. 12), although some dolomite is coincident with lagoonal rocks. Two types of dolomite present in the Edwards have been designated stratal and massive dolomite.

Stratal dolomite.--This type is generally very thinly laminated to thinly bedded, contains rip-up clasts, and some beds display desiccation cracks. Individual dolomite beds are commonly less than 2 feet thick. This type of dolomite is commonly interbedded with micrite and biosparite. Tightly packed and generally anhedral dolomite crystals are in the 10-micron range. Stratal dolomite is interpreted as early, pre-lithification replacement which formed in an environment analogous to recent supratidal flats (Shinn, 1964; Shinn et al., 1965; Purdy and Imbrie, 1964; Deffeyes et al., 1965; Ebanks, 1967; Alderman and Skinner, 1957; and Alderman, 1965).

Massive dolomite.--This type is interpreted to reflect post-lithification replacement. Dolomite preferentially replaces grainstones, and rarely replaces micrite and reef core, even where grainstones are extensively dolomitized, indicating an underlying permeability control. Dolomitized sequences are on the order of 10 feet thick. Crystals are fine to coarse (commonly greater than 30 microns), normally euhedral, and have an open fabric. This type of dolomite commonly underlies, or is marginal to, parts of the lagoonal fill.

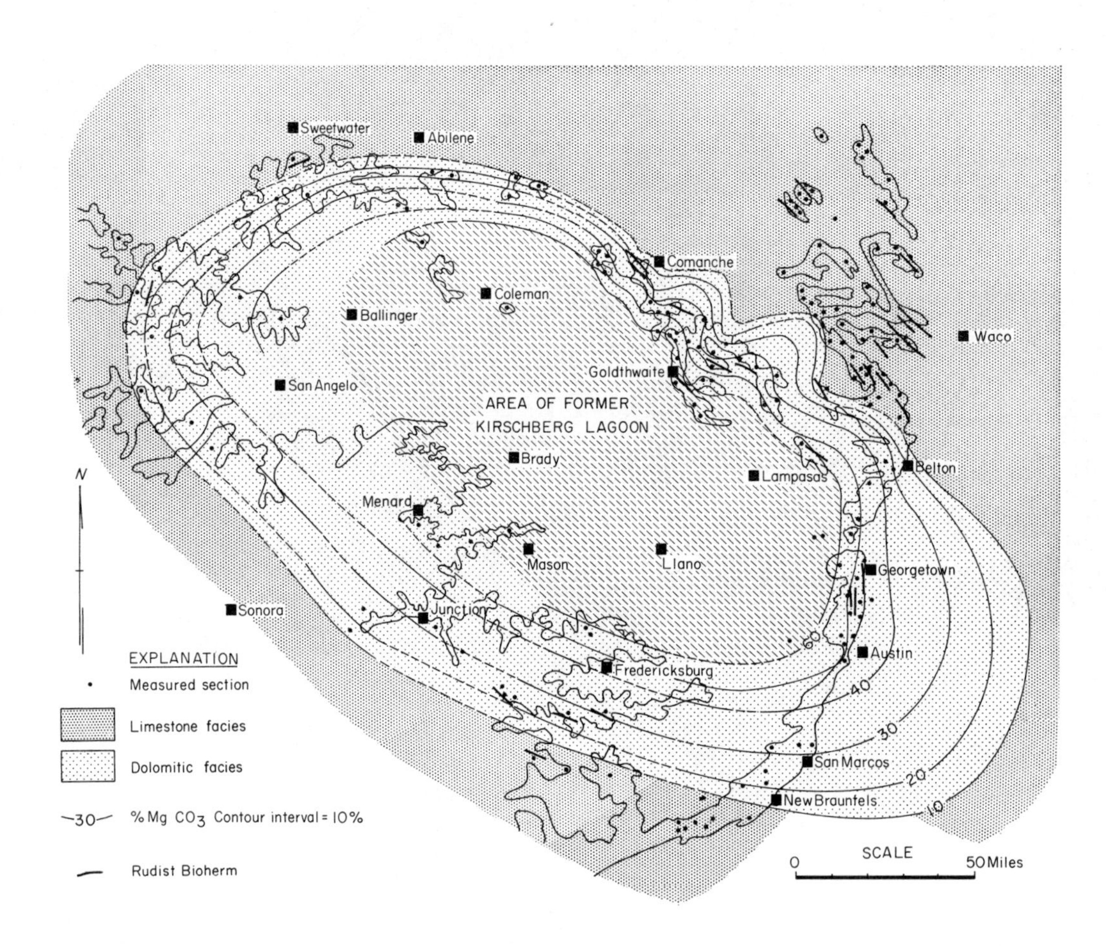

FIG. 12. Dolomite distribution, Edwards Formation.

Origin of Dolomite

The two types of Edwards dolomite are related to specific depositional environments (Fig. 13). Stratal dolomite formed in an area of low physical energy to the south of the lagoonal system, whereas massive reflux dolomite replaced grainstones formed in the high physical energy area asso-

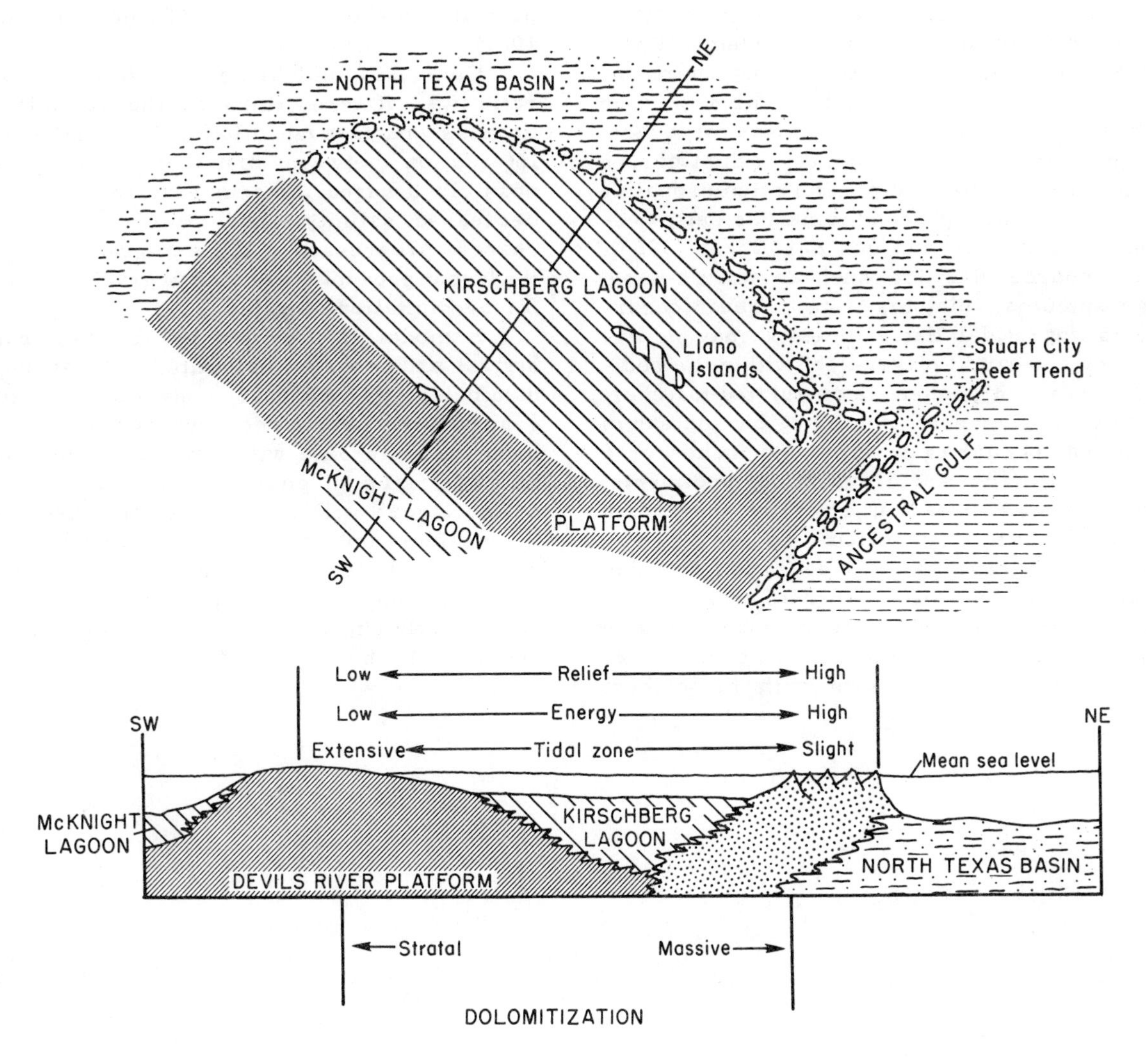

FIG. 13. Physiographic control of stratal and massive dolomite.

ciated with reefs. In the lagoon and on the platform, astronomical tides probably were not effective sediment transporting agents. Dominance of micrite over grainstones attests to low turbulence. Under these relatively low energy conditions, early replacement dolomite formed in supratidal areas by concentration of brine in local depressions in which gypsum was deposited. The resulting Mg-rich brine percolated downward replacing the calcium ion in the carbonate sediment. Some replacement may have occurred through upward movement of water as described by Shinn et al. (1965).

Surface water salinity increased toward the platform, and brine flowed toward the lows of the lagoon floor, where evaporites accumulated. Brine escaped through permeable lime sand that floored parts of the lagoon near the rudistid reefs (Fig. 13). In this area grainstones are normally dolomitized, whereas associated micrite is commonly unaffected. Micrite, in many instances, shielded underlying permeable but undolomitized grainstones. Reef cores composed of micrite are commonly not dolomitized, a fact which further demonstrates a primary permeability control of massive reflux dolomite. If permeable rock were brought into contact with Mg-rich brines, dolomitization was essentially complete, and rocks which had little permeability are those which now have low magnesium content. The mechanism for this type of dolomitization was by reflux as postulated by Adams and Rhodes (1960).

Origin of Chert

Brines were rich in both magnesium and silicon as shown by close association of massive dolomite and replacement chert (Fig. 13). Replacement or diagenetic origin of chert is shown by preservation of host rock structures and fabric. The greatest concentration of chert occurs to the east and northeast of the lagoon system, and is coincident with reef development.

It is postulated that the mechanisms which

concentrated magnesium and silicon were related. Although the source of these two elements was not the same, massive dolomite and chert apparently were emplaced by reflux of brine through the grainstone facies. The process of quartz solution in the high pH environment of the Coorong area (South Australia) and precipitation of opaline silica under lower pH conditions has been described by Peterson and von der Borch (1965). Given a source of silica such as quartz sand, sponge spicules, diatoms, etc., chert occurrence in supratidal flat sediments may be explained by a vertical pH gradient within the sediment. A similar gradient is postulated for chert associated with massive dolomite, but instead of the gradient occurring within a few inches or a few feet vertically, as in supratidal sediments, the gradient was spread out laterally, perhaps a few tens of feet to a few miles. Source for chert is postulated to have been siliceous sponges which grew in the current swept interreef areas.

Chert extends farther north than massive dolomite (Fig. 14), but maximum chert occurrence is generally coincident with the highest percent massive dolomite (Fisher and Rodda, 1967).

Composition of brine was altered as Mg/Ca ratio was lowered through the dolomitization process and through dilution by connate water. Interaction between brine and connate water possibly lowered the temperature of the brine, decreased salinity, and lowered pH. These changes progressed as brine migrated away from the lagoon producing a chemical gradient favorable for precipitation of silica.

A model is provided by the Edwards for stratal and massive dolomite. It can be predicted that the low energy stratal dolomite will generally be chert free, but will occur in thin beds that alternate with limestone, and that the massive reflux dolomite on the high energy side of the lagoon will occur as thick units high in magnesium content, but will generally contain chert. Reef flank sediments, on the other hand, are commonly completely dolomitized and are consistently chert free. The best dolomite prospects are in the massive facies.

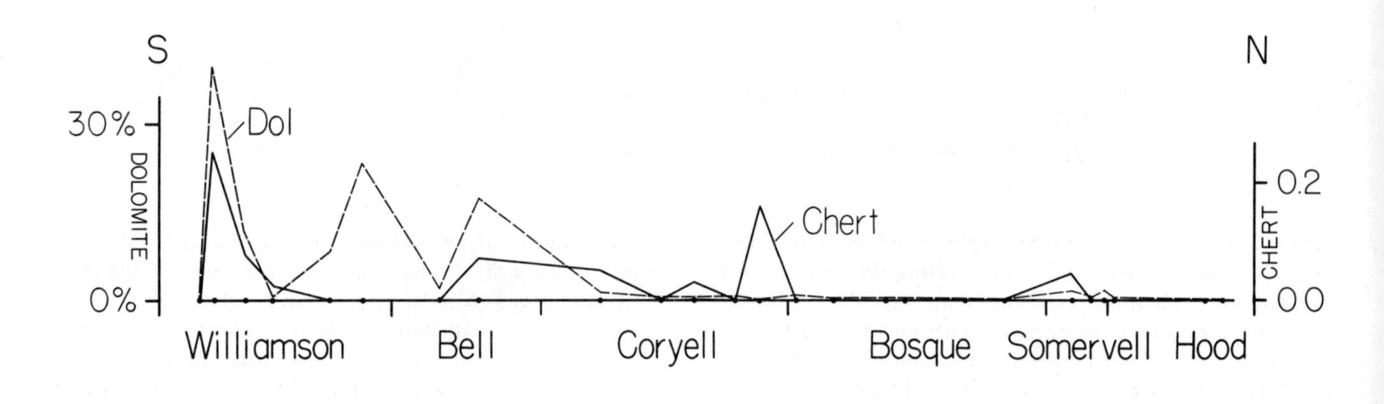

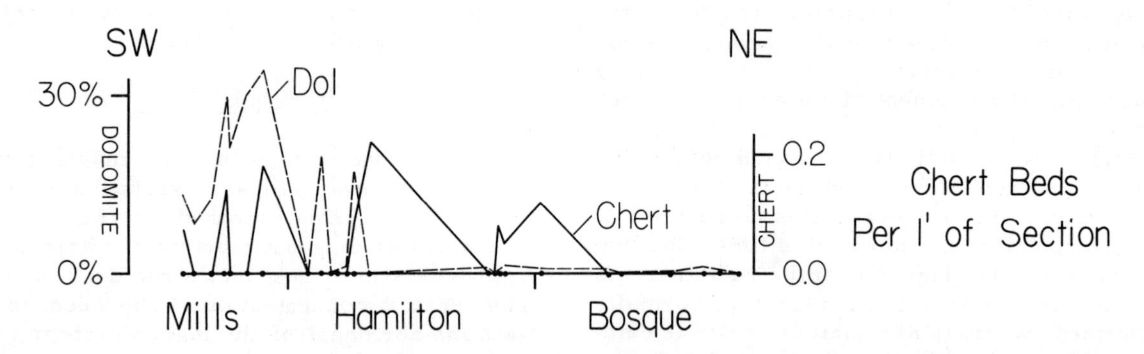

FIG. 14. Chert and dolomite distribution, Edwards Formation.

SUMMARY

The use of depositional systems in mineral exploration will help delineate favorable areas and to a certain degree allow prediction of quality and magnitude of the deposit.

Kaolinite can be expected to occur in muds and clays immediately adjacent to tributary and slightly meandering channel sands, and within highly meandering channel sand, especially in upper point bar and overbank sediments.

Recognition of fluvial-delta relationships should indicate areas of the most persistent and best quality lignite occurrences; recognition of the reef-lagoonal control for stratal dolomite and the massive dolomite-chert association should be of value in exploration for chert-free dolomite.

REFERENCES

Adams, J. E., and Rhodes, M. L. (1960) Dolomitization by seepage refluxion: Bull. Amer. Assoc. Petrol. Geol., vol. 44, pp. 1912-1932.

Alderman, A. R. (1965) Dolomitic sediments and their environment in the southeast of south Australia: Geochim. et Cosmochim. Acta, vol. 29, pp. 1355-1365.

__________ and Skinner, H. C. (1957) Dolomite sedimentation in the south-east of Australia: Amer. Jour. Sci., vol. 255, pp. 561-567.

Allen, J. R. L. (1963) The classification of cross-stratified units, with notes on their origin: Sedimentology, vol. 2, pp. 93-114.

Beutner, E. C., et al. (1967) Bedding geometry in a Pennsylvanian channel sandstone: Bull. Geol. Soc. America, vol. 78, pp. 911-916.

Bishop, B. A. (1966) Stratigraphy and carbonate petrography of the Sierra de Picachos and vicinity, Nuevo Leon, Mexico: Unpub. Ph. D. dissertation, Univ. Texas at Austin, 451 pp.

Brown, L. F., Jr. (in press) Virgil-lower Wolfcamp repetitive depositional environments in north-central Texas, in Cyclic sedimentation in the Permian Basin: West Texas Geological Society, Midland, Texas.

Clark, O. M., Jr. (1964) Clay deposits of the Tuscaloosa Group in Alabama: Clays and Clay Minerals, Proc., 12th Natl. Conf., Pergamon Press, pp. 495-507.

Deffeyes, K. S., et al. (1965) Dolomitization of Recent and Plio-Pleistocene sediments by marine evaporite waters on Bonaire, Netherlands Antilles, in Dolomitization and limestone diagenesis: Soc. Econ. Paleontologists and Mineralogists Spec. Pub. 13, pp. 71-88.

Ebanks, W. J. (1967) Recent carbonate sedimentation and diagenesis, Ambergris Cay, British Honduras: Unpub. Ph. D. dissertation, Rice Univ., 189 pp.

Fisher, W. L. (1963) Lignites of the Texas Gulf Coastal Plain: Univ. Texas, Bur. Econ. Geology Rept. Inv. 50, 164 pp.

__________ and McGowen, J. H. (1967) Depositional systems in the Wilcox Group of Texas and their relationship to occurrence of oil and gas: Gulf Coast Assoc. Geol. Socs. Trans., vol. 17, pp. 105-125.

__________ and Rodda, P. U. (1967) Stratigraphy and genesis of dolomite, Edwards Formation (Lower Cretaceous) of Texas, in Proceedings of the 3d Forum on Geology of Industrial Minerals: State Geol. Survey of Kansas Spec. Dist. Pub. 34, pp. 52-75.

__________ et al. (1965) Rock and mineral resources of east Texas: Univ. Texas, Bur. Econ. Geology Rept. Inv. 54, 439 pp.

Fisk, H. N. (1958) Recent Mississippi River sedimentation and peat accumulation, in Van Aelst, Ernest, ed., Congres pour l'Avancement des Etudes de Stratigraphie et de Geologie du Carbonifere, 4th, Heerlen, 1958, Compte Rendu, vol. 1, pp. 187-199.

Folk, R. L. (1968) Petrology of sedimentary rocks: Hemphill's, Austin, Texas, 170 pp.

Frazier, D. E. (1967) Recent deltaic deposits of the Mississippi River: their development and chronology: Gulf Coast Assoc. Geol. Socs. Trans., vol. 17, pp. 287-315.

__________ and Osanik, A. (1961) Point-bar deposits, Old River Locksite, Louisiana: Gulf Coast Assoc. Geol. Socs. Trans., vol. 11, pp. 121-137.

Freas, D. H. (1962) Occurrence, mineralogy, and origin of the lower Golden Valley kaolinitic clay deposits near Dickinson, North Dakota: Bull. Geol. Soc. America, vol. 73, pp. 1341-1364.

Frost, J. G. (1967) The Edwards Limestone of Central Texas, in Comanchean (Lower Cretaceous) stratigraphy and paleontology of Texas: Soc. Econ. Paleontologists and Mineralogists, Permian Basin Sec., Pub. 67-8, pp. 133-156.

Glass, H. D. (1956) Clay mineralogy of Pennsylvanian sediments in southern Illinois, in Swineford, A., ed., Clays and clay minerals, Natl. Research Council Pub. 566, pp. 227-241.

Harms, J. C., et al. (1963) Stratification in modern sands of the Red River, Louisiana: Jour. Geology, vol. 71, pp. 566-580.

Kesler, T. L. (1956) Environment and origin of the Cretaceous kaolin deposits of Georgia: Econ. Geology, vol. 51, pp. 541-554.

Kohls, D. W. (1967) Petrology of the Simsboro Formation of northeast-central Texas: Jour. Sed. Petrology, vol. 37, pp. 184-204.

Kolb, C. R., and van Lopik, J. R. (1966) Depositional environments of the Mississippi River deltaic plain, southeastern Louisiana, in Deltas in their geologic framework: Houston Geol. Soc., Houston, Texas, pp. 17-61.

Koperina, V. V. (1958) Facies and types of coal accumulation in the Donets coal measures: Bull. Acad. Sciences U.S.S.R., Geol. Ser. (English trans.), vol. 1, pp. 8-17.

Lytle, S. A., and Driskell, B. N. (1954) Physical and chemical characteristics of the peats, mucks, and clays of the coastal marsh area of St. Mary Parish, Louisiana: Louisiana Agr. Expt. Sta. Bull. 484, 37 pp.

Moore, C. H., Jr. (1967) Stratigraphy of the Edwards and associated formations, west-central Texas: Gulf Coast Assoc. Geol. Socs. Trans., vol. 17, pp. 61-75.

Moore, Derek (1958) The Yoredale Series of Upper Wensleydale and adjacent parts of the northwest Yorkshire: Yorkshire Geol. Soc., Proc., vol. 31, pp. 91-146.

Pask, J. A., and Turner, M. D. (1952) Geology and ceramic properties of the Ione Formation, Buena Vista area, Amador County, California: California Dept. Nat. Res., Div. Mines Spec. Rept. 19, 39 pp.

Peterson, M. N. A., and von der Borch, C. C. (1965) Chert: modern inorganic deposition in a carbonate-precipitating locality: Science, vol. 149, pp. 1501-1503.

Potter, P. E., and Siever, R. (1956) Sources of basal Pennsylvanian sediments in the eastern interior basin, I, Cross-bedding: Jour. Geology, vol. 64, pp. 225-244.

Purdy, E. G., and Imbrie, J. (1964) Carbonate sediments, Great Bahama Bank, in Guidebook, Field Trip No. 2: Geol. Soc. America, pp. 1-58.

Rodda, P. U., et al. (1966) Limestone and dolomite resources, Lower Cretaceous rocks, Texas: Univ. Texas, Bur. Econ. Geology Rept. Inv. 56, 286 pp.

Shinn, E. A. (1964) Recent dolomite, Sugarloaf Key, in Guidebook, Field Trip No. 1, South Florida carbonate sediments: Geol. Soc. America, pp. 62-67.

__________ et al. (1965) Recent supratidal dolomite from Andros Island, Bahamas, in Dolomitization and limestone diagenesis: Soc. Econ. Paleontologists and Mineralogists Spec. Pub. 13, pp. 112-123.

Weaver, C. E. (1960) Possible uses of clay minerals in the search for oil, in Swineford, A., ed., Clays and clay minerals, vol. 8: Natl. Conf. Clays and Clay Minerals, 8th, Norman, Okla., Oct. 1959, Proc., pp. 214-227.

Wisler, C. O., and Brater, E. F. (1963) Hydrology: John Wiley and Sons, New York, 408 pp.

PREVIOUS FORUMS

1. A Symposium on Geology of Industrial Limestone and Dolomite (1965) published in Ohio Journal of Science, 1966, vol. 66, pp. 97-224.

2. A Symposium on Geology of Cement Raw Materials (1966) published by Indiana Geological Survey, 1966, 197 pp.

3. A Symposium on Industrial Mineral Exploration and Development (1967) published by Kansas Geological Survey, Special Distribution 34, 1967, 183 pp.